The
DNA
PROVIRUS
Howard Temin's
Scientific Legacy

The DNA PROVIRUS

Howard Temin's Scientific Legacy

Edited by

Geoffrey M. Cooper
Dana-Farber Cancer Institute and
Harvard Medical School
Boston, Massachusetts

Rayla Greenberg Temin
Department of Genetics
University of Wisconsin
Madison, Wisconsin

and

Bill Sugden
McArdle Laboratory for Cancer Research
University of Wisconsin Medical School
Madison, Wisconsin

ASM Press • *Washington, D.C.*

Copyright © 1995 American Society for Microbiology
1325 Massachusetts Ave., N.W.
Washington, DC 20005

Library of Congress Cataloging-in-Publication Data

The DNA provirus : Howard Temin's scientific legacy / edited by
  Geoffrey M. Cooper, Rayla Greenberg Temin, and Bill Sugden.
    p.   cm.
  Includes index.
  Includes selected papers of Howard M. Temin and proceedings of Howard
Temin Commemorative Symposium, held in Madison, Wisconsin, Oct.
13–15, 1994.
    ISBN 1-55581-098-5
    1. Oncogenes—Congresses. 2. Retroviruses—Congresses. 3. Temin,
Howard M.—Congresses. 4. Temin, Howard M.   I. Temin, Howard M.
II. Cooper, Geoffrey M.   III. Temin, Rayla Greenberg.   IV. Sugden,
Bill.   V. Howard Temin Commemorative Symposium (1994 : Madison,
Wis.)
    [DNLM: 1. Proviruses—genetics—collected works. 2. DNA, Viral—
collected works. 3. HIV-1—genetics–congresses. 4. Neoplasms—
genetics—congresses. 5. Oncogenes—genetics—congresses.
6. Proviruses—genetics—collected works.   QW 168.5.R18 D629 1995]
RC268.42.D63   1995
616′.0194—dc20
DNLM/DLC
for Library of Congress   95-12355
                                                                      CIP

Cover photos: The electron micrograph illustrates virus particles budding from an
HIV-infected cell. Magnification ×128,000. The micrograph is courtesy of Dr.
Henry S. Slater, Dana-Farber Cancer Institute. (*Inset*) Model of HIV reverse transcriptase complexed with an inhibitor (nevirapine). The α-carbon of the polymerase (purple) is shown bound to DNA (blue and yellow). Active site aspartic acid
residues 185 and 186 are shown in red. Nevirapine (green) binds near tyrosine residues 181 and 188 (white). Reprinted with permission from L. A. Kohlstaedt, J.
Wang, J. M. Friedman, P. A. Rice, and T. A. Steitz (Science **256:**1783–1790,
1992). Copyright 1992 American Association for the Advancement of Science.
Courtesy of Dr. Thomas A. Steitz, Yale University.

Contents

I. SELECTED PAPERS OF HOWARD M. TEMIN

II. CANCER

III. VIROLOGY

IV. HIV, AIDS, AND THE IMMUNE SYSTEM

Participants at the Howard Temin Commemorative Symposium

CANCER

GEOFFREY M. COOPER
Dana Farber Cancer Center
Boston, Mass.

On the Origin of Oncogenes

THOMAS D. GILMORE
Boston University
Boston, Mass.

Malignant Transformation by the
v-Rel Oncoprotein

WILLIAM F. DOVE
University of Wisconsin
Madison, Wis.

The Adenomatous Polyposis Coli
(*apc*) Gene of the Mouse in
Development and Neoplasia

J. MICHAEL BISHOP
University of California
San Francisco, Calif.

Protooncogenes and Cell Signaling

GENOME REPLICATION

WEI-SHAU HU
West Virginia University
Morgantown, W.V.

Plus Strand Transfer of
Retroviruses

JIAYOU ZHANG
University of Wisconsin
Madison, Wis.

Analysis of Minus-Strand Strong
Stop DNA Transfer during
Reverse Transcription

VINAY K. PATHAK
West Virginia University
Morgantown, W.V.

Aberrant Provirus Structure of a
Promoter-Trap Shuttle Vector

SANDRA WELLER
University of Connecticut
Farmington, Conn.

DNA Replication and Genome
Maturation in Herpes Simplex
Virus

VIROLOGY AND GENETICS

DAWN P. W. BURNS
University of Wisconsin
Madison, Wis.

Rate and Nature of Mutations
within Long Runs of Identical
Nucleotides during SNV
Replication

ERIC O. FREED
*National Institute of Allergy and
  Infectious Diseases*
Bethesda, Md.

HIV-1 Matrix: Characterization of
Diverse Functions in the Virus
Life Cycle

CESTMIR ALTANER
Cancer Research Institute
Bratislava, Slovak Republic

Cell Receptor for Bovine
Leukemia Virus

MOSHE KOTLER
Hebrew University
Jerusalem, Israel

Regulation of Retroviral Protease

KOUICHI IWASAKI
University of Washington
Seattle, Wash.

Genetic Analysis of Defecation
Behavior in the Nematode
Caenorhabditis elegans

HUMAN VIRUSES AND DISEASE

IRVIN S. Y. CHEN
UCLA School of Medicine
Los Angeles, Calif.

The SCID/Human Chimera Mouse
as a Model for Human Retrovirus
Pathogenesis and Gene Therapy

DANI BOLOGNESI
Duke University
Durham, N.C.

The Dilemma of Developing and
Testing AIDS Vaccines

SAMUEL BRODER
National Cancer Institute
Bethesda, Md.

National Cancer Program—Past,
Present, and Future

ROBERT C. GALLO
National Cancer Institute
Bethesda, Md.

Select Aspects of HIV
Pathogenesis and Some New
Attempts at Controlling HIV
Replication

DAVID BALTIMORE
Rockefeller University
New York, N.Y.

Control of HIV Replication

HIV AND ANTIVIRAL STRATEGIES

KATHLEEN A. BORIS-LAWRIE
University of Wisconsin
Madison, Wis.

Replication of Hybrid Retrovirus
Genomes

MICHAEL EMERMAN
*Fred Hutchinson Cancer Research
  Center*
Seattle, Wash.

The Mechanism of HIV Cell
Death and Latency

ANTONIO T. PANGANIBAN   Encapsidation of HIV-1 RNA
University of Wisconsin
Madison, Wis.

C. YONG KANG   HIV *gag-env* Chimera: a Novel
University of Western Ontario   Strategy of AIDS Vaccine
London, Ontario, Canada

W. GARY TARPLEY   The BHAP HIV-1 RT Inhibitors
Upjohn Laboratories
Kalamazoo, Mich.

TAK W. MAK   T-Cell Recognition and
Ontario Cancer Institute   Development
Toronto, Ontario, Canada

ONCOGENES AND CANCER

GEORG W. BAUER   TGF-β-Treated Normal
University of Freiburg   Fibroblasts Induce Apoptosis in
Freiburg, Germany   Transformed Cells

THOMAS C. MITCHELL   Stimulation of NFκB Activity by
University of Wisconsin   Mutant Derivatives of the Latent
Madison, Wis.   Membrane Protein of Epstein-
                                         Barr Virus

DAVID E. BOETTIGER   Analysis of β1 Integrin Function
University of Pennsylvania   during Transformation of Cells by
Philadelphia, Pa.   RSV

MARK HANNINK   Protein-Protein Interactions That
University of Missouri   Mediate Transformation by v-Rel
Columbia, Mo.

ELI KESHET   VEGF-Mediated Tumor
Hebrew University   Angiogenesis
Jerusalem, Israel

VIROLOGY

JOHN M. COFFIN   Retrovirus Variation and
Tufts University   Evolution
Boston, Mass.

HAROLD E. VARMUS   Proviral Insertion Mutations and
National Institutes of Health   Activation of Proto-Oncogenes
Bethesda, Md.

Roland R. Rueckert
University of Wisconsin
Madison, Wis.

Common Cold Viruses: Virus
Structure and Drug Design. How
Are We Doing?

Bill Sugden
University of Wisconsin
Madison, Wis.

Epstein-Barr Virus: the
Prototypical Human Tumor Virus

Contributors

J. MICHAEL BISHOP
  The G.W. Hooper Research Foundation and Department of Microbiology and Immunology, University of California, San Francisco, San Francisco, California 94143

DANI P. BOLOGNESI
  Duke University Medical Center, Durham, North Carolina 27710

SAMUEL BRODER
  National Cancer Institute, Bethesda, Maryland 20892

IRVIN S. Y. CHEN
  Division of Hematology-Oncology, School of Medicine, University of California, Los Angeles, and Jonsson Comprehensive Cancer Center, Los Angeles, California 90024

JOHN M. COFFIN
  Department of Molecular Biology and Microbiology, Tufts University School of Medicine, Boston, Massachusetts 02111

GEOFFREY M. COOPER
  Division of Molecular Genetics, Dana-Farber Cancer Institute, and Department of Pathology, Harvard Medical School, Boston, Massachusetts 02115

WILLIAM F. DOVE
  McArdle Laboratory for Cancer Research, University of Wisconsin, Madison, Wisconsin 53706

ROBERT C. GALLO
  Laboratory of Tumor Cell Biology, National Cancer Institute, National Institutes of Health, Bethesda, Maryland 20892-4255

THOMAS D. GILMORE
  Biology Department, Boston University, Boston, Massachusetts 02215

TAK WAH MAK
  The Amgen Institute, The Ontario Cancer Institute, and Departments of Medical Biophysics and Immunology, University of Toronto, Toronto, Ontario, Canada M4X 1K9

SUGATA SARKAR
  Biology Department, Boston University, Boston, Massachusetts 02215

SAÏD SIF
  Department of Molecular Biology, Massachusetts General Hospital, Boston, Massachusetts 02114

BILL SUGDEN
McArdle Laboratory for Cancer Research, University of Wisconsin, Madison, Wisconsin 53706

HAROLD VARMUS
National Institutes of Health, Bethesda, Maryland 20892-0148

SANDRA K. WELLER
Department of Microbiology, University of Connecticut Health Center, Farmington, Connecticut 06030

DAVID W. WHITE
Biology Department, Boston University, Boston, Massachusetts 02215

ELIZABETH S. WITHERS-WARD
Department of Microbiology and Immunology, School of Medicine, University of California, Los Angeles, and Jonsson Comprehensive Cancer Center, Los Angeles, California 90024

Preface

On 9 February 1994, Howard M. Temin, age 59, died after a year-and-a-half-long battle with lung cancer. His loss was deeply felt not only by his devoted family but also by his friends and colleagues throughout the scientific community. His scientific contributions have been central to the development of molecular biology, particularly with respect to the major advances that have been made in our understanding of cancer and retroviruses over the last 25 years. In Howard, these accomplishments were joined with a clear moral sense that guided his approach to both science and life, making him a rare combination of a great scientist and a good man.

Howard's early work as a graduate student pioneered the use of cell transformation in culture as an experimental model for studying the induction of cancer by viruses, in particular Rous sarcoma virus. As Howard proceeded with these studies, he made a series of observations, based principally on genetic analysis and experiments with metabolic inhibitors, indicating that the replication of Rous sarcoma virus involved the synthesis of a DNA provirus in infected cells. This proposal—the DNA provirus hypothesis—required the synthesis of DNA from a viral RNA template: a seemingly heretical reversal of the "central dogma" of molecular biology. Throughout the 1960s, the DNA provirus hypothesis was given little credence in the scientific community, instead being met with scorn and even anger by many of Howard's colleagues. Nonetheless, Howard believed his data and continued to test his hypothesis, obtaining increasingly more convincing evidence in its support. These efforts culminated in 1970, with the discovery of a viral enzyme, now known as reverse transcriptase, that synthesized DNA from an RNA template—an unambiguous biochemical demonstration that the "central dogma" could be reversed. Howard concluded his paper with the thought that "this result would have strong implications for theories of viral carcinogenesis and, possibly, for theories of information transfer in other biological systems." As he predicted, the discovery of RNA-directed DNA synthesis has had a fundamental impact on virtually all areas of cell and molecular biology, as well as leading directly to subsequent discoveries of oncogenes and human retroviruses.

The acceptance of the DNA provirus hypothesis was a striking personal vindication for Howard. It was followed by many accolades and honors, including the Nobel prize, which he shared with David Baltimore and Renato Dulbecco in 1975. Howard fully understood the implications of awards like the Nobel, but he refused to let this or other forms of external recognition change his life or priorities. While using his power and influence for causes that he thought appropriate, Howard's top priorities remained his family, his research, and his students. He continued to be actively involved in research and teaching, always interacting closely with the students and postdoctoral fellows in his laboratory. Over the last 20 years, Howard made many important and novel contributions to our understanding of retrovirus replication and the mechanisms responsible for genetic variation of these viruses, including their unique ability to acquire cellular oncogenes. Much of the current interest in retroviruses is related to the AIDS epidemic, and Howard was deeply involved in dealing with human immunodeficiency virus as a critical public health issue.

Howard's impact on science has also been felt through his influence on his students and colleagues. To many of Howard's trainees, he remained a mentor and friend throughout his life. His approach to science was marked by undaunted enthusiasm and absolute integrity. For Howard, it was always the data that counted, not the politics. His purity of purpose made Howard a "scientist's scientist" and a role model for those of us honored by his friendship.

When Howard was diagnosed with lung cancer in the fall of 1992, several of his former trainees wanted to organize a symposium in his honor. Howard, who was undergoing intensive chemotherapy at the time, asked that the meeting be held on the occasion of his 60th birthday, which would have been on 10 December 1994. He then proceeded to plan the symposium, formulating a program that included not only his former trainees and colleagues from the McArdle Laboratory and University of Wisconsin but also colleagues from the broader scientific community who had made key contributions based on Howard's work, such as the discoveries of proto-oncogenes and human retroviruses. Major sessions of the symposium thus focused on three areas of current research whose foundations rest on Howard's discoveries: the molecular biology of cancer, HIV, and other animal viruses. The meeting took place as planned, though sadly as a commemorative symposium held several months after Howard's death. This volume is the proceedings of that meeting, together with reprints of some of Howard's major papers.

A major goal of this book is to integrate Howard Temin's contributions as a scientist and mentor with the body of science he affected. The foreword is a biographical portrayal of Howard by Rayla Greenberg Temin. The volume is then divided into four parts. Part I contains selected reprints of Howard's contributions leading to the discovery of the DNA provirus and discussing the biological implications of reverse transcription. Each of these papers is prefaced by introductions that review the scientific context and impact of his work. The next three parts of the book reflect current research in areas which then grew from Howard's discoveries and to which he, his students, and his colleagues have made seminal contributions. Most of these papers are from participants at the symposium, written not only to discuss current science but also to provide personal reflections on Howard's contributions as a teacher, mentor, and colleague. Three of Howard's more recent papers are also included in the Virology and AIDS sections, illustrating his continuing contributions to these areas.

While we are all mortal, some of us are able to achieve a kind of immortality in the hearts of family and friends. Howard Temin's immortality encompasses a greater sphere, as did the accomplishments of his life. He lives on in the hearts of a larger scientific family, including his trainees and colleagues, as well as in the minds of the many scientists whose work rests on his discoveries. It is our hope that this book will help to commemorate his unique insights and experiments, which have opened so many new doors in biology and medicine.

Geoffrey M. Cooper
Bill Sugden

Foreword

A tenacious memory, a quick perception, and other mental equipments enable one to acquire information, but to be a sage one must be in possession of a great soul.
　　　　　　　—L. Ginsberg, *Students, Scholars, and Saints* (5)

Howard Temin's roots as a sage emerged early in his life, when he set out as a scholarly, public spirited young fellow who spoke up for what he thought was right. His insatiable desire for knowledge led him, in graduate school, to a discovery that ultimately created a new body of knowledge. With penetrating insight he breached the then-current boundaries of thought and pursued what he deeply sensed had to be. Generous of spirit and global in his intellect and interests, Howard enriched the lives of all with whom he interacted. He guided not only his own children but also his trainees with a combination of rigor and loving-kindness. This volume will exemplify how he lives on through all of his students.

YOUTH

Born on 10 December 1934 in Philadelphia, Pa., to Annette and Henry Temin, Howard was the middle of three sons. Michael, the eldest, is an attorney in Philadelphia, and Peter is a professor of economics at the Massachusetts Institute of Technology. Henry, an attorney, and Annette, committed to the educational sector, were both active in many community and civic organizations. Annette kept scrapbooks of her sons during their boyhood years, and some of the information on Howard's youth and pre-University of Wisconsin (UW) period is gleaned from these. Howard attended the Charles W. Henry School at Carpenter and Greene Streets, a small public elementary school founded in 1908. Henry School prided itself on "educating the youth of Germantown for lives of usefulness and service" and as an influence in the neighborhood for beauty, establishing Carpenter's Woods, the first bird sanctuary in that section of the country. A voracious reader throughout his entire life, Howard began during his Henry School days to keep a log and synopses of the numerous books he read, ranging from history, especially battle stories, to science to adventure to science fiction. Elected editor-in-chief of *The Henry Crier,* the student-run newspaper, Howard published editorials that reflected a concern for social justice and for safety. "What does charity mean to you?" he exhorted his classmates (Dec. 19, 1947). "Does it mean going to your mother or father and asking them for some money to give? Remember, your parents make contributions of their own. Or do you refrain from the ice cream cone or movies to contribute . . . You cannot have peace without contentment and happiness. Hungry people are not contented and happy; worried people are not contented and happy; neither are sick people or uncared-for people." Later, in what was perhaps a forerunner of his legendary antismoking speech at the Nobel Banquet, Howard alerted his schoolmates to the dangers of snowball throwing and careless sledding: "In winter great emphasis is placed on winter safety. Everyone knows not to throw snowballs at cars and people. How many people stop throwing snowballs? Very few! How many people know why they should not throw snow-balls? Practically everyone!" (Jan. 26, 1948). A sense of public service was instilled in a concrete way at Howard's Bar Mitzvah on 20 December 1947, when he and his parents decided not to have a celebratory reception. Instead, in view of the desperate post-World War II situation for Jews in Europe, they sent the funds that such a reception would entail to the Displaced Persons camp overseas. These were used to buy clothing for children in the British zone of occupation in Germany who were preparing to emigrate to Palestine under Hadassah's Youth Aliyah program.

As president of his eighth grade class of 13 boys and 11 girls, Howard gave the address at the closing program in January 1948. It was the first of many public orations, including the salutatory address at graduation from Hebrew High School on 16 June 1949, following the full course of after-school study at the Germantown Jewish Center. At Philadelphia's Central High School, Howard was president of the debating club, the beginning of a lifelong skill in vigorous debate on a wide range of topics. In his Valedictory address at high school graduation in June 1951, Howard spoke about the scientific and ethical challenges posed by the recent explosion of the hydrogen bomb on the one hand and the possibility of sending a man to the moon on the other.

JACKSON LABORATORY AND SWARTHMORE

Formative in Howard's scientific career was his participation in 1950 in the Jackson Laboratory summer school for high school students at Bar Harbor, Maine, directed by Frederick R. Avis, known as "Doc." The 10 students selected for the program, then in its second year, paid $20.00 a week for board and lodging and supplied their own dissection kit, microscope, and lab coat. Howard investigated gonadotropic hormones from the pituitary gland as they related to ovulation in the rabbit. In an interview with the *Philadelphia Inquirer,* Howard, age 15, reported, "Most of our time is spent working and that is really my favorite pastime . . . Scientists aren't an eccentric bunch at all, the way they're pictured in some books and movies. They're perfectly normal, hard-working men and women. This is the first time I've come into close contact with working scientists, and I've found them democratic, friendly, and helpful. Now more than ever I plan to go into science myself, tho' I'm not sure yet in what field . . . I'm enjoying the experience tremendously. I've never done any of this kind of science—actual research—before. In fact, some of the work we're doing has never been done by anyone before! The scientists here are really interested in young people, and enthusiastic about this summer program. The other students are swell people. We all have lots of group fun, combined with scientific thrills." (7).

Howard returned to the Jackson Laboratory the next two summers and in 1955 as an instructor. His first scientific paper, in 1953, "Genetic Determinants of Hypoxia-Induced Congenital Anomalies," (6) was from this period. In 1952,

Fred Avis wrote to Annette and Henry: "It is difficult for me to evaluate Howard without using flowery terms. The boy has been a tremendous help to both Patty and myself . . . Howie is unquestionably the finest scientist of the fifty-seven students who have attended the program since its beginning. In my own experience I would throw in another three thousand, whom I have had the pleasure of teaching in my regular school work. I can't help but feel this boy is destined to become a really great man in the field of science."

At Swarthmore College, 1951–1955, Howard majored in biology, but he felt limited by the available courses, all of the classical mode. There was little of molecular biology, a field then newly burgeoning with excitement because of Watson and Crick's 1953 discovery of the structure of DNA. Howard wanted to learn more about this. His notes while at Swarthmore indicate that he contemplated for his career "something which is demonstrable and solely controlled by lawful nature as opposed to sentient life, thus not a doctor, who can do some good, but for whom more depends on luck and the patient." Howard's bent was to "set up and run an experiment. Nature supplies the answers. If unsuccessful, it's up to me," he wrote, "because nature . . . would always answer well-designed experiments." During his senior year, Howard conducted a 7-page, 114-question detailed survey among Swarthmore students concerning their education. The yearbook cites him as Dean Howard, overseer of Martin [biology] Library, and one of the future giants in experimental biology. Because of a disagreement with the zoology honors program and the external examiners (it was said that he "fought his examiners to a draw"), Howard rebelled at graduation by not donning cap and gown nor marching in the processional. He attended the ceremony with his family and heard his name read in absentia. Years later he did return for Swarthmore graduation and, in full regalia, receive an honorary degree.

CAL TECH

In selecting among graduate schools, Howard made a careful comparison of courses at four schools strong in modern biology: University of California, University of Chicago, The California Institute of Technology, and Washington University, seeking a physical biochemistry approach in experimental morphogenesis, physicochemical biology and cytology, virus studies, genetics, and cell organization. He selected Cal Tech, with its extraordinary tradition in genetics and phage work and the addition of molecular biology. Howard lived in the so-called "Prufrock house," with Matt Meselson, Frank Stahl, Jan Drake, and John Cairns. The house itself, over and beyond even that of the Division of Biology, had an exciting and stimulating scientific atmosphere, with daily discussions of now historic experiments, then in the making. The household was organized in a systematic way, and Howard did a good bit of the cooking. When he arrived in Madison, Wis., he brought with him several heavy pieces of cookware from his Cal Tech days, and he continued to enjoy cooking. Howard became part of Max and Manny Delbruck's extended family, joining camping trips to Joshua Tree National Monument and the California Desert, weekends filled with story telling and other diversions.

At Cal Tech, Howard started out in experimental embryology, studying fertilization in the lab of A. Tyler, the last student of T. H. Morgan. But embryological development was in a descriptive phase, and Howard sought a discipline where tools and approaches were available to dissect a biological process more analytically. He joined the laboratory of Renato Dulbecco, who in 1951 had developed the plaque assay for animal viruses. Harry Rubin, the postdoctoral fellow, was extending the cell culture approach to tumor viruses. Plaque production could not be used as an assay here, because tumor viruses did not detectably kill the cells they infected; rather they caused cells to proliferate aberrantly. Howard worked with Harry on inventing a reproducible and workable assay for Rous sarcoma virus, based on this virus' capacity to transform chicken cells in culture. The focus assay, a quantitative method based on the formation of foci of transformed cells, was Howard's first important work, reported in 1958 in *Virology* (13).

Shortly after this publication Howard communicated to his parents (22 January 1959): "I have been working on the nature and origin of variation in cell resistance to infection with Rous sarcoma virus and in cell type after infection. I have established that there are 'genetic factors' in the cell and in the virus for these characters and that the variations arise in a mutation like way, with a high rate. I have not elucidated the nature of the process causing the mutations."

After receiving his Ph.D. degree in 1959, Howard stayed at Cal Tech for another year. In "Cancer and Viruses," a review article published in January 1960 (8), Howard described how his field of research could bring together two seemingly opposed theories about human cancer. One held that cancer results from cellular mutations and the other held that viruses cause human cancer, as they do cause certain animal tumors. He posited that "from a functional point of view, there's little difference between chromosomal or gene mutation and infection by Rous sarcoma virus. Both sets of events cause genetic changes in the cell." He had observed that Rous sarcoma virus was similar to a cellular gene in determining, for example, whether an infected cell is round or fusiform and in being able to mutate. His experiments revealed that there was a very small number of genetic copies of the virus in the cell (less than two) and that these were regularly inherited. In an early formulation of his views, he stated, "the virus, in some structural sense, as well as functional sense becomes a part of the genome of the cell. Probably it does not attach to a chromosome and may not even be in the nucleus, but becomes part of the general apparatus of the cell which controls what a cell is." This was a stepping stone that laid the groundwork for his ultimately revolutionary views on the transfer of genetic instructions.

Throughout 1960 Howard kept his parents regularly informed as he probed the intricacies of the life cycle of Rous sarcoma virus and its replication. On 4 May 1960 he notified them, "I have made some very exciting discoveries. They indicated that the Rous sarcoma virus is not a virus as such but is composed of two separable parts. In a few weeks I'll know the nature of the parts and so can describe what I have found. This will be important for giving a cancer without complicating virus production." On 8 July he communicated, "I have found a new type of virus-cell complex in which the information for the virus is carried in a cell, passed on without virus being produced unless whole new virus is added to the cells. (This is also another way I do not understand). This differs from previously discovered bacterial cases in that the cells are not resistant to superinfecting virus, but produce the original virus after such superinfection. The cells carrying but not producing virus appear to be tumor cells and not to have viral antigen. This rules out most hypotheses for the way in which the virus causes the cell to become malignant." On 18 July he stated, "I am getting ready for a large project to find the site of the virus genome (set of genes)." On 26 July he related, "I've found that cells can be infected by a virus and become tumor cells in the absence of reproduction of that virus (the virus is masked, latent, or hidden). However, addition of a yet undefined second factor causes (induces) the production of the first factor (virus)." Then, on 27 December he communicated, "I am working on which parts of infection are necessary for carcinogenesis: conversion of morphology and/or virus production. I have virus producing, non-converted and non-virus producing, converted. I am studying what is the nature of the virus-cell complex and whether or not these cells are neoplastic." Howard's mother kept carefully typed notes of these communications.

UNIVERSITY OF WISCONSIN: LAUNCHING CAREER AND FAMILY

In the fall of 1960 Howard joined the faculty of the McArdle Laboratory for Cancer Research at the University of Wisconsin, where he worked for the rest of his life. His first grant proposal was nearly rejected because he was considered too young, at age 25, to have an independent laboratory, but Harold Rusch, the chairman of McArdle and highly respected in the cancer field vouched for Howard, and the research was funded. I met Howard in 1961 at the UW while I was a graduate student in the *Drosophila* population genetics group of James F. Crow. Howard was a newly arrived young assistant professor, and I had just returned from a year in Edinburgh at the Institute of Animal Genetics in the laboratory of Charlotte Auerbach. We were married on 27 May 1962. From the beginning, Howard and I shared a life in many realms including the scientific, demonstrated by the fact that we even spent part of our wedding trip at Cold Spring Harbor, N.Y., in June 1962. There Howard presented his paper, "The Separation of Morphological Conversion and Virus Production in Rous Sarcoma Virus Infection" (9). In the summer of 1970 we returned to Cold Spring Harbor as a complete family, with our daughters, Sarah, born in 1966, and Miriam, born in 1968. Howard taught the tumor virology course that summer with Tom Benjamin. This was but one of numerous trips to Cold Spring Harbor for Howard, especially for the annual tumor virus meetings. In 1993 the cover of the abstracts book displayed Howard's 1963 paper (10) and stated "The DNA Provirus Hypothesis on its 30th Anniversary." Unable to attend because of his cancer therapies, Howard was moved by the many messages of affection and hope inscribed on his copy of the book by the participants at the tumor virus meeting that year.

DNA PROVIRUS HYPOTHESIS

Earlier, Howard had proposed that an RNA tumor virus, such as Rous sarcoma virus, causes cancer by actually adding new genes to the cell, in the form of a regularly inherited information-bearing structure that he had named a "provirus." In 1963, solidifying what he had previously intuited, he had evidence for the provirus being DNA. It was known that certain DNA viruses could cause cells to become malignant by attaching viral DNA to chromosomal DNA, incorporating new genetic instructions for how the cell grows (an application of the model for lysogeny by phage). The intriguing puzzle was: how could an RNA virus manage to act like a gene in altering the cell? No known mechanism existed. Howard believed that the RNA virus directs the formation of a DNA copy, which then becomes part of the chromosomal master blueprint of the cell. The method was still mysterious. But in looking for the key to how certain RNA viruses could convert a cell into a tumor cell, Howard discovered a new and fundamental secret in biology: RNA can direct the synthesis of DNA!

The notion of RNA as an informational template for DNA was heretical in that it ran counter to the "central dogma" of molecular biology: that genetic information flows just one way, from DNA into RNA and then from RNA into protein. When Howard presented this unorthodox view at the April 1965 meeting of the American Association for Cancer Research (AACR), the *Philadelphia Inquirer* reported, "It was met with some skepticism by virus experts." A Sunday *New York Times* article on viruses and cancer featured Howard's report at the AACR, noting, "If further research confirms this idea some of the important current concepts of biology may be turned topsy-turvy." (May 2, 1965).

REVERSE TRANSCRIPTASE AND THE NOBEL PRIZE

Howard believed his theory to be correct, and he continued experimenting to rule out alternative explanations. Once he had convinced himself that it had to be true, he was able to pursue the story in spite of widespread incredulity. He knew that it was not enough, however, to satisfy himself; he had to get evidence that would convince the scientific world at large, and so he did. Howard and Dr. Satoshi Mizutani, his postdoctoral fellow, had sought and found in the virions the special viral enzyme that can make a DNA copy from an RNA template, precisely the chemical reaction hypothesized. He announced the discovery on 28 May 1970 at the 10th International Cancer Congress in Houston, Tex., in a 15-min. talk entitled "Role of DNA in the Replication of RNA Viruses." The session was chaired by Howard's good friend and colleague from Prague, Jan Svoboda. The announcement of the enzyme, which Howard preferred to call "RNA-directed DNA polymerase," but which was named reverse transcriptase, was riveting.

On 10 December 1975, the day of his 41st birthday, Howard received the Nobel Prize, with Renato Dulbecco and David Baltimore. In preparation for the festivities, Howard read the biography of Alfred Nobel, the story of the prizes, the history of Sweden, the poetry of Montale (the laureate in Literature that year), and biographical material about former prize winners, for that year was to be the ingathering of all the laureates of years past. During the Nobel ceremony Howard whispered something to King Carl Gustav as he received his medal. He had thought for a long time about how he would use that moment when he had the ear of the king, and it was to be for his children: "Your Majesty," he said, "I would like to request an audience for my daughters." The private audience, with Sarah and Miriam, ages 9 and 7, in long dresses, was granted and captured by the Swedish press. At the banquets and receptions Howard made a point of chatting with, indeed interviewing, each of the biologists, given the unique opportunity of meeting so many historical figures in one place, and he taped these conversations. At the splendid Nobel banquet in Stockholm City Hall, Howard gave the brief response on behalf of the three biology laureates, to the royalty and 1,200 assembled guests, many of whom were smoking. He turned heads when he proclaimed that, as he had come to Stockholm to receive a prize for cancer research, he was "outraged that the one major measure available to prevent much cancer, namely the cessation of smoking, has not been more widely adopted." Prior to our trip, we had carefully studied Muriel Beadle's description of Nobel Week in her book *These Ruins Are Inhabited* (3), about the year when George Beadle received the prize. That led to some overpreparation on our part, such as the purchase of long white gloves and practicing bowing and curtseying, customs which actually had gone out with the former king, Gustav Adolph. On the other hand Muriel Beadle was helpful in warning us that we would be awakened very early on Santa Lucia Day by a beautiful chorus of young women, adorned with crowns of lighted candles, entering our hotel rooms unannounced, singing, with a bevy of photographers in tow. One magical occasion followed another. After the formal ceremonies were over, Howard wanted to view the warship Vasa, the huge wooden ship which sank to the bottom of Stockholm Harbour in 1628 and which was recently dredged up and in the process of being restored. Anders Franzen, the marine archeologist who in 1956 had located the ill-fated but beautifully preserved Vasa and led the salvage project, gave us a private tour. Franzen, who had a mild cold, presented Howard with his book on the history of the Vasa (4), inscribing it, "best wishes from Anders Franzen who wants to see you soon again receiving another Nobel Prize for curing the common cold." Howard relished everything during that special week, so much so that after returning to Madison he organized many evenings at our home to share with friends the pictures and stories of the Nobel Award.

UW—MADISON

In the 18 years following the Nobel Prize, Howard continued to make important contributions to the understanding of retrovirus replication and the mechanisms responsible for their genetic variation. He kept his research active and his laboratory small, closely supervising students and postdocs with great thought and effort. His guiding principles were that you must look for the important problem, you must know intimately the biology of your system, and you must tell a whole story. From the specifics of the experiments, you frame the results into an overview true to the biology that underlies it. Howard treated his students and postdocs as professionals and set high standards for them. First, he introduced them to the fundamentals of virology and cell culture with a simple project. He then advanced the complexity of the question being addressed experimentally as the student grew intellectually. He guided his students and postdocs with a Friday afternoon so-called (by them, not Howard) "confession-session" in which they met one-on-one and discussed experiments in detail. Howard always had the lab on his mind, making notes for experiments in a little notebook that he kept in the inside pocket of his Harris Tweed. Any evening when we had occasion to drive on University Avenue past the McArdle building, Howard would count up to the 5th floor to see how many windows in his lab shone with light and felt satisfied if someone was working! Howard formed close personal ties with the group and kept in touch with his trainees and their careers long after they left the UW. Every July 4th the entire lab gathered for the annual picnic at our home, with Howard leading croquet games and grilling a turkey, carving it with expertise.

Highly organized himself, Howard was adept at creating structure. He established an informal Tuesday lunch with Bill Sugden, Roland Rueckert, and Paul Kaesberg, and he initiated the Tuesday tumor virology group. Monday was lunch with Bernie Weisblum, Masayasu Nomura, Millard Susman, Bill Dove, and others, starting in 1964. By the force of his personality Howard kept these weekly commitments going regularly for decades, whatever else was going on. Mentoring his trainees was a high priority, as was classroom teaching; he taught the graduate Virology course for more than 30 years, first with Dave Pratt and for the past 20 years with Bill McClain. Howard enjoyed explaining science to journalists and to the public, in addition to lecturing. Once, he was invited to Van Hise Elementary School to talk to second graders about the immune system and AIDS. He had a cold at the time and put it to good use to explain about viruses and the body's "police department" that "shoots them down." Howard was charmed by the whimsically illustrated and written reports from the children that showed how well they learned what he taught.

Howard felt a responsibility to apply his science for the public good. He was actively involved in critical public health issues at local, national, and global levels, from antismoking platforms to gene therapy to cancer to AIDS. He used his prestige to try to influence public policy at every opportunity. Howard was consulted on many matters, for the myriad of facts at his command and for his good sense and fairness. The NCAB (National Cancer Advisory Board) said that his wit often relieved the tension in a dispute. Indeed, Howard was often brought in to resolve and bring light to conflictual situations. His thoughtful advice was sought not only in the world of science but also by friends and colleagues as well. "Let's ask Howard," everyone said. With regard to funding science, he urged support for the individual investigator, knowing that "the trail of discovery will be blazed by painstaking individual research . . . Science is not a eureka sort of thing" (2). His own work illustrates that you cannot predict what is going to happen with basic research and a vision but that the combination can be wonderfully illuminating. The practical applications are not necessarily the goal but may be the fallout of the quest for new knowledge. Reverse transcriptase has turned out to be a mainstay for research in gene cloning and biotechnology. Retroviruses are an important vector for gene therapy. None of this was in Howard's mind when he set out at Cal Tech to understand the life cycle of RNA tumor viruses.

OUTSIDE OF MADISON: TRAVEL

Howard made several trips abroad, although in general he limited his travel in duration and frequency. Devoted to family and laboratory, he generally preferred to be at home. He never took a sabbatical leave. For him the day-to-day life was central, but this was punctuated by special journeys. Two of these stand out, one to Russia and one to China. In October 1976 Howard travelled to Moscow and Leningrad at the invitation of The Academy of Medical Sciences of the Soviet Union. He read the history of Russia, he conferred with professional scholars at the UW, and he learned the Cyrillic alphabet, so that, among other things, he could rewrite into Russian the street addresses he had been given in English. In addition to official talks and visits to laboratories, he made plans to seek out refusenicks and dissidents, gathering names from the Federation of American Scientists, the Committee of Concerned Scientists, and other contacts. The refusenicks, so named because they had been refused exit visas to emigrate to Israel, were ousted from their labs as a result of their applying to leave the Soviet Union. Howard, armed with a pocket flashlight to sight apartment numbers in the dark, 35 mm and 16 mm cameras, and a small tape recorder, set forth with a great deal of nervousness, for he was bringing in articles from *Science* and *Nature* as well as Hebrew books, not allowed in the Soviet Union at that time. Following his lectures at various research institutions, he embarked on his unofficial itinerary. The first night he went out to find Evgeny Abezgauz, one of the courageous underground Soviet artists, the designated recipient of a letter, two Hebrew primers, and five Hebrew reading books that Howard had carried, with deep concern, from the United States. Standing outside his hotel, Howard held up a slip of paper with the address in Russian. One of the private cars waiting there took him to the apartment house, where he managed to find the correct door and gain entry. A class was studying the Hebrew language, using a book like the ones he had brought. Abezgauz drew a self-portrait for Howard; it is hanging in our home to this day, a reminder of that grim and perilous era. Some nights Howard skipped dinner in order to go out again on his clandestine missions, so he ate from the stash of nuts and raisins tucked in his suitcase. Once he went to the unofficial seminar of refusenick scientists at the home of Dr. Mark Azbel, a meeting known as the Sunday Seminar. He was invited to come back again on Monday to give a second lecture. When he arrived he learned that most of the members of the seminar had been arrested in the street that morning by the KGB. One of the witnesses described it on tape for Howard. Later that evening Howard visited Andrei Sakharov, the physicist awarded the 1975 Nobel Peace Prize for his work in human rights. While they awaited news together, Sakharov gave Howard messages to take back to the United States. Eventually word came that the refusenick scientists were released, after being held for 12 hours without charge. At almost midnight, Azbel joined them at Sakharov's apartment and made a tape reviewing the events. They asked Howard to publicize the cutting off of the scientific seminar, which was the first time this had ever happened. Howard was never sure whether the arrest had anything to do with his intended lecture, but he was told that in the Soviet system even giving a lecture to the Sunday Seminar could be defined as espionage. His hosts were concerned that the tapes would be confiscated and told Howard to keep them on his person. Fortunately he was not inspected. When he got back, Howard notified the newspapers, and he shared the tapes and his detailed, chilling diary (11). He pressed the cause of the dissidents to several political and scientific groups and gave information on Soviet prisoners to groups such as Amnesty International. After a trip testifying to his own committment and courage, Howard told the local newspaper, "one can't do too much for such a cause." (1).

In April 1977 Howard travelled to the People's Republic of China as part of an economic trade mission led by Governor Lucey of Wisconsin, one of a handful of missions allowed into the PRC at that time. The mission consisted mostly of business leaders and included U.S. District Judge John Reynolds, Jr., and the UW Regent, Dr. Ben Lawton of the Marshfield Clinic. Along with the official tour and sightseeing, Howard visited and lectured at many biological and medical institutes. He provided his appraisal and technical summaries of education and science, including research on the cancers common in China—liver, esophageal, and nasopharyngeal—and their epidemiology. Friends and scholars read his informal journal and reports (12) of the country and the science with interest; these write-ups were among the earliest of that genre available after the opening of China. Howard loved Chinese landscape painting and brought back several beautiful scrolls to remind him of that trip. He held soirees at our home after Russia and China, relating his observations and insights about his extraordinary experiences.

Whenever he travelled, near and far, Howard found a way to get the most out of the trip and to contribute. In Cleveland he pursued his interest in Asian art and he arranged to see not only the museum but also the extensive collection not on display, and he held animated discussions with the curator on the various art periods. In Israel and elsewhere, including at national parks in the United States, he challenged the guides on archeological treks or nature hikes with questions, always wanting to know more. In Thailand in 1984 he tried to convince the health officials that an AIDS epidemic was coming to their country. In 1990 during a family trip in Kenya, Howard visited medical laboratories in Nairobi to discuss the AIDS epidemic.

Our last international journey together was in March 1992 to the Andalusia region in Spain. We were called back just upon arriving in Granada, because Howard's mother, Annette, had died. Howard was close to his mother and was especially attentive to her in the final difficult years of her long and fruitful life, phoning on a daily basis, helping in the supervision of her care as she became frail, and scheduling frequent trips to the East Coast so he could visit her in Philadelphia. Six months after her death he was diagnosed with the very disease to which he had dedicated his life's research.

THE CRUCIBLE OF CANCER

When diagnosed with cancer, Howard had to figure out how to distill the core essence of his life. Fearing that he might not attain the full measure of his days, he asked for and helped plan an ingathering of his trainees for his 60th birthday, to have been 10 December 1994. This book emanates from that wish. The way Howard summoned his amazing courage and fierce determination to live with hope and spirit in the face of grave illness was inspiring to all those around him. He came to know firsthand the realities of the clinical problems of cancer and the limits of current knowledge; trying experimental regimens gave him some measure of hope. Even while undergoing procedures, his curiosity extended to details concerning, for instance, the physics of the radiation equipment. He willingly answered questions from medical personnel about science, especially about AIDS, and he was thanked by the staff more than once for what he had taught them during a sojourn in the hospital. He went to work as regularly as possible within the confines and arduous demands of his numerous therapies and he continued to be productive throughout, even while at the edge of his life. He published 15 papers during this period, he helped students and postdocs prepare talks for the tumor virus meetings at Cold Spring Harbor, he applied for patents on a model for an AIDS vaccine, and he participated in NCAB meetings in Washington, D.C., in February 1993 and by speaker phone the next fall. In October 1993, Howard presented his current research on the genetics of retroviruses at a conference in Chicago on the 40th anniversary of the discovery of the structure of DNA. This was to be his last public address.

Outside of science and medical involvements, Howard lived beyond limits as well. He tried to make each day as full as it could be and to hold on to what is precious. It did not take cancer to teach these principles; these were extensions of the way Howard always lived, making the best use of time. He said that we cannot choose what happens to us, only how we respond. Family was uppermost; he stayed in close touch with Sarah and Miriam throughout, visiting parks and gardens with them during their frequent sojourns in Madison. He read diligently, especially interested now in biographies and autobiographies, of Darwin (commenting on Darwin's lengthy illness), Fisher, Wright, Muller, Jacob, Feinman, and McClintock. He was captivated by the Patrick O'Brian series, recommended to him by one of his oldest friends, Dr. Beppino Giovanella. Visits with his brothers, Michael and Peter, and with out-of-town colleagues often took place in his garden, where he liked to sit or stroll.

In December 1993 Howard gave several lectures on retroviruses in the virology course. He supervised the completion of the Ph.D. theses of his three final graduate students, Tom Mitchell; Shaio-Lan Yang, whom he escorted at her graduation in December 1993; and Gary Pulsinelli, whose final oral exam Howard conducted on 10 January 1994. This was but one month before he died. Howard had just returned from San Diego where our family gathered for a winter vacation. Howard loved the sunshine, the eucalyptus trees, the multicolored flowers in bloom, the orange trees, the ocean's roar. He read four books that week and had lists of more that he was going to read.

A BALANCED LIFE

Howard's zest for life was boundless. Nothing was carried out halfway, whether mapping his garden with notebooks and computer records of the succession of blooms or delving into the history and culture of China. He loved nature and the outdoors, watching the migrating ducks and geese in University Bay and Goose Pond in spring and autumn. For years he made the long daily walk from home to office, past Picnic Point, even in the coldest Wisconsin winters, and in the other seasons, he biked (with helmet, of course). He was a familiar figure at concert series and was fond of opera; Beethoven's *Fidelio* was a favorite. He enjoyed films, viewing *Alexander Nevsky* and *Casablanca* each many times over. Always enjoying a hearty laugh, he especially liked early comedies starring Harold Lloyd and, later, Alec Guiness. Howard brought a spiritual dimension into his life by attending Sabbath services regularly, which provided for him a weekly sanctuary in time and place.

In a full and busy life, Howard knew what was important, and above all it was his family. A caring, devoted, and loyal family man, Howard cherished his daughters, following their lives and peregrinations with enthusiam and interest. Sarah, in Berkeley, Calif., and Miriam, in Cambridge, Mass., are both involved in health-related activities. Sarah volunteers with a nonprofit organization for women with cancer, while studying biology, chemistry, and art. Miriam is at the Harvard School of Public Health. From them, Howard learned about feminist issues and he became a strong supporter of women in science. In June 1992, when Howard was awarded the National Medal of Science at the White House by President Bush, Sarah and Miriam wanted to use the occasion to make a statement. They brought a long length of red ribbon and from it cut and fashioned ten loops, one for each of us in attendance, including Howard, to wear in support of those striken with AIDS and for the President to view as he walked into the Rose Garden. Howard took pride in the strong social consciences of his daughters. His legacy will be carried on by them and by his numerous "scientific children" represented in this volume.

Rayla Greenberg Temin

REFERENCES

1. **Anonymous.** 1976. Temin loses his audience to KGB. *Capital Times*, Madison, Wis. (16 December).
2. **Anonymous.** 1977. In hospitals and labs, 9 researchers wage war on the elusive enemy, cancer. *People*, vol. 8, no. 7.
3. **Beadle, M.** 1961. *These Ruins Are Inhabited.* Doubleday, New York.
4. **Franzen, A.** 1974. *The Warship Vasa.* Norstedts, Stockholm.
5. **Ginsberg, L.** 1928. *Students, Scholars, and Saints*, p. 242. Jewish Publication Society of America, Philadelphia.
6. **Ingalls, T. H., F. R. Avis, F. J. Curley, and H. M. Temin.** 1953. Genetic determinants of hypoxia-induced congential anomalies. *J. Hered.* 5:185–194.
7. **Segal, A.** 1950. Teen-age scientist in the making. *Today* Magazine, *The Philadelphia Inquirer* (27 August).
8. **Temin, H. M.** 1960. Cancer and viruses. *Engineering and Science,* (Cal. Tech) 23(4):21–24.

9. **Temin, H. M.** 1962. The separation of morphological conversion and virus production in Rous sarcoma virus infection. *Cold Spring Harbor Symp. Quant. Biol.* **27:**404–417.

10. **Temin, H. M.** 1963. The effects of actinomycin D on growth of Rous sarcoma virus in vitro. *Virology* **20:**577–582.

11. **Temin, H. M.** 1977. Moscow diary. *The Sciences* (NY) **17**(4):26–27.

12. **Temin, H. M.** 1978. China diary. *The Sciences* (NY) **18**(4):27–28.

13. **Temin, H. M., and H. Rubin.** 1958. Characteristics of an assay for Rous sarcoma virus and Rous sarcoma cells in tissue culture. *Virology* **6:**669–688.

(*Left*) Howard Temin at 8th grade closing program, Charles W. Henry School, Philadelphia, Pa., February 1948.

(*Below*) Howard Temin with other high school students at Jackson Laboratory summer research program, Bar Harbor, Maine. Summer 1950.

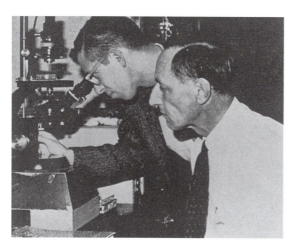

(*Left*) Howard Temin with Boris Ephrussi, Cal Tech, 1960.

(*Right*) The faculty of the McArdle Laboratory for Cancer Research, taken in August 1964 in front of the original McArdle Laboratory building, just prior to the move to the present location. Front row: Van R. Potter, Charles Heidelberger, Harold P. Rusch (Director), James A. Miller, and Elizabeth C. Miller. Back row: Howard M. Temin, Waclaw Szybalski, Gerald C. Mueller, Henry C. Pitot, and Roswell K. Boutwell.

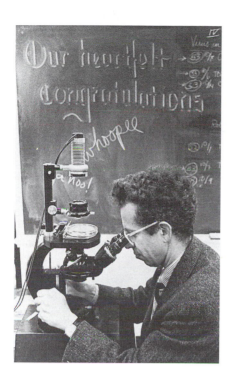

(*Left*) Howard Temin in his laboratory on the day the Nobel Prize was announced, 16 October 1975.

(*Below*) Howard Temin at the banquet in the Stadtshalle in Stockholm, Sweden, after the Nobel Award Ceremony, 10 December 1975, giving the response on behalf of the laureates in Physiology or Medicine, during which Howard made his famous anti-smoking plea.

(*Right*) The Temin family at a private audience with the king of Sweden, in the Stadtshalle of Stockholm after the Nobel Banquet, 10 December 1975. Sarah, Howard, King Carl Gustav, Rayla, Miriam.

(*Left*) Rayla and Howard Temin at a banquet for Laureates given by the Swedish King Carl Gustav at the Palace, Stockholm, 11 December 1975, the day following the Nobel Award Ceremony.

(*Right*) Ph.D. candidate Shiaolan Yang at graduation, escorted by her professor, Howard Temin. 19 December 1993, University of Wisconsin, Madison.

(*Left*) Some members of the Temin research group from the 1970s. From left: Norman Dulak, C. Yong Kang, Theresa Lewandowski, Geoffrey Cooper, Susan Hellenbrand, John Coffin, Narayana Battula, Victoria Kassner, David Boettiger, Atsumi Kato, Sandra Weller, and Cestmir Altaner.

(*Right*) A few of the Temin group from the 1980s. From left: Antonito Panganiban, Thomas Gilmore, Michael Emerman, Bakary Sylla, Susan Hellenbrand, Virginia Goiffon, Irvin Chen.

(*Left*) Recent members of the Temin group, from the 1990s. Front row: Cestmir Altaner, Rebecca Wisniewski, Kathleen Boris-Lawrie, Gary Pulsinelli, Shiaolan Yang, and Wei-Shau Hu. Back row: Jeffrey Jones, Eric Freed, Louis Mansky, Kouichi Iwasaki, Vinay Pathak, Dawn Burns, and Thomas Mitchell.

The Temin family—Sarah, Rayla, Miriam, and Howard—June 1992 on the lawn of the White House when Howard received the National Medal of Science.

Howard Temin in his cell culture room in the McArdle laboratory.

Acknowledgments

At the end of the symposium, the participants gathered to thank Dr. Ilse Reigel for all the care-filled help she provided to them while they were at McArdle. She helped edit their papers and their dissertations in addition to giving them her personal support, and she also helped organize the symposium. We too thank Dr. Reigel for helping assemble and edit this volume, as she did for so many of Howard Temin's manuscripts.

PART I

SELECTED PAPERS OF
HOWARD M. TEMIN

The DNA Provirus: Howard Temin's Scientific Legacy
Edited by G. M. Cooper, R. Greenberg Temin, and B. Sugden
© 1995 American Society for Microbiology, Washington, DC 20005

Chapter 1

Reprint of Temin and Rubin's 1958 Paper Describing the Focus Assay for Transformation by Rous Sarcoma Virus

In 1958, Howard Temin (then a graduate student) and Harry Rubin (a postdoctoral fellow) developed the first quantitative assay for the transformation of cultured cells by tumor viruses. Prior to their work, Rous sarcoma virus (isolated in 1911 by Peyton Rous) had become the first generally accepted tumor virus because of its ability to cause sarcomas following inoculation of chickens. Such in vivo experiments, however, did not provide an experimental system suitable for studies of either virus replication or the molecular events responsible for the conversion of a normal cell to a cancer cell. In order to address these fundamental questions of tumor virology, an in vitro assay that allowed virus-induced transformation to be analyzed and quantitated under controlled and reproducible experimental conditions was needed. This was provided by Temin and Rubin, whose assay for cell transformation by Rous sarcoma virus remains the standard assay used in studies of transformation induced not only by tumor viruses but also by chemicals, radiation, and isolated oncogenes of both viral and cellular origins.

The focus assay developed by Temin and Rubin is based on the morphological changes characteristic of transformed cells. Chick embryo fibroblasts grown in tissue culture are infected with Rous sarcoma virus, and morphologically altered cells appear within 1 or 2 days. These cells continue to grow, giving rise to characteristic foci of transformed cells within a week after virus infection. The assay is not only rapid but also readily quantitated by counting the number of foci in a culture dish. Temin and Rubin demonstrated that the number of foci obtained was proportional to the concentration of virus used to initially infect the cells, thus providing a quantitative basis for the assay as well as showing that cell transformation results from infection by a single virus particle.

This simple assay for cell transformation in vitro has formed the basis for subsequent studies of tumor viruses and oncogenes. By permitting quantitative studies of virus replication, it provided the fundamental tool needed for analysis of the virus life cycle, including Temin's subsequent experiments from which the provirus hypothesis was derived. The availability of the focus assay also allowed other workers to isolate Rous sarcoma mutants defective in transformation, leading to the identification of the *src* oncogene. The focus assay has similarly been applicable to studies not only of other tumor viruses but also of cellular oncogenes. For example, mutated oncogenes in human cancers were first identified by using

3

the focus assay to detect their biological activities. Since a variety of oncogenes have now been isolated from both viral and cellular genomes, the focus assay continues to be used to analyze their roles in cell transformation. Developed over 35 years ago, Howard Temin's assay for Rous sarcoma virus thus remains a standard tool in virology and molecular oncology.

Geoffrey M. Cooper

Copyright 1958, Academic Press

VIROLOGY **6**, 669–688 (1958)

Characteristics of an Assay for Rous Sarcoma Virus and Rous Sarcoma Cells in Tissue Culture[1]

HOWARD M. TEMIN[2] AND HARRY RUBIN

Division of Biology, California Institute of Technology, Pasadena, California

Accepted July 30, 1958

An accurate tissue culture assay for Rous sarcoma virus (RSV) and Rous sarcoma cells is described. The Rous sarcoma virus changes a chick fibroblast into a morphologically new and stable cell type with the same chromosomal complement as ordinary chick embryo cells. One virus particle is enough to change one cell, but at any one time 90% of the cells in a culture are not affected by RSV. The cellular resistance is the same in a clonal population. The physiological state of the cell is of some importance in deciding whether or not it is competent to be infected by RSV but so far attempts to infect all the cells in a chick embryo culture by altering the physiological condition have failed.

INTRODUCTION

The customary technique for assaying the Rous sarcoma virus (RSV) has been the production of tumors in chickens (Rous, 1911; Bryan, 1946). More recently, the virus has been assayed by infecting the chorioallantoic membrane of the developing chicken embryo (Keogh, 1938; Rubin, 1955; Prince, 1957). Although these techniques are satisfactory for many experiments, they do not provide the accuracy nor the ease of manipulation required for a diversified study of virus-host interactions at the level of the individual cell. An assay technique comparable to the methods used for assaying bacteriophages and cytopathogenic animal viruses has now been developed and is described in the present paper. The assay depends upon the morphological change caused by RSV in chick embryo cells.

The first clear demonstration that chick embryo cells grown in tissue culture were changed into Rous sarcoma cells by infection with RSV was

[1] Supported by grants from the American Cancer Society and the United States Public Health Service, grant E 1531.

[2] United States Public Health Service predoctoral fellow.

669

given by Halberstaedter *et al.* (1941), who exposed chick fibroblast cultures to pieces of intensely irradiated Rous sarcoma cultures and observed gross changes in the culture and cytological changes similar to those present in cultures of Rous sarcoma cells derived from a chicken sarcoma. Lo and associates (1955) succeeded in changing fibroblasts *in vitro* with partially purified RSV. They also showed that the culture containing changed cells produced virus for several months. Following this lead Manaker and Groupé (1956) succeeded in producing discrete foci of changed cells upon treatment of monolayer cultures of chick embryo cells with RSV. In the range tested the number of foci per culture was proportional to the concentration of virus added.

In the assay method to be described here an agar overlay has been introduced to decrease the chance of formation of additional foci by virus liberated from infected cells and by detachment and reattachment of infected cells. In addition, a technique will be described for determining the number of infected cells in a mixed population. The properties and nature of the changed cells are also discussed.

MATERIALS AND METHODS

Solutions and Media

"Standard medium":
    8 parts Eagle's medium (Eagle, 1955) with double concentration of amino acids and vitamins and 4.2 g $NaHCO_3$ per liter.
    1 part Difco Bacto-Tryptose phosphate.
    1 part serum (usually 0.8 calf and 0.2 chicken serum).
Trypsin:
    0.25 % Bacto-Trypsin in tris buffer.
Eagle's minus:
    Eagle's medium in which the salts of Ca^{++} and Mg^{++} have been omitted.

Incubation

All cultures were made in petri dishes. They were incubated at 37° in a water-saturated atmosphere. The pH was regulated at about 7.3 by a bicarbonate buffer controlled by CO_2 injected into the incubator.

Examination of Cultures

Cultures were examined at 25- and 100-fold magnifications of a Zeiss plankton (inverted) microscope.

Chick Embryo Cultures

Primary cultures of 9- to 13-day-old chicken embryo cells were made by the method of Dulbecco as modified by Rubin (1957). Four million cells in 10 ml of standard medium were placed in a 100-mm petri dish. After 3–5 days of incubation secondary cultures were made. The primary cultures were washed twice with Eagle's minus, and 2 ml of 0.05% trypsin in Eagle's minus were added. After 15 minutes' incubation 2 ml of standard medium were added to stop the action of the trypsin. The cells were pipetted twice and centrifuged at 1500 rpm for 1 minute. The pellet was resuspended in standard medium and the cells counted. Two or three hundred thousand cells in 3 or 5 ml of standard medium were placed in 50-mm petri dishes. After incubation overnight these plates were used for the assay of Rous sarcoma virus and Rous sarcoma cells. The cultures consisted chiefly of fibroblasts.

Assay for Virus

The secondary culture was washed once with Eagle's medium and then virus in a volume varying from 0.1 to 0.8 ml was added. After an adsorption period of from 10 minutes to 1 hour 5 ml of standard medium containing 0.6% agar was added. (The concentration of agar is not critical for focus development.) Three days later the culture was fed by adding 2 ml of agar medium on top of the first agar layer. Five to seven days after infection 3 ml of a 1/20,000 solution of neutral red in Eagle's medium was added to allow easier counting of the Rous sarcoma foci. After 2 hours' incubation the neutral red was removed and the plate was placed on a piece of glass with a rectangular grid of 2-mm squares and was scanned for foci at a 25-fold magnification of the inverted microscope.

Assay for Rous Sarcoma Cells or Foci Formers

The cells to be assayed were trypsinized, counted, diluted, and added to a secondary plate. After incubation for 8–16 hours to allow attachment of the cells, the medium was removed and standard medium with 0.6% agar was added. Three days later the culture was fed by adding 2 ml of agar medium on top of the first agar layer. After 5–7 days 3 ml of a 1/20,000 solution of neutral red in Eagle's medium was added and the plate scanned for foci at a 25-fold magnification.

Virus Stocks

The original virus used in this work was generously supplied by Dr. Bryan of the National Cancer Institute. This preparation contained about 2×10^6 focus-forming units (FFU) per milliliter and was derived from tumors in chickens. Tissue culture virus was obtained by infecting secondary chick embryo cultures with 10^5 FFU of RSV and growing the cultures in standard medium without chicken serum for 2 weeks with four transfers. The supernatants were then collected. They contained between 10^5 and 10^6 FFU per milliliter.

Chromosome Counts

Cells were transferred to dishes containing sterile 22×22-mm No. 0 cover slips and incubated in standard medium. Eighteen hours later the cover slips were put through the following modification of the procedure of Hsu and co-workers (1957):

1 part Eagle's:9 parts distilled water	15 minutes
Dry rapidly under blower	
100% Methyl alcohol	1 hour
Aceto-orcein	2 hours
Ethyl alcohol (absolute)	Rinse
Ethyl alcohol (absolute)	10 minutes
Mount in Euparal	

Drawings of well-spread chromosome figures were made with a camera lucida at 1350 magnification. The actual counts were made from the camera lucida drawings.

Cloning Technique

Cells were cloned by the feeder layer technique of Puck *et al.* (1957). Twenty-four hours before cloning cultures containing 5×10^4 HeLa cells or chick fibroblasts were X-rayed with a dose of 5000 r. The cells to be cloned were dispersed with trypsin, counted, and 2×10^3 cells added to each plate. At sixteen hours 5 ml of standard medium containing 0.3% agar was added. At 8 days and every 4 days thereafter the cultures were fed by adding agar medium on top of the first agar layer. When the clones were large enough to be picked, the agar was removed and trypsin added. Under a dissecting microscope the clones were picked with a micropipette and transferred to a new dish.

X-Irradiation

Cultures were X-irradiated from a Machlett OEG.60 tube with tungsten target and beryllium end window operated at 50 kilovolts and

30 milliamperes. The outlet of the tube was covered with a 0.38-mm Al filter. The irradiation of the cells was carried out at a distance of 6.6 cm from the target in a covered petri dish at an intensity of 2640 roentgens per minute.

Ultraviolet Irradiation

Ultraviolet irradiation was given from a Westinghouse germicidal lamp. The virus was in 1 ml of medium, 11¾ inches from the lamp.

Abbreviations

RSV—Rous sarcoma virus
FFU—Focus-forming unit.

EXPERIMENTAL

Morphological Changes in Chick Embryo Cells following Infection with Rous Sarcoma Virus

One or two days after addition of Rous sarcoma virus to a chick embryo culture rounded refractile cells appear. As single cells these can be confused with cells in mitosis. However, these cells and their progeny retain their changed appearance throughout interphase, giving rise after 2 or 3 days to groups of rounded refractile cells. These groups or foci can easily be distinguished from the background of fibroblasts (Plate I, Fig. A). The cells differ markedly in their colonial characteristics from the fibroblasts. They grow in a grapelike cluster only loosely attached to one another and to the glass. Often a focus becomes multilayered and the rounded cells migrate out on top of the other cells (Plate I, Fig. B). The cell sheet may also tear and retract in the region of a focus leaving a hole surrounded by the round, refractile cells (Plate I, Fig. C). As long as the culture is fed every 3 days, the number of cells in a focus doubles about every 18 hours (Fig. 1). If the culture is not fed, the cells in a focus lose their refractility and become difficult to distinguish from the background of fibroblasts. Upon addition of fresh medium the cells in a focus may regain their refractility.

Mechanism of Growth of a Focus

There are two processes which could be concerned in the growth of a focus: division of the cells in a focus; and infection of normal cells by the virus released from cells in the focus. Several experiments were carried out to determine whether both of these processes actually occur. Cells

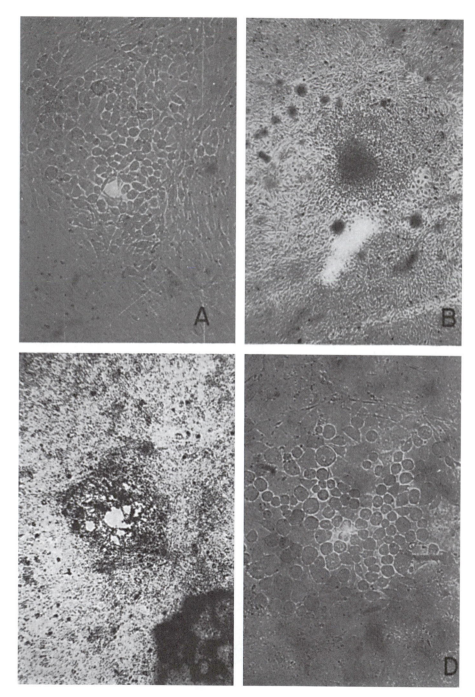

PLATE I

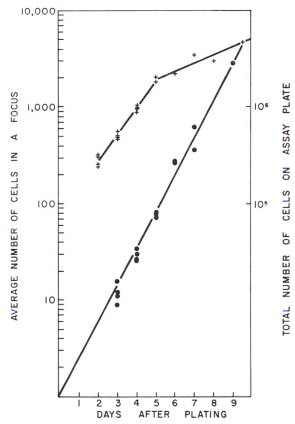

FIG. 1. Increase in number of cells in a focus. Rous cells were plated on assay plates. Twenty hours later agar was added to the plates. Every day the number of cells in 10–20 foci was counted. The total number of cells on the plate was also counted: first, by directly counting a fraction of the cells on the plate; and after the third day, by removing the agar, trypsinizing the cells, and counting in a hemocytometer. ● = Average numer of cells in a focus. + = Number of cells on an assay plate.

FIG. A. Focus of Rous sarcoma cells formed on a secondary chick embryo culture. Unstained; magnification: × 80.

FIG. B. Focus of Rous sarcoma cells formed on a secondary chick embryo culture. Lightly stained with neutral red; magnification: × 20.

FIG. C. Focus of Rous sarcoma cells formed on a secondary chick embryo culture. Heavily stained with neutral red; magnification: × 20.

FIG. D. Focus of Rous sarcoma cells formed on a clonal population of chick embryo fibroblasts. Unstained; magnification: × 80.

which had been infected more than 2 days before with RSV were plated in the following ways:

1. The cells were plated as described under methods for assay of Rous sarcoma cells.
2. The cells were plated as in (1) but on a secondary culture of cells 1/40 as sensitive to virus infection as the strain regularly used.
3. The cells were plated as in (1), but the secondary culture was X-rayed previously.
4. The cells were X-rayed to prevent their multiplication and plated as in (1).

Under (1) both reinfection and division can occur. Under (2) reinfection is 1/40 as likely and under (3) impossible. (Previous work had shown that X-rayed fibroblasts cannot be infected by RSV.) Under (4) infection would be necessary for initiation of a focus.

The relative number of foci resulting from the four methods of plating are presented in Table 1. The table shows that the number of foci was the same in all cases. It is concluded that a focus can be formed by dividing Rous sarcoma cells in the absence of reinfection but that nondividing Rous sarcoma cells can initiate formation of a focus, showing that virus released from a cell on the glass can infect the surrounding cells. Since there was little difference in the size of the foci obtained in the different ways even when reinfection was impossible, (3), the major role in the development of the foci must be assigned to division of changed cells.

TABLE 1
INITIATION OF A FOCUS[a]

Number of foci after standard plating of cells (1)	Treatment	Number of foci after designated treatment
105	2	111
23, 44, 79	3	40, 46, 66
197, 212	4	180, 204

[a] Rous sarcoma cells were plated in the following ways: (1) The cells were plated on a secondary plate as described under methods for assaying Rous sarcoma cells. (2) The cells were plated on a secondary plate composed of chick embryo cells $\frac{1}{40}$ as sensitive to RSV as the strain usually used. (3) The cells were plated on a X-rayed feeder layer. (4) The Rous sarcoma cells were X-rayed and then plated as in (1). The number of foci resulting by each procedure are given in relation to (1).

Properties and Nature of Rous Sarcoma Cells

After the agar overlay is removed from an infected culture, the cells can be transferred and grown in fluid medium. Chick serum is omitted to allow further infection. After 2 weeks the culture becomes composed chiefly of round, refractile cells. These cells are producing RSV. They resemble the basophilic, round cells of Doljanski and Tenenbaum (1943).

The Rous sarcoma cell is a stable cell type which can be grown in pure culture. Rous sarcoma cells were cloned and grown for twenty generations in the absence of ordinary chick embryo cells. After more prolonged culture the cells became giant and diverse in appearance.

To check whether or not the morphological change was associated with a gross change in the chromosome complement, chromosome counts were done on Rous sarcoma cells and on chick embryo fibroblasts. The number of chromosomes found in less than 1-month-old Rous sarcoma cells and in normal fibroblasts was the same (Fig. 2). The mode was in the middle seventies for both types of cells and the spread was similar.

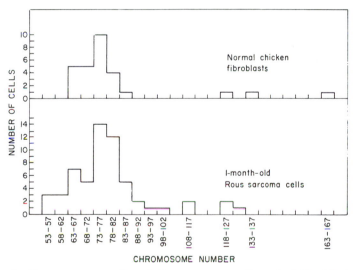

Fig. 2. Chromosome number of Rous sarcoma cells and of fibroblasts. Rous cells or normal cells were transferred to cover slips and fixed and stained by a modification of the procedure of Hsu. Drawings of well-spread chromosome figures were made with a camera lucida at 1350 magnifications, and counts were made from the drawings.

After more prolonged culture of the Rous sarcoma cells at a time when many giants were present, over half of the cells were polyploid.

The nature of the chick embryo cell which is changed by RSV to a Rous sarcoma cell has been controversial since Carrel's claim that only macrophages could be infected (Carrel, 1924, 1925). Several workers have disputed this contention, claiming the fibroblast as the cell of origin of the Rous sarcoma cell (quoted in Lo *et al.*, 1955). To determine unequivocally whether the fibroblast, which is the major cell type of the secondary chick embryo culture, is indeed the cell which is changed by infection with RSV, clonal lines of fibroblasts were exposed to the virus. The foci which resulted were similar in number and appearance to those found on the usual assay plates (Plate I, D).

Quantitative Features of the Assay

Adsorption volume. The following experiment was carried out to determine the relative efficiency of virus adsorption as a function of the volume of virus inoculum placed on the secondary culture. Virus was diluted in Eagle's medium. A constant amount of virus was suspended in different volumes of fluid and placed on separate assay plates. After 30 minutes the supernatant was removed and an agar overlay added. The

TABLE 2

ADSORPTION VOLUME[a]

Volume of original virus suspension	Volume of inoculum	Number of foci
1/5000 ml	0.1 ml	177
	.2	200
	.4	135
	.8	55
1/20,000	.1	41
	.2	40
	.4	23
	.8	11
1/80,000	.1	11
	.2	9
	.4	6
	.8	3

[a] Virus was diluted in Eagle's medium. A constant amount of virus was suspended in different volumes of fluid and placed on separate assay plates. After 30 minutes the supernatant was removed and an agar overlay added.

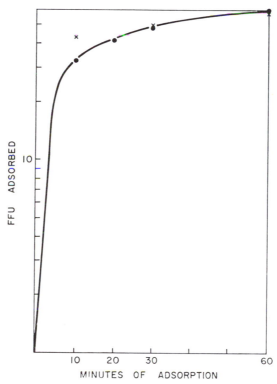

FIG. 3. Kinetics of adsorption. An inoculum of virus containing 50 FFU in 0.2 ml was added to each of several secondary cultures. After adsorption for varying lengths of time the supernatant was removed and an agar overlay added. Two separate experiments are shown. Each point is the average of two assay plates.

results are in Table 2. The number of foci per plate for the same amount of virus doubles as the volume of the inoculum was lowered from 0.8 ml to 0.4 ml and from 0.4 ml to 0.2 ml. No difference was found between 0.2 ml and 0.1 ml. Two-tenths of a milliliter is therefore used as a standard inoculum.

Kinetics of adsorption. An inoculum of virus containing 50 FFU in 0.2 ml was added to each of several cultures. After adsorption for varying lengths of time the supernatant was removed and an agar overlay added. The number of foci per plate is plotted as a function of the time of adsorption (Fig. 3). For practical reasons 30 minutes has been used as a standard adsorption time.

Relationship between virus concentration and number of foci. A series of
twofold dilutions between 1/80 and 1/2560 of a virus stock containing
3.5×10^5 FFU/ml were placed on secondary culture for 30 minutes and
agar added. The number of foci per plate was found to be proportional
to the virus concentration over a thousand fold range (Fig. 4). This
linear dose response confirms earlier results, using *in vivo* assay techniques, which showed that one virus particle is sufficient to cause an
infection (Keogh, 1938; Rubin, 1955; Prince, 1957).

Relationship between virus concentration and number of infected cells. A
series of twofold dilutions from undiluted virus to 1/512 of a virus stock

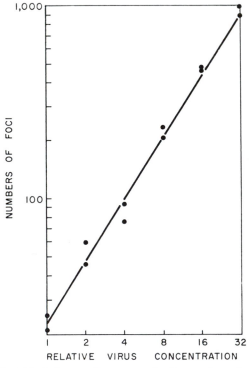

Fig. 4. Relationship between virus concentration and number of foci. A series
of twofold virus dilutions between 1/80 and 1/2560 of a virus stock containing
7×10^4 FFU/ml were placed on secondary cultures for 30 minutes and agar added.
To enable counting of as many as 1000 foci per plate the cultures were counted
on the fifth day after infection.

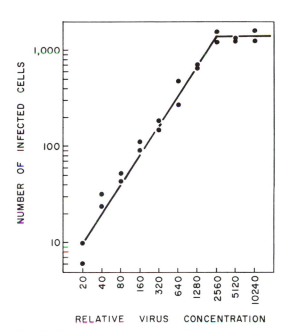

FIG. 5. Relationship between virus concentration and number of infected cells. Virus dilutions from undiluted to 1/512 of a stock containing 10^5 FFU/ml were placed on secondary cultures for 30 minutes. The cultures were washed, trypsinized, and various dilutions of the infected cell suspension were placed on each of two secondary cultures. Sixteen hours later agar was added. The plates were counted on the sixth day after infection.

containing 10^5 FFU/ml were placed on secondary cultures for 30 minutes. The cultures were washed, trypsinized, and various dilutions of the infected cell suspension were placed on each of two secondary cultures. Sixteen hours later agar was added. The number of foci per plate on the sixth day is plotted in Fig. 5. The number of Rous sarcoma cells per plate was a linear function of virus concentration until 0.5 % of the cells were infected. Further increases in virus concentration failed to increase the fraction of cells infected. In other experiments the maximal fraction of cells infected usually varied from 1.0 to 10 %.

The efficiency of the technique for detecting the number of infected cells by plating them in suspension as focus-formers was tested in the following manner. Three plates were infected with 100–600 FFU of RSV. After a 30-minute adsorption period the plates were washed. Agar

was added to two plates and they were incubated as in the standard
virus assay. The number of foci showed the number of cells infected. The
cells on the third plate were suspended with trypsin and plated on
secondary cultures. The results of several experiments are presented in
Table 3. The efficiency of the technique to determine the number of
infected cells by plating them in suspension varied between 10 and 100 %
when compared with the number of foci in the direct assay. This variation
in efficiency of plating is partly responsible for the fluctuation in the
maximum fraction of cells infected measured above, but does not explain
why no more than 10 % of the cells can be infected in *any* experiment.

Factors limiting the fraction of cells which can be infected at any given time.
There are several possibilities to explain the restriction in the proportion
of cells infected at a given time by RSV. The culture may be heterogeneous with regard to the genetic or embryological origin of the cells; with
regard to their physiological state, either in general or in reference to the
division cycle; or there may be interfering substances, such as an inactive
virus, in the inoculum. Investigation of these possibilities was carried out.

1. Variation in genetic or embryological origin of the cells: Two lines
of clonal fibroblasts were obtained. They were infected with serial threefold dilutions at high virus concentrations and the number of infected
cells determined as above. The results of such an experiment are presented in Fig. 6. A plateau in the number of infected cells was found at

TABLE 3

EFFICIENCY OF PLATING INFECTED CELLS IN SUSPENSION[a]

Number of foci resulting		Efficiency of plating infected cells in suspension (%)
From direct assay	From plating of infected cell in suspensions	
555	48, 54	9.2
280, 210	80	33
135, 123	55, 75	50
855, 815	645, 510	69
101, 102	84, 104	93
595, 610	653, 662	109

[a] Three plates were infected with from 100 to 600 FFU of RSV. After a 30-
minute adsorption period the plates were washed. Agar was added to two plates
and they were incubated as in the standard virus assay. The number of foci showed
the number of cells infected. The cells on the third plate were suspended with
trypsin and plated at a dilution of 1/1 to 1/4. A number of different experiments
are listed.

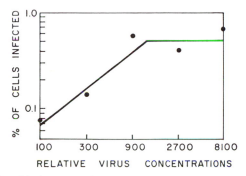

FIG. 6. Relationship between virus concentration and number of infected cells in a cloned population. Cultures of 2×10^5 cells were made from a clonal line of chick embryo fibroblasts. They were infected with serial threefold dilutions at high virus concentrations and the number of infected cells determined.

the same level as was found with secondary chick embryo cultures of varied cell origin. Since the clonal population was from a single cell, the plateau could not be due to heterogeneity in the origin of the cell.

2. Variation in physiological susceptibility: The fraction of infectible cells on secondary cultures was determined under different conditions as a function of the time which had elapsed since the plates were made. One group of cultures was kept in standard conditions; a second was placed in Eagle's medium without glutamine or glucose for 24 hours (Vogt, personal communication); a third group was kept at 14° for 24 hours in an attempt at obtaining synchronization (Chèvremont et al., 1957). At 3-hour intervals after the cells were returned to standard conditions, different cultures were infected with 10^5 FFU of RSV and the number of infected cells determined as above. The results presented in Fig. 7 show there is a considerable fluctuation in the fraction of susceptible cells in a culture with time and under different environmental conditions.

3. Interference: The possible occurrence of interference by inactive virus particles was tested by the following experiment. Stock virus was killed by irradiation with ultraviolet-light or by incubation at 37°. Two-tenths milliliter of killed virus was placed on a secondary plate. After 30 minutes the supernatant was removed and live virus added as in the direct assay. The results presented in Table 4 indicate that such killed virus preparations interfere with infection by live virus.

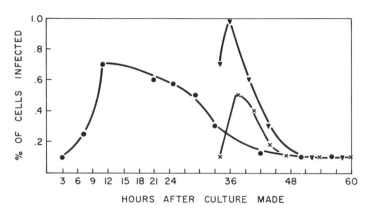

FIG. 7. Physiological state and per cent cells infected. Secondary cultures were made at time 0. One group of cultures was incubated in the usual manner; a second group was placed in Eagle's medium without glucose or glutamine for 24 hours and then standard medium was replaced; a third group in sealed dishes was placed at 14° for 24 hours and then returned to 37°. At the indicated times the cultures were infected with a high concentration of virus. After adsorption they were trypsinized, the cells counted, and 10^4 cells placed on each of two secondary cultures. Each point represents an average of two plates. ● = Control; × = starved for 24 hours; ▼ = 14° for 24 hours.

TABLE 4

INTERFERENCE[a]

Treatment	Untreated virus	Treated virus (diluted 1:10)	Treated virus (undiluted) followed by untreated
UV 10 min	533	11	121, 120
37° 72 hours	1005, 1100	0, 1	85, 110

[a] Stock virus was inactivated by UV irradiation or by prolonged incubation at 37°C. The undiluted inactivated virus was placed on secondary cultures for thirty minutes. The supernatants were then removed and live virus at a concentration of about 1/500 was added.

Since the standard virus stocks would be expected to contain significant amounts of 37°-inactivated virus, such virus could cause autointerference and restrict the number of cells infected. To minimize the amount of inactivated virus, fluid was collected from a culture of Rous sarcoma cells 6 hours after washing. The maximum number of infectible cells, using this virus stock for infection, was only slightly higher than when

TABLE 5
INFECTION WITH 6-HOUR VIRUS[a]

Virus stock (hours after washing)	Titer (FFU/ml)	Per cent cells infected
6	5×10^5	8.5
48	2.5×10^5	5.6

[a] Fluid was collected from a culture of Rous sarcoma cells 6 hours after washing. Secondary cultures were infected with this virus stock. The percentage of cells infected is compared with the percentage of cells infected using a standard virus stock.

standard virus was used for infection (Table 5). This finding indicates that autointerference by 37°-inactivated virus plays a minor role in limiting the number of cells infected.

4. Repeated infection with RSV: The results in the previous sections suggest that the physiological condition of the cells is an important factor in limiting their susceptibility to infection. An experiment was carried out to determine whether or not the physiological variation of chick embryo cells to RSV infection was due to a transient period of competence in individual cells. Several cultures were infected with a saturating dose (10^5 FFU) of RSV for 1 hour. Supernatants were removed from the cultures and medium added. Each hour thereafter for 4 hours, virus was added to a different culture for 1 hour to determine whether any more cells had become susceptible in the interval. One culture was not re-exposed to virus. At the end of 5 hours the number of infected cells in each culture was determined. The results of two such experiments are given in Fig. 8. The number of infected cells in a culture could be increased beyond the usual plateau level by re-exposing the cells to virus at a time after the initial exposure. The maximum increase was obtained when the interval between the first and second inocula was 2–3 hours; longer intervals did not increase it further, suggesting that cells undergo a 2–3-hour transient period of competence.

Relative efficiency of in vitro assay and chorioallantoic membrane assay. Comparative titrations of a virus stock sent by Bryan were carried out on the chorioallantoic membrane of the developing chick embryo and by the standard tissue culture technique. The titer on various occasions varied from 1 to 4×10^6 FFU/ml in both assays.

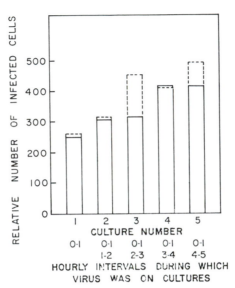

FIG. 8. Repeated infection with RSV. Five cultures were infected with 10^5 FFU of RSV for 1 hour. Supernatants were removed from the cultures and medium added. Each hour thereafter for 4 hours virus was added to one culture for 1 hour. One culture was not re-treated. At the end of 5 hours the number of infected cells in each culture was determined. Each line is the average of two plates. Two separate experiments are shown.

DISCUSSION

Use of the tissue culture assay for RSV which has been described here has shown that at a given time over 90 % of the cells in at chick embryo culture are not suscetpible to being infected by RSV. This resistance is not due to the diverse genetic or embryological nature of the cells. There is some interference by material in the virus stock which influences the number of infectible cells but does not appear to be the major limiting factor. The physiological state of the cells does affect their susceptibility and experiments with repeated infection suggest that the cells pass through a transient state of competence to infection and change, similar to that described by Hotchkiss (1954) for transformation of pneumococci. Among temperate phage, the physiological condition of the cells has also been shown to play an important role in determining the proportion of cells which will undergo the lysogenic cycle (Bertani, 1957).

Attempts to synchronize cultures of chick embryo cells with prolonged starvation has given results indicative of cyclical changes in competence (Temin, unpublished). At the time of maximum competence, however, 90 % of the cells still are not infected.

It is possible that methods of synchronization which require less drastic treatment of the cells will enable all cells to be infected with RSV simultaneously.

REFERENCES

BERTANI, L. E. (1957). The effect of the inhibition of protein synthesis on the establishment of lysogeny. *Virology* **4,** 53–71.

BRYAN, W. R. (1946). Quantitative studies on the latent period of tumors induced with subcutaneous injections of the agent of chicken tumor. I. 1. Curve relating dosage of agent and chicken response. *J. Natl. Cancer Inst.* **6,** 225–237.

CARREL, A. (1924). Action de l'extrait filtre du sarcome de Rous sur les macrophages du sang. *Compt. rend. soc. biol.* **91,** 1069–1071.

CARREL, A. (1925). Effets de l'extrait de sarcomes fusocellulaires sur des cultures pures de fibroblastes. *Compt. rend. soc. biol.* **92,** 477–479.

CHÈVREMONT, S., FIRKET, H., CHÈVREMONT, M., and FREDERIC, J. (1957). Contribution à l'étude de la préparation à la mitose. *Acta Anat.* **30,** 175–193.

DOLJANSKI, L., and TENENBAUM, E. (1943). Studies on Rous sarcoma cells cultivated *in vitro.* 2. Morphologic properties of Rous cells. *Cancer Research* **3,** 585–603.

EAGLE, H. (1955). The specific amino acid requirements of a human carcinoma (cell strain HeLa) in tissue culture. *J. Exptl. Med.* **102,** 37–48.

HALBERSTAEDTER, L., DOLJANSKI, L., and TENENBAUM, E. (1941). Experiments on the cancerization of cells *in vitro* by means of Rous sarcoma agent. *Brit. J. Exptl. Pathol.* **22,** 179–187.

HOTCHKISS, R. D. (1954). Cyclical behaviour in pneumococcal growth and transformability occasioned by environmental changes. *Proc. Natl. Acad. Sci. U. S.* **40,** 49–54.

HSU, T. C., POMERAT, C. M., and MOORHEAD, P. S. (1957). Mammalian chromosomes *in vitro.* VIII. Heteroploid transformation in the human cell strain Majes. *J. Natl. Cancer Inst.* **19,** 867–872.

KEOGH, E. V. (1938). Ectodermal lesions produced by the virus of Rous sarcoma. *Brit. J. Exptl. Pathol.* **19,** 1–8.

LO, W. H. Y., GEY, G. O., and SHAPRAS, P. (1955). The cytopathogenic effect of the Rous sarcoma virus on chicken fibroblasts in tissue cultures. *Bull. Johns Hopkins Hosp.* **97,** 248–256.

MANAKER, R. A., and GROUPÉ, V. (1956). Discrete foci of altered chicken embryo cells associated with Rous sarcoma virus in tissue culture. *Virology* **2,** 838–840.

PRINCE, A. M. (1957). Quantitative studies in Rous sarcoma virus. I. The titration of Rous sarcoma virus on the chorioallantoic membrane of the chick embryo. *J. Natl. Cancer Inst.* **20,** 147–158.

PUCK, T. T., CIECIURA, S. J., and FISHER, H. W. (1957). Clonal growth *in vitro* of human cells with fibroblastic morphology. *J. Exptl. Med.* **106,** 145–158.

ROUS, P. (1911). A sarcoma of the fowl transmissible by an agent separable from the tumor cells. *J. Exptl. Med.* **13,** 397–411.

RUBIN, H. (1955). Quantitative relations between causative virus and cell in the Rous No. 1 chicken sarcoma. *Virology* **1,** 445–473.

RUBIN, H. (1957). Interactions between Newcastle disease virus (NDV), antibody and cell. *Virology* **4,** 533–562.

The DNA Provirus: Howard Temin's Scientific Legacy
Edited by G. M. Cooper, R. Greenberg Temin, and B. Sugden
© 1995 American Society for Microbiology, Washington, DC 20005

Chapter 2

Reprint of Temin's 1964 Paper Proposing the DNA Provirus Hypothesis

The development of the focus assay provided the means for studying both cell transformation and the replication of Rous sarcoma virus (RSV). As Temin proceeded with these studies, he made a series of unexpected observations indicating that the replication of RSV was fundamentally different from that of other RNA-containing viruses. On the basis of these findings, he proposed the DNA provirus hypothesis, which stated that the viral RNA was copied into DNA in infected cells. The resulting DNA provirus was then replicated and stably inherited as part of the infected cell's genome.

The DNA provirus hypothesis was based on several different types of experimental evidence, which are presented in this 1964 paper. First, studies of cell transformation using mutants of RSV indicated that the morphology of transformed cells was determined by genetic information from the virus. Importantly, this information was regularly transmitted to daughter cells following cell division, even in the absence of virus replication or a large number of viral genomes in infected cells. On the basis of these observations, Temin proposed that the viral genome was present in infected cells in a stably inherited form, which he called a provirus.

Evidence that the provirus is DNA was then derived from experiments with metabolic inhibitors. First, actinomycin D, which inhibits DNA-directed RNA synthesis, was found to inhibit virus production by RSV-infected cells. Second, inhibitors of DNA synthesis were found to inhibit early stages of cell infection by RSV. It thus appeared that DNA synthesis was required early in infection and that DNA-directed RNA synthesis was subsequently needed for production of progeny virus. These findings led to the proposal that the provirus was a DNA copy of the viral RNA genome. Temin sought further evidence for this proposal by using nucleic acid hybridization to detect viral sequences in infected-cell DNA, but the sensitivities of the available techniques were limited, and the data were unconvincing.

The DNA provirus hypothesis was thus proposed on the basis of genetic experiments and the effects of metabolic inhibitors. It was a radical proposal that ran counter to the generally accepted ''central dogma'' of molecular biology: that there is a unidirectional flow of genetic information from DNA to RNA to protein. In this setting, Temin's hypothesis that RSV replicates by transfer of information from RNA to DNA not only failed to win acceptance from the scientific community but was met with general derision. Howard Temin's proposal of the DNA provirus and his continuing pursuit of experimental evidence to support the proposal are thus a testimony to his intellectual courage as well as to his scientific originality.

Geoffrey M. Cooper

Nature of the Provirus of Rous Sarcoma [1,2]

HOWARD M. TEMIN, *Ph.D., McArdle Laboratory, University of Wisconsin, Madison, Wisconsin*

MANY viruses have been shown to cause both tumors in animals and parallel morphological alterations in cell culture systems. The Rous sarcoma viruses are unique among tumor viruses. Their action is efficient and direct, both *in vivo* and *in vitro*. With the other tumor viruses, the infection of a sensitive cell does not always bring about a rapid morphological alteration of the infected cell. Baluda (*1*) has shown that, although the virus of avian myeloblastosis will infect many cells, only some of these become tumor cells. Weisberg (*2*) has shown that the alteration of mouse embryo cells by polyoma virus is a multistage process taking some weeks to complete. For the action of avian myeloblastosis virus, a special cell type is needed; for the action of polyoma virus, a series of events subsequent to infection is needed.

The efficiency and directness of action of the Rous sarcoma viruses are due to their method of infection which adds to the cell genome sufficient information to make it a tumor cell. The nature of this information, the provirus (*3, 4*), will be the subject of this discussion.

The analysis of the nature of the provirus has been facilitated by the availability of genetically marked strains of virus controlling different morphological alterations of infected cells. Two strains of virus, *morph*[r] and *morph*[f], mutationally related, have been isolated by cloning from the standard Bryan strain of Rous sarcoma virus (RSV) (*5*). (The two strains differ in host range from the standard Bryan RSV.)

The different responses of the same cell type to infection by these two virus strains gave the first evidence that the morphological alterations in the cells were not a nonspecific response to virus production. Both fibroblasts (*5*) and pigmented cells from the iris epithelium (*6*) responded to infection with alterations that were specific for the strain of virus used for infection rather than for the cell type infected. The separation of virus production and morphological alteration was

[1] Presented at the International Conference on Avian Tumor Viruses, Duke University, Durham, N.C., March 31 to April 3, 1964.
[2] This investigation was supported by grants C-5250 and CA-07175 from the National Cancer Institute, National Institutes of Health, Public Health Service.

557

achieved with the discovery of converted, non-virus-producing (CNVP) cells (7). Infection with *morph^r* virus produced CNVP cells of *r* type, and infection with *morph^f* virus produced CNVP cells of *f* type (7). These results suggested that there were separate determinants in the virus for virus production and for alteration in morphology of infected cells.

The morphological alterations produced in infected cells by genetic determinants in the virus have been called conversion (5). Superinfection experiments (8) implicate the continued role of the virus genome in the maintenance of the specific morphological alteration of an infected cell.

When cells infected with one strain of RSV are exposed to a second strain of RSV several alternative responses are observed. The most common is immunity (7–9). However, the immunity is not absolute. The second virus can cause infection of a previously infected cell and replacement of the original virus. When the two viruses differ in the type of conversion they cause, there is then a change in cell morphology from that controlled by the original virus to that controlled by the superinfecting virus. In such experiments, a cell and its descendant cells have different morphologies due to being infected with different viruses.

The second virus can also cause infection of a previously infected cell with the appearance of a doubly infected cell. Doubly infected cells were found having *r* character and releasing both *morph^r* and *morph^f* RSV. They were also found having *f* character and releasing both *morph^r* and *morph^f* RSV. Therefore, either virus may be dominant in the conversion process; the factors determining which will be dominant are not clear (10).

The existence of doubly infected cells enables us to ask whether the two viruses are inherited regularly together or whether they are inherited irregularly so that they segregate to separate progeny cells. Cells were infected with *morph^f* RSV, cloned, at 3 days exposed to *morph^r* RSV, at 9 days subcloned, and at 17 days virus yields from single cells determined in microdrops. The totals of results from 5 experiments in which superinfection led only to addition, and not to replacement, are presented in text-figure 1. Only 2 of 67 cells which produced *morph^f* virus in the microdrop were producing *morph^f* virus alone. The other 65 produced both *morph^f* and *morph^r* virus. Since about 15 generations had occurred since superinfection, and since there is no selection in cell culture against cells carrying *morph^f* virus, these results show that the 2 viruses are inherited regularly together and that segregation to separate progeny cells is a rare event.

A similar question can be asked about CNVP cells. Is the virus information transmitted to all progeny cells? The data in table 1 indicate that there is also regular inheritance in the absence of virus production.

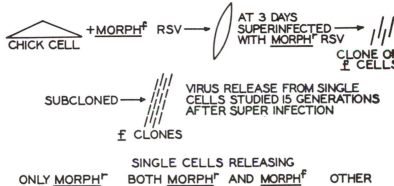

TEXT-FIGURE 1.—Inheritance of virus in doubly infected cells. This experiment is taken from (8).

These experiments indicate that at division of an infected cell both progeny cells receive the same viral information. This regular inheritance could be due to random inheritance of a large number of proviruses (or vegetative viruses) or ordered inheritance of a small number of proviruses.

There are two independent ways to determine the genetic characters carried by the proviruses of Rous cells. In one, the morphology of the cells is examined; in the other, the genotype of the released virus is studied. The morphology of a cell will presumably reflect the genotype of a majority of the proviruses present. The frequency of mutation of this morphology will be proportional to the rate of mutation per provirus raised to a power of the number of proviruses. The frequency of mutation of the genotype of the released virus will be proportional to the rate of mutation per provirus.

Text-figure 2 presents data comparing the frequency of mutation of cell morphology and genotype of released virus. The results indicate that the frequencies of mutation are similar in both cases and, therefore, that the number of proviruses per cell is one or two. The existence of such a small number of proviruses is inconsistent with regular inheritance due to random processes and suggests that the proviruses segregate to daughter cells in an ordered fashion.

The mutation rate per provirus per cell division appears to be high— of the order of 1 in 10^4 cell divisions. This high rate can be due to the character, the morphology of infected cells, being controlled by any of several genes on the provirus, or to some mutation-generating process.

The experiments summarized here establish that genetically the provirus is a regularly inherited information-bearing structure. Other experiments are required to determine its location in the cell and its chemical composition.

AVIAN TUMOR VIRUSES

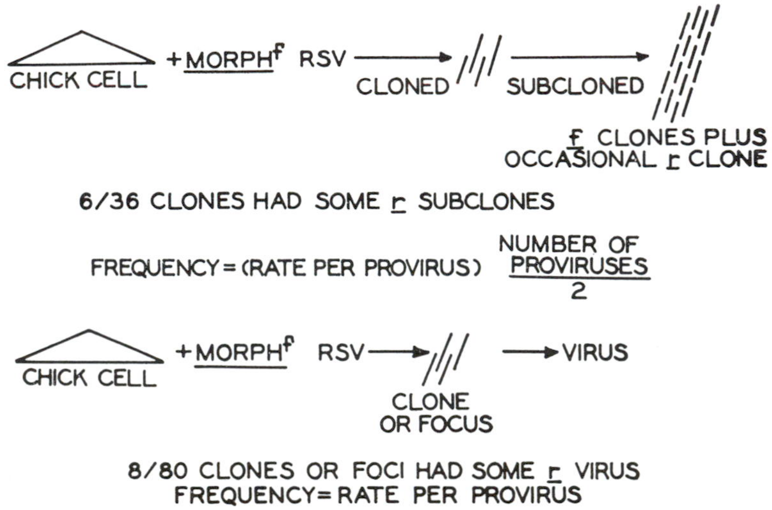

TEXT-FIGURE 2.—Mutation frequency of cell morphology and viral genotype. This experiment is taken from (8).

The isolated virions of the strains of avian tumor viruses that have been studied appear to contain single-stranded RNA (11–13). The strain of RSV used by us also seems to contain single-stranded RNA (14).

To tell that the information in the virion responsible for conversion is RNA, further observations are needed. Chick cells infected with Fujinami virus produce little free virus. However, all cells converted by Fujinami virus are capable of infecting sensitive chick cells when the cells converted by Fujinami virus are plated as infective centers on the sensitive chick cells. This process of infection by the converted cells is inhibited by the presence of 2 µg per ml RNase in the medium (15). Control experiments indicate that the RNase must be destroying a subviral infectious agent. Therefore, in this case the converting agent is RNA.

Numerous studies (14, 16–18) with compounds which specifically block DNA synthesis have shown that stoppage of DNA synthesis does not inhibit production of converting virus by Rous cells. These results cannot be interpreted as due to the presence of a large pool of preformed viral nucleic acid (3). Therefore, all of these studies agree that the virus information is in RNA when it is extracellular.

However, the effect of actinomycin D on production of virus and nucleic acid by Rous cells appears to be inconsistent with the provirus being RNA. When virus-producing Rous cells were exposed to 0.1 µg per ml of actinomycin, virus production was inhibited (text-fig. 3). This inhibition was removed once the antibiotic was removed. Similar results, inhibition of RSV production by actinomycin, have been reported by others (18–21).

NATIONAL CANCER INSTITUTE MONOGRAPH NO. 17

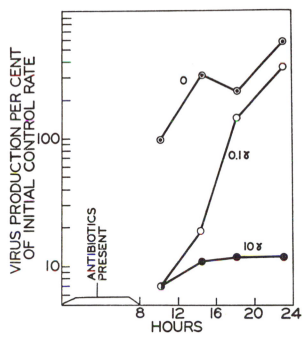

TEXT-FIGURE 3.—Effect of actinomycin on virus production by Rous cells. Experiment taken from (3).

To determine whether this effect of actinomycin is to inhibit production of nucleic acid for the virion, the extent of inhibition by actinomycin of RNA synthesis in parallel cultures of uninfected and infected cells was studied. No difference in the amount of synthesis resistant to actinomycin treatment in the Rous and the uninfected cells was found. However, this method is not too sensitive.

To check this result, an experiment was performed to see if Rous cells treated with actinomycin produce any high molecular weight RNA. Parallel cultures of normal and Rous cells were exposed to actinomycin. They were labeled with tritiated uridine for 4 hours, the cells harvested, RNA extracted, and then studied by sucrose gradient centrifugation. There was no difference between the pattern of labeling in the Rous and in the uninfected cells (text-fig. 4).

Two additional figures are needed to tell the significance of this result. Cells treated with 4 μg per ml of actinomycin incorporate about 0.6 percent of the label incorporated into untreated cells, and about 0.1 percent of the uridine label in untreated cells goes into virus. Therefore, about 15 percent of the counts in the experiment of text-figure 4 would have been due to virus material if it had been made. Since no difference was found between the pattern of label of the Rous and the normal cells, it is suggested that no virus nucleic acid was synthesized.

The results of these experiments with actinomycin suggested that, although the virion of RSV contained RNA, the provirus consisted

AVIAN TUMOR VIRUSES

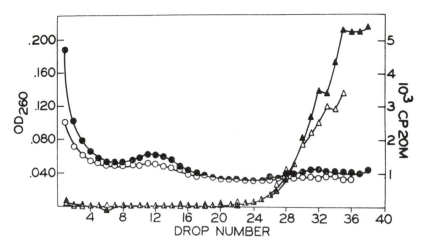

TEXT-FIGURE 4.—Effect of actinomycin on type of RNA made in Rous and uninfected
cells.   Tertiary cultures containing 10^7 chick fibroblasts or Rous cells were prepared.
Five ml of medium containing 4 μg per ml actinomycin was added.   After 10
minutes, 10μc of tritiated uridine was added.   Four hours later the cells were
harvested, RNA was extracted with Duponol and hot phenol, and they were
centrifuged at 4° C for 8 hours at 35,000 rpm in the SW 39 head of the Spinco
Model L ultracentrifuge.   Drops were collected, optical density at 260 mμ was
measured, and the radioactivity in each fraction was determined.   Experiment
taken from (*14*).

O = Optical density at 260 mμ of RNA of fibroblasts.
● = Optical density at 260 mμ RNA of Rous cells.
△ = Counts of RNA of fibroblasts.
▲ = Counts of RNA of Rous cells.

of DNA.   To see if synthesis of new DNA at infection was required
for virus production, experiments were carried out on the effects of
amethopterin and 5-fluorodeoxyuridine (FUDR) on infection of chick
cells by RSV.

Secondary cultures of chick cells were infected with *morph*r RSV at a
multiplicity of infection (moi) of about 1.   Four hours after infection,
the medium on the cultures was replaced by medium made with dialyzed
serum.   In two cultures the medium contained no amethopterin; in two
more, 8×10^{-7} M amethopterin, 10^{-5} M adenosine, and all amino acids;
in two more, 8×10^{-7} M amethopterin, 10^{-5} M adenosine, all amino
acids, and 10^{-5} M thymidine.   Twenty hours after infection these media
were replaced by normal medium and virus production was studied.

The results of such an experiment (text-fig. 5) show that amethopterin
caused a pronounced inhibition of virus infection, which was prevented
by thymidine.

A similar inhibition of virus infection, which also was prevented by
the presence of thymidine, was found when FUDR was used to inhibit
DNA synthesis at infection.   Cultures containing 8×10^5 chick cells
were made.   From $7\frac{1}{2}$ hours before infection with a moi of 1 with

NATIONAL CANCER INSTITUTE MONOGRAPH NO. 17

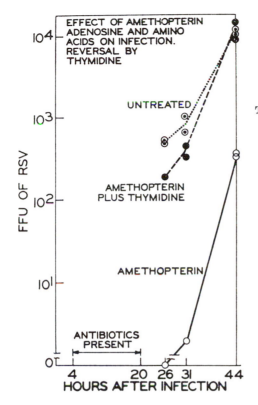

EFFECT OF AMETHOPTERIN ADENOSINE AND AMINO ACIDS ON INFECTION. REVERSAL BY THYMIDINE

UNTREATED

AMETHOPTERIN PLUS THYMIDINE

AMETHOPTERIN

ANTIBIOTICS PRESENT

FFU OF RSV

HOURS AFTER INFECTION

TEXT-FIGURE 5.—Effect of amethopterin on infection. Secondary cultures containing 8×10^5 chick embryo cells were infected at a multiplicity of infection of about 1 with *morph^r* RSV. After 40 minutes' incubation regular medium was added. Four to 20 hours after infection, medium made with dialyzed serum and containing the indicated other components was added. The number of focus-forming units (FFU) of RSV produced in the indicated intervals was determined. Experiment taken from (*14*).

TABLE 1.—Inheritance of provirus in converted non-virus-producing cells *

Experiment 1	Number of cells plated	Number of foci appearing
a	5	31
b	5	4
c	700	275

Experiment 2	Number of subclones	Number of virus-producing subclones
a	12	11
b	8	8
c	9	9

*Lines of CNVP cells carrying *morph^r* RSV were grown for at least 15 generations. In experiment 1, they were exposed to a multiplicity of infection (moi) of about 4 of *morph^f* RSV. A known number of cells were tested for ability to produce *morph^r* RSV. In experiment 2, the CNVP cells were subcloned, the subclones were exposed to a moi of about 4 of *morph^f* RSV and the cells tested for release of *morph^r* RSV. Data taken from (7).

morph^r RSV to 24 hours after infection, 2 cultures each were exposed to medium made with dialyzed serum containing no FUDR; 0.5 μg per ml FUDR; 0.5 μg per ml FUDR; and 10^{-5} M uridine; 0.5 μg per ml FUDR and 10^{-5} M thymidine; or 0.5 μg per ml FUDR, 10^{-5} M thymi-

AVIAN TUMOR VIRUSES

dine, and 10^{-5} M uridine. Virus was harvested at 41 hours after infection. The results (table 2) show that lack of thymidine caused a large inhibition of infection.

TABLE 2.—Effect of FUDR on infection*

	Cultures treated with				
FUDR 0.5 μg per ml	0	+	+	+	+
Thymidine 10^{-5}M	0	0	0	+	+
Uridine 10^{-5}M	0	0	+	0	+
FFU's of RSV	90, 150	30, 30	45, 45	240, 465	660, 945

*Cultures containing 8×10^5 chick cells were exposed to medium, made with dialyzed serum, and containing the indicated compounds from 7½ hours before to 24 hours after infection with a multiplicity of infection (moi) 1 of *morph*r RSV. Virus yields were harvested 41 hours after infection.

The experiments with amethopterin and FUDR indicate that there is a requirement for thymidine, presumably for DNA synthesis, during infection of chick embryo cells by RSV. These results support the results with actinomycin and suggest that the provirus is DNA.

The presence of provirus in the cell is the essential difference between Rous cells and normal chick cells (7, 22). If, as the above experiments have suggested, the provirus is DNA, Rous cells should have an extra piece of DNA which is not found in normal cells. Such DNA might be found by use of the hybridization techniques developed for the study of complementarity in RNA (23, 24). These experiments were carried out.

Cells from parallel cultures of chick cells and Rous cells were collected and frozen (25). Nuclei were isolated from about 2×10^8 cells, with 80 percent glycerol as an isolation medium (26). DNA was then extracted from the isolated nuclei. The DNA was trapped in agar by the methods of Bolton and McCarthy (24).

Virus was labeled with tritiated uridine and purified by two cycles of potassium tartrate density gradient centrifugation. Control experiments (14) have shown that only 15 percent of the label in this preparation is from cellular material. RNA was isolated from the purified virus by extraction with Duponol and hot phenol (27) after the virus was shaken with ether. Rous cells were labeled with tritiated uridine (28) for 19 hours and harvested 24 hours after the label was removed. RNA was extracted as for the experiment in text-figure 4. It contained 3×10^3 counts per minute (cpm) per μg RNA.

These RNA's were then tested for complementarity to the DNA from Rous and from normal cells by the methods of Bolton and McCarthy (24). For each run 10 fractions were collected at 60° C after washing with 0.3 M salt, and 5 fractions were collected at 75° C after washing with 0.0015 M salt. The samples were counted in a liquid scintillation counter for 20 to 60 minutes. The cpm in the last 3 fractions were taken as background. The percent of the counts above background in each fraction was then calculated. The results (table 3) are given for

TABLE 3.—Hybridization of labeled RNA from Rous cells and Rous virus with DNA from Rous cells and uninfected cells*

DNA from	RNA from	Percent of counts in fraction						
		6	7	8	9	10	11	12
Uninfected cells	Rous cells	0.68	0.47	0.60	0.44	0.58	0.23	−0.12
				Avg. of 6-10 = 0.55				
Uninfected cells	Rous cells	0.65	0.73	0.58	0.23	0.52	0.40	−0.01
				Avg. of 6-10 = 0.54				
Rous cells	Rous cells	0.26	0.17	0.22	0.24	0.19	3.76	2.46
				Avg. of 6-10 = 0.22				
Rous cells	Rous cells	1.03	0.06	0.31	0.21	0.24	0.91	0.00
				Avg. of 7-10 = 0.20				
Uninfected cells	RSV	−0.23	−0.34	0.85	−0.34	1.93	−0.79	0.28
				Avg. of 6-10 = 0.37				
Uninfected cells	RSV	1.1	0.59	0.98	0.90	1.23	0.55	0.06
				Avg. of 6-10 = 0.23				
Rous cells	RSV	1.74	0.75	1.01	0.23	1.22	2.56	0.33
				Avg. of 7-10 = 0.80				
Rous cells	RSV	2.1	0.32	0.32	−0.19	0.32	3.30	0.78
				Avg. of 7-10 = 0.19				

*DNA was extracted from isolated nuclei of parallel Rous and normal cells. RNA labeled with tritiated uridine was isolated from Rous cells and purified RSV. Hybridization was carried out by the agar-column method of Bolton and McCarthy (24). The column of DNA from uninfected cells contained approximately 15 percent more DNA than that from Rous cells. The specific activity of the nucleic acid from RSV was approximately three times that from Rous cells.

the last 5 fractions collected at 60° C with high salt, nonspecifically bound RNA, and the first 2 fractions collected at 75° C with low salt, specifically bound RNA. The average value for the last of the non-specifically bound fractions is also given.

It appears that there is a new DNA in Rous cells, not present in normal cells, and that this DNA is complementary to RNA isolated from purified RSV. Further experiments using another technique of hybridization have been published (*42*).

Infection of a normal cell by RSV and conversion of that cell into a tumor cell appears to involve the addition of new nuclear DNA genes to the cell genome. This hypothesis gives no information about the biochemical mechanisms responsible for conversion. Presumably the genes on the provirus somehow cause the conversion.

The genes responsible for production of virus or of virus-related material are not active in CNVP cells (*29, 30*) (table 4). These genes may be present in the provirus with their function repressed (*7, 22*) or absent (*30*).

However, certain other metabolic properties of Rous cells appear to be altered by infection (*31–33*). To gain some insight into the mechanism of these alterations, studies of the relative rate of synthesis of total cell protein and of hexokinases or of acid mucopolysaccharides (AMPS) in parallel populations of Rous and of normal cells were undertaken. Total cell protein was measured by the Lowry technique (*34*); hexokinases were measured by the techniques of Sols (*35*) with a Gilford multiple

TABLE 4.—Lack of serum-blocking power of material from CNVP cells*

	Virus incubated with:		
	Medium	Antiserum	Antiserum adsorbed with material from CNVP cells
FFU surviving	830, 1010	0, 1	0, 0

Virus incubated with antiserum adsorbed with:

	OD of RSV				OD of CNVP material
FFU surviving	0. 1 4100	0. 01 305	0. 001 137	0. 0 137	0. 025 117

*Antiserum was turkey anti-RSV kindly supplied by Dr. Rauscher. RSV was incubated with antiserum adsorbed by material from cultures of CNVP cells which banded in a potassium tartrate gradient in a position similar to that of RSV. Antiserum was also adsorbed by heat-killed purified RSV. The relative optical densities at 260 mμ of the material used for adsorption is given.

NATIONAL CANCER INSTITUTE MONOGRAPH NO. 17

sample absorbance recorder (*36*); and AMPS were measured by the Dische method, as modified by Bitter and Ewins (*37*), on samples purified by the method of Bollet, as modified by Morris and Davidson (*38*).

Rous cells broken up by sonification always had more hexokinase activity per mg protein than uninfected cells similarly treated. However, as seen in text-figure 6, the rate of synthesis of hexokinases was similar in the two types of cells. The difference in specific amount of enzyme was therefore probably due to the presence in uninfected cells of some nonhexokinase material not present in Rous cells, although it could also be due to a longer life of the enzyme. Therefore, the change in specific amount of hexokinase is probably not a primary change, but due to changes of synthesis of nonhexokinase material in Rous cells.

Morgan had previously demonstrated (*39*) that foci of Rous cells stained positively for mucopolysaccharides. However, it is possible that the rate of synthesis of AMPS, a specific fibroblast function, might have been changed after infection.

The results of experiments to test this possibility (text-fig. 7) indicate that the rate of synthesis of free AMPS is somewhat higher in Rous than in uninfected cells. (There is little Dische positive material in the cells.) Further experiments with Fujinami virus and with avian myeloblastosis virus suggests an increase in rate of synthesis of AMPS of 5 to 10 times the rate of normal cells. Information for changes in AMPS synthesis may, therefore, be carried in the provirus. Changes in AMPS in polyoma virus-infected cells have also been reported (*40, 41*). Further

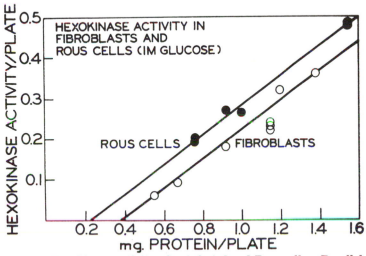

TEXT-FIGURE 6.—Hexokinase activity of uninfected and Rous cells. Parallel cultures of uninfected and of Rous cells were made. Each day some of each were taken and broken up by sonification, and total protein and hexokinase activity at 10^{-3} M and 1 M glucose were measured. Results for 1 M glucose are presented. The results for 10^{-3} M glucose were similar.

AVIAN TUMOR VIRUSES

work on characterizing the AMPS of avian tumor virus-infected cells is in progress.

The results presented here, taken together, signify that the virus acts as a carcinogenic agent by adding some new genetic information to the cell. Since this genetic information for carcinogenesis is not the same as that required for virus production, it is clear that Koch's postulates cannot be expected to apply to virus-induced tumors. Methods which can look for pieces of virus nucleic acid or virus-carried information must be applied to a search for agents in tumors of unknown etiology.

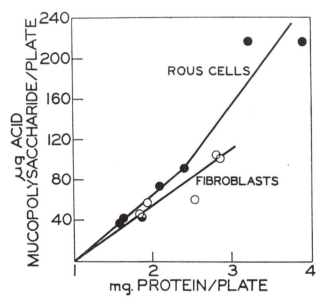

TEXT-FIGURE 7.—Production of free acid mucopolysaccharide by uninfected and Rous cells. Parallel cultures of uninfected and of Rous cells were made. Each day the supernatants of some of each were taken and centrifuged, and AMPS was prepared. The amount of AMPS was measured by the Dische carbazole reaction. Each day some of each culture was trypsinized and the total protein measured. All determinations were done in duplicate.

REFERENCES

(1) BALUDA, M. A., MOSCOVICI, C., and GOETZ, I. E.: Specificity of the *in vitro* inductive effect of avian myeloblastosis virus. Nat Cancer Inst Monogr 17: 449–458, 1964.

(2) WEISBERG, R. A.: Delayed appearance of transformed cells in polyoma virus-infected mouse embryo cultures. Virology 21: 669–671, 1963.

(3) TEMIN, H. M.: The effects of actinomycin D on growth of Rous sarcoma virus *in vitro*. Virology 20: 577–582, 1963.

(4) ———: Malignant transformation by viruses *in vitro*. Health Lab Sci 1: 79–83, 1964.

(5) ———: The control of cellular morphology in embryonic cells infected with Rous sarcoma virus *in vitro*. Virology 10: 182–197, 1960.

(6) TEMIN, H. M. and EPHRUSSI, B.: Infection of chick iris epithelium with the Rous sarcoma virus *in vitro*. Virology 11: 547–552, 1960.

(7) TEMIN, H. M.: Separation of morphological conversion and virus production in Rous sarcoma virus infection. Sympos Quant Biol 27: 407–414, 1962.

(8) ———: Mixed infection with two types of Rous sarcoma virus. Virology 13: 159–163, 1961.

(9) TING, R. C.: Studies of the early stages of Rous sarcoma virus infection *in vitro*. Virology 22: 568–574, 1964.

(10) PRINCE, A. M.: Factors influencing the determination of cellular morphology in cells infected with Rous sarcoma virus. Virology 18: 524–534, 1962.

(11) EPSTEIN, M. A., and HOLT, S. J.: Observations on the Rous virus; integrated electron microscopical and cytochemical studies of fluorocarbon purified preparations. Brit J Cancer 12: 363–369, 1958.

(12) CRAWFORD, L. V., and CRAWFORD, E. M.: The properties of Rous sarcoma virus purified by density gradient centrifugation. Virology 13: 227–232, 1961.

(13) BONAR, R. A., PURCELL, R. H., BEARD, D., and BEARD, J. W.: Virus of avian myeloblastosis (BAI strain A). XXIV. Nucleotide composition of the pentose nucleic acid and comparison with strain R (erythroblastosis). J Nat Cancer Inst 31: 705–716, 1963.

(14) TEMIN, H. M.: The participation of DNA in Rous sarcoma virus production. Virology 23: 486–494, 1964.

(15) ———: Unpublished studies.

(16) RICH, M. A., PEREZ, A. G., and EIDINOFF, M. L.: Effect of 5-fluorodeoxyuridine on the proliferation of Rous sarcoma virus. Virology 16: 98–99, 1962.

(17) GOLDÉ, A., and VIGIER, P.: Inhibition du developpement du virus du sarcome de Rous par le 5-fluorouracile en culture de tissus. C R Acad Sci (Paris) 252: 1693–1695, 1961.

(18) BATHER, R.: Influence of 5-fluorodeoxyuridine and actinomycin D on the production of Rous sarcoma virus *in vitro*. Proc Amer Ass Cancer Res 4: 4, 1963.

(19) EIDINOFF, M. L., BATES, B., PEREZ, A., and DE LA SIERRA, A.: Effect of inhibitors of nucleic acid synthesis on the cell culture system: Chick embryo cell-Rous sarcoma virus. Proc Amer Ass Cancer Res 4: 18, 1963.

(20) VIGIER, P.: Investigations on the replication of Rous sarcoma virus. Nat Cancer Inst Monogr 17: 407–419, 1964.

(21) BADER, J. P.: Nucleic acids of Rous sarcoma virus and infected cells. Nat Cancer Inst Monogr 17: 781–790, 1964.

(22) TEMIN, H. M.: Further evidence for a converted, non-virus-producing state of Rous sarcoma virus-infected cells. Virology 20: 235–242, 1963.

(23) HALL, B. D., and SPIEGELMAN, S.: Sequence complementarity of T2-DNA and T2-specific RNA. Proc Nat Acad Sci USA 47: 137–146, 1961.

(24) BOLTON, E. T., and McCARTHY, B. J.: A general method for the isolation of RNA complementary to DNA. Proc Nat Acad Sci USA 48: 1390–1397, 1962.

(25) DOUGHERTY, R. M.: Use of dimethylsulphoxide for preservation of tissue culture cells by freezing. Nature (London) 193: 550–552, 1962.

(26) ANTONI, F., HIDVEGI, E. J., and LONAI, P.: Isolation of cell nuclei from Lettre Ehrlich ascites tumor cells in glycerol medium. Acta Physiol Acad Sci Hung 21: 325–334, 1962.

(27) WECKER, E.: The extraction of infectious virus nucleic acid with hot phenol. Virology 7: 241–243, 1959.

(28) TEMIN, H. M.: The effects of actinomycin D on growth of Rous sarcoma virus *in vitro*. Virology 20: 577–582, 1963.

(29) VOGT, P. K.: Fluorescence microscopic observations on the defectiveness of Rous sarcoma virus. Nat Cancer Inst Monogr 17: 523–541, 1964.

AVIAN TUMOR VIRUSES

(30) HANAFUSA, H.: The nature of defectiveness of Rous sarcoma virus. Nat Cancer Inst Monogr 17: 543–556, 1964.

(31) MORGAN, H. R.: The biologic properties of cells infected with Rous sarcoma virus *in vitro*. Nat Cancer Inst Monogr 17: 395–406, 1964.

(32) WAGLE, S. R., ASHMORE, J., LOVE, W. C., AHMED, M., and LEVINE, A. S.: Biochemical studies of Rous sarcoma virus-induced tumors. Nat Cancer Inst Monogr 17: 769–779, 1964.

(33) SMIDA, J., THURZO, V., and SMIDOVA, V.: Comparison of some avian tumor viruses. Nat Cancer Inst Monogr 17: 231–236, 1964.

(34) LOWRY, O. H., ROSEBROUGH, N. J., FAIR, A. L., and RANDALL, R. J.: Protein measurements with the Folin phenol reagent. J Biol Chem 193: 265–277, 1951.

(35) VINUELA, E., SALAS, M., and SOLS, A.: Glucokinase and hexokinase in liver in relation to glycogen synthesis. J Biol Chem 238: 1175–1177, 1963.

(36) PITOT, H. C., PERAINO, C., PRIES, N., and KENNAN, A. L.: Glucose repression and induction of enzyme synthesis in rat liver. Adv in Enzyme Regulation 2: 237–247, 1964.

(37) BITTER, T., and EWINS, R.: A modified carbazole reaction for uronic acids. Biochem J 31: 43P, 1961.

(38) DAVIDSON, E. H.: Heritability and control of differentiated function in cultured cells. J Gen Physiol 46: 983–998, 1963.

(39) ERICKSEN, S., ENG, J., and MORGAN, H. R.: Comparative studies in Rous sarcoma with virus, tumor cells, and chick embryo cells transformed *in vitro* by virus. I. Production of mucopolysaccharides. J Exp Med 114: 435–440, 1961.

(40) FORRESTER, J. A., AMBROSE, E. J., and MACPHERSON, J. A.: Electrophoretic investigations of a clone of hamster fibroblasts and polyoma-transformed cells from the same population. Nature (London) 196: 1068–1070, 1962.

(41) DEFENDI, V., and GASIC, G.: Surface mucopolysaccharides of polyoma virus transformed cells. J Cell Comp Physiol 62: 23–31, 1963.

(42) TEMIN, H. M.: Homology between RNA from Rous sarcoma virus and DNA from Rous sarcoma virus-infected cells. Proc Nat Acad Sci USA 52: 323–329, 1964.

DISCUSSION

Dr. Bonar: Did you say that the RNA in the mature avian virus particles exists as a single strand, that is, as one long unit?

Dr. Temin: No, a single as opposed to a double strand, not a single molecule.

Dr. Bonar: And what evidence supports that?

Dr. Temin: Your evidence on the base ratios, and our evidence on the RNase susceptibility of the viral RNA in 0.4 M salt.

Dr. Vigier: When one studies base composition of RNA of purified RSV, the results are not at all indicative of double-strandedness. Yet, if we have a mixture of RSV and RAV, with predominance of the latter, single-strandedness is certain for RAV.

Dr. Temin: In our stocks of RSV, RAV is present as a minor component, if at all.

Chapter 3

Reprint of Boettiger and Temin's 1970 Paper on Inactivation of Bromodeoxyuridine-Containing Proviral DNA by Visible Light

During the 1960s, Howard Temin pioneered different routes for testing his hypothesis of the DNA provirus. The techniques available today for characterizing specific nucleic acids were then either unavailable or primitive. Site-specific endonucleases were just being discovered at the end of the decade and were not yet used to analyze DNAs, the procedures and enzymes needed to label DNAs in vitro to high specific activity were not available, and the α-^{32}P-labeled nucleoside triphosphates required for these procedures had yet to be synthesized commercially. Howard did use nucleic acid hybridization to search for DNA sequences homologous to retroviral RNA in infected cells, but the sensitivity of this approach in 1964 rendered it less than compelling. In the latter part of the decade, he and David Boettiger perfected an experimental protocol to test whether or not 5'-bromodeoxyuridine (5BUdR) was incorporated into retroviral replicative intermediates. This analog of thymidine had previously been shown to be incorporated into DNA, and being brominated, it rendered the DNA that contained it more sensitive to visible light than unhalogenated DNA would be. Boettiger and Temin induced chicken cells to become stationary by withdrawing serum from their cultures. These nonproliferating cells did not efficiently incorporate 5BUdR into their chromosomal DNA and thus were not sensitive to exposure to visible light. Temin and Boettiger infected the stationary cells, treated them with 5BUdR, exposed them to increasing doses of visible light, plated the treated chicken cells on a feeder layer of rat cells resistant to infection, and provided serum. They found that the number of retrovirus-induced foci that survived was inversely proportional to the length of the exposure to light. This experiment, along with its described controls, indicated that in stationary cells, infecting retroviruses do synthesize DNA intermediates that can be tagged by the 5BUdR they incorporate. This finding was the most direct proof of the existence of a DNA provirus until that proviral DNA was isolated and detected 5 years later. However, it was not as emotionally satisfying as detection of the viral enzyme that mediated the synthesis of DNA from a viral RNA template. This discovery followed immediately.

One unemphasized finding of the work with 5BUdR was that a DNA intermediate of retroviral replication is synthesized in stationary cells but decays unless these cells are induced to re-enter the cell cycle and proliferate. This finding formed the basis for later studies in which Howard Temin and his colleagues explored the structure of the replicative intermediate and found that newly infected cells must pass through mitosis for infection to be completed. We know today that the lentiviruses, which include human immunodeficiency virus, avoid this potential block to replication faced by other retroviruses and do in fact successfully infect stationary cells.

Bill Sugden

(Reprinted from Nature, Vol. 228, No. 5272, pp. 622–624, November 14, 1970)

Light Inactivation of Focus Formation by Chicken Embryo Fibroblasts infected with Avian Sarcoma Virus in the Presence of 5-Bromodeoxyuridine

by

DAVID BOETTIGER
HOWARD M. TEMIN

McArdle Laboratory,
University of Wisconsin,
Madison, Wisconsin 53706

When RNA tumour viruses infect cells growing in the presence of 5BUdR, exposure to visible light inactivates the virus. This confirms that synthesis of DNA from viral RNA is necessary for viral transformation.

THE RNA tumour viruses (leucoviruses) seem to have a different mode of replication from other RNA-containing viruses. The replication of RNA tumour viruses is sensitive to inhibitors of DNA synthesis early in infection[1,2]; virus production by converted cells is sensitive to actinomycin D (refs. 2 and 3); and viral RNA seems to have a greater homology with DNA from converted cells than with DNA from uninfected cells[4,5]. These observations are summarized in the DNA provirus hypothesis[6]: following infection, the viral RNA is transcribed into DNA and maintained as DNA in subsequent cell generations.

Cells grown in the presence of 5-bromodeoxyuridine (5BUdR) incorporate this analogue into their DNA. This substituted DNA is more sensitive to inactivation by ultraviolet and short wavelength visible light than is unsubstituted DNA[7,8]. Kao and Puck[9] used this sensitization to select against animal cells which could undergo DNA synthesis in deficient media.

The DNA provirus hypothesis predicts that a new DNA is formed soon after infection. If this new DNA could be labelled with 5BUdR, the infection could be aborted by visible light. In addition, the 5BUdR might provide a means to separate the new DNA from cellular DNA, and thus aid in its isolation. To avoid killing cells by the exposure to light, the incorporation of 5BUdR into essential cellular DNA must be reduced. This reduction can be accomplished by maintaining the cells for several days in medium without serum, which collects the cells in the G_1-phase of the cell cycle[10–12] and reduces the number of cells entering S-phase (nuclear DNA synthesis) to 1 to 5 per cent over a 24 h period. (A similar approach is being used by Balduzzi and Morgan, personal communication.)

In preliminary experiments, stationary cultures of chicken embryo fibroblasts were infected with 100 to 1,000 focus-forming units (f.f.u.) of the avian sarcoma virus, B77, in the presence of 5BUdR (ref. 6). Focus formation was inactivated in proportion to the amount of subsequent exposure to visible light following 5BUdR treatment. (Earlier experiments suggested that treatment of stationary cultures of chicken cells with 5BUdR immediately after infection inhibited focus formation[12] but we did not observe this. The two experiments differed

because here the serum-containing media were added to the cultures immediately after the removal of the 5BUdR. In the earlier experiments, there was a 24 h delay before addition of serum. It has been shown (unpublished results of D. B.) that there is an exponential loss of focus formation ($t_{\frac{1}{2}} = 9 \cdot 5$ h) during the interval between removal of 5BUdR and addition of serum.)

In these experiments, this technique was modified. Stationary cultures of chicken embryo fibroblasts were exposed to virus, treated with 5BUdR, exposed to light, and replated on a feeder layer of rat cells which are resistant to infection. This assay system gives a linear relationship between the number of foci produced and the multiplicity of infection (at multiplicities of infection of less than 0·1 f.f.u./cell (unpublished results of D. B.)[13]. For a cell to register as a focus in this assay, it must (a) contain non-inactivated viral genetic information, and (b) be capable of division and conversion. At a multiplicity of infection of 0·1 f.f.u./cell, the inactivation of focus formation by light followed single hit kinetics (Fig. 1). Essentially the same curve of inactivation by light was obtained using concentrations of 5BUdR from 2–500 µg/ml. in the post-infection incubation medium. Similar curves were obtained for cells infected with the B77 or the Schmidt–Ruppin strains of avian sarcoma virus.

In all systems where it has been examined, the effects of 5BUdR result from its incorporation into DNA in place of thymidine[8,14,15]. [3]H-5BUdR was used to measure the incorporation of 5BUdR into DNA. Cultures of stationary or dividing chicken cells were incubated for 20 h in medium with or without serum containing 1 µCi of [3]H-5BUdR. Nucleic acid was separated by a modified Schmidt–Thannhauser procedure. Less than 1 per cent of the perchloric acid precipitable counts were hydrolysed after incubation for 20 h at 37° C in 0·3 M KOH. Thus less than 1 per cent of the incorporated 5BUdR seems to be associated with RNA. Bader has reported a similar result for dividing chicken cells[16]. Also the light sensitization of focus formation induced by 5BUdR was inhibited when thymidine was added to the 5BUdR-containing medium. The 5BUdR sensitization observed in these experiments is therefore probably a direct result of incorporation into DNA.

Table 1. EFFECTS OF 5BUdR

Experiment	5BUdR (per cent)	5BUdR +30 min exposure to light (per cent)
1. Initiation of DNA synthesis*	80·8	78·2
2. Initiation of cell division†	25	25
3. Relative plating efficiency of cells infected after 5BUdR treatment ‡	100	88

* Cultures of stationary cells, infected with 3 f.f.u./cell, treated with 5 μg/ml. 5BUdR and exposed to light were incubated in ³H-thymidine and serum 0–20 h following exposure to light; autoradiograms were made and labelled nuclei counted.

† Cultures of stationary cells were infected, treated with 5BUdR, and exposed to light. Serum was added. 17 h later 5×10⁻⁷ M colcemid was added; after an additional 8 h, cells arrested in metaphase were counted by phase-contrast microscopy.

‡ As described in the text except that the cells are infected after removal of 5BUdR and before exposure to light.

There are several explanations to account for the inactivation of focus formation by visible light after infection in the presence of 5BUdR: first, the infected cells incorporate 5BUdR into essential cellular DNA and subsequent exposure to light prevents them from dividing; second, some particular fraction of cellular DNA which is essential for cell conversion is replicated on infection, incorporates 5BUdR, and is inactivated by the exposure to light; or third, the virus genetic information is transcribed to a 5BUdR containing DNA which is inactivated on exposure to visible light. Experiments were done to test these hypotheses.

The first hypothesis was examined as shown in Table 1. The ability of cells to respond to the addition of serum by initiating DNA synthesis as examined by autoradiography or to divide as examined by counting cells arrested in metaphase by colcemid was not significantly altered by the exposure to light. When the infection is delayed until after 5BUdR treatment there is no significant change in plating efficiency after exposure to light. The lack of cell killing in these experiments (Table 1) results from the use of cultures of stationary cells in which only 1–5 per cent of the cells pass through S-phase DNA synthesis and incorporate enough 5BUdR to become sensitive to visible light. Thus cell killing and differences in plating efficiency between the cells exposed to light and those unexposed to light could account for only a small portion of the light inactivation of focus formation observed in Fig. 1.

To test the possibility that only potential focus-forming cells incorporated 5BUdR and were killed by light, the

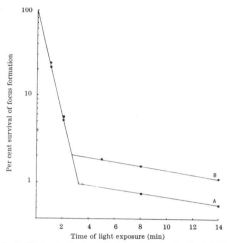

Fig. 2. The inactivation by visible light of the capacity of cells labelled with 5BUdR to produce foci. Cultures of stationary cells were either infected with 8 × 10⁴ f.f.u. of B77 virus and incubated with 5 μg/ml. 5BUdR and 4 per cent foetal calf serum for 22 h (A), or incubated in the same medium and infected after incubation (B). The same protocol described in the methods section was followed from this point.

sensitivity to light of cells containing 5BUdR substituted cellular DNA was measured. Cells were exposed to 5BUdR in the presence of serum so that 5BUdR would be incorporated into cellular DNA. These cells were infected either before or after the exposure to 5BUdR. The fraction of cells able to produce foci after exposure to light was measured (Fig. 2). The inactivation followed single hit kinetics with a resistant fraction of 1–2 per cent. In another experiment cells were exposed to 5BUdR and serum for 3 days preceding infection. The resistant fraction was reduced to less than 0·1 per cent. The resistant fraction represents those cells which did not incorporate 5BUdR. The inactivation measured is the result of cell killing. The initial slope of the cell killing curve is about twenty-fold greater than that for focus formation (Fig. 1). This difference in slope demonstrates that, when stationary cultures are used, the inactivation of focus formation is not a result of cell killing. In addition the difference in slope means that there is no significant stimulation of S-phase DNA synthesis by infection with avian sarcoma viruses. The stimulation of DNA synthesis which has been reported[17] is probably an early effect of cell conversion.

The difference between the second and third hypotheses is whether the new DNA is replicated from cellular DNA or whether it is transcribed from viral RNA. To distinguish between these possibilities, cultures of stationary chicken cells were infected at multiplicities of greater than 1 f.f.u./cell. If the new DNA is transcribed from viral genes, increasing the multiplicity of viral genes present in a cell might increase the number of copies of the new DNA; and the cell's focus-forming potential might be more resistant to light inactivation than that for a cell with a single copy of the new DNA. If the new DNA is cellular, the number of copies of the genes per cell would be independent of the multiplicity of infection, and the rate of focus inactivation by light would be the same at all multiplicities of infection. Parallel cultures of stationary chicken cells were infected with Schmidt–Ruppin virus at 3 f.f.u./cell or 0·15 f.f.u./cell in the presence of 5BUdR and exposed to light (Fig. 3.) Focus production in cells infected at the higher multiplicity was more resistant to inactivation by light than it was for cells infected at the low multiplicity. Thus it seems that the new DNA is transcribed from viral genes.

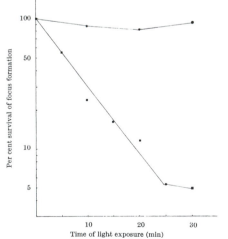

Fig. 1. Inactivation of focus formation by visible light. The protocol is explained in the text. Infection was carried out with about 8 × 10⁴ f.f.u./culture of B77 virus and the cultures were overlaid with Eagle's medium with (■) or without (●) 5BUdR (5 μg/ml.). The cultures were exposed to light and replated on rat cells. In all cases, points represent the average of counts of duplicate cultures.

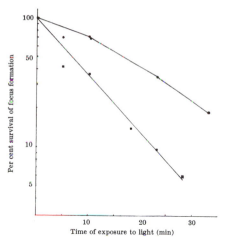

Fig. 3. The effect of multiplicity of infection on light sensitization caused by 5BUdR. Methods are described in the text. Cultures were infected with either 3 f.f.u./ml. (●) or 0·15 f.f.u./ml. (■) of the Schmidt-Ruppin virus and overlaid with medium containing 50 μg/ml. 5BUdR.

This experiment also serves as a further control for cell killing. If the light inactivation curves were the result of cell killing, increasing the multiplicity of infection should either increase the amount of cell killing or it should remain the same. But inactivation of focus formation is actually reduced by increasing the multiplicity of infection. This result is incompatible with the cell killing hypothesis.

The requirement for DNA synthesis early in infection of chicken cells by avian sarcoma viruses has been well documented[1,2,6]. There seem to be two parts to this requirement for DNA synthesis. In addition to the requirement for normal S-phase DNA synthesis and cell division[18], there is a requirement for non-S phase DNA synthesis. This non-S phase DNA synthesized following infection can be selectively labelled with 5BUdR using cultures of stationary cells. Subsequent exposure to visible light demonstrates that the non-S phase DNA required for focus formation is less sensitive to light than 5BUdR labelled S-phase DNA. The controls show that the inactivation of focus formation by light exposure is not the result of cell killing when the 5BUdR treatment is done in cultures of stationary cells.

The non-S phase DNA synthesis could involve the selective replication of cellular genes required for infection by avian sarcoma viruses, but the dependence of this non-S phase DNA synthesis on the multiplicity of infection makes this hypothesis less likely. Furthermore, because the DNA synthesized following infection by avian sarcoma viruses is specific for the viral strain used in the infection[6,19], a different set of cellular genes would be required for each strain of avian sarcoma virus.

These results are consistent with the model that the required S-phase DNA synthesized contains the genetic information for continued cell growth and division, and the required non-S phase DNA synthesized represents viral genetic information transcribed from viral RNA to a DNA intermediate.

Cultures and Irradiation

A modified Eagle's minimum essential medium (E) (obtained as a powder from Schwarz BioResearch, Orangeburg) was used. ET medium was made by adding 20 per cent tryptose phosphate broth (Difco Laboratories; Detroit, Michigan). The medium was supplemented with calf serum (Microbiological Associates) and/or foetal calf serum (Grand Island Biological Co.) as noted.
Primary cultures were made from 11–13 day old chicken embryos[20] (White Leghorn eggs were obtained from Sunnyside

Hatchery, Oregon). Cultures containing $7·5 \times 10^5$ cells were prepared in 60 mm Petri dishes (Falcon Plastics) in ET medium using cells from the first to fifth transfer generation. The medium was changed to fresh ET one day after preparation and the cultures were infected after 2 days of further incubation. Rat cell primary cultures were prepared from 12–14 day old embryos. Feeder cultures containing 5×10^5 cells were prepared in 60 mm Petri dishes in ET media supplemented with 4 per cent foetal calf serum.

B77 and Schmidt-Ruppin stocks have been described. High multiplicity of infection experiments used virus concentrated $100 \times$ by centrifugation[21]. Chicken cell cultures were exposed to 0·2 ml. virus, incubated for 40 min at 38° C, and overlaid with Eagle's medium containing 5BUdR (Calbiochem) or other analogues as indicated. After incubation for 18–22 h, the cultures were washed and overlaid with Tris buffered saline (0·14 M NaCl, 0·005 M KCl, 0·00033 M Na_2HPO_4, 0·025 M Tris, 0·0005 M $MgCl_2$, 0·001 M $CaCl_2$), exposed to the light source, resuspended with the aid of trypsin, diluted, and replated on previously prepared rat cell feeder cultures. 12–18 h later the medium was replaced with ET medium containing 0·6 per cent agar, 4 per cent calf serum and 4 per cent foetal calf serum. Foci were counted 6–8 days later. Counts on duplicate cultures differed by less than 20 per cent.
Cultures in Tris buffer were exposed from the underside at 6 cm distance to a light source consisting of two Westinghouse FS20 fluorescent sun lamps. This lamp emits a peak efficiency at 310 nm. Because irradiation from this source was found to cause considerable damage to the cells, filters consisting of 225 ml. tissue culture flasks (Falcon Plastics) filled with 0·3 mg/ml. thymidine (path length = 3 cm) were interposed. When this filter was used very little damage was observed to control cells which contained no 5BUdR.
Following incubation with ^{3}H-thymidine the cells were fixed with ethanol-acetic acid (3 : 1), washed with water, air-dried, coated with 'NTB3' nuclear track emulsion (Kodak) and exposed for 7 days at 4° C. The cultures were developed and stained with Harris's haematoxylin. For determination of 5BUdR incorporation into RNA, cells incubated in medium without serum containing 1 μCi of [3]H-5BUdR (Schwarz BioResearch, 10·6 Ci/mmole) were sonicated for 45 s and divided into two aliquots. Total incorporation was measured by precipitating twice with 1·5 per cent perchloric acid (PCA), resuspending in 0·3 M KOH and counting in 10 ml. of 'Scintisol' (Isolab). RNA incorporation was measured by precipitating with 1·5 per cent PCA + 100 μg bovine serum albumin, filtering on Whatman 'GF/A' filters, incubating at 37° C for 20 h in 0·3 M KOH, reprecipitating the supernatant with 1·5 per cent PCA and counting the remaining supernatant in 10 ml. of 'Scintisol'.

This work was supported by a US Public Health Service research grant from the National Cancer Institute. D. B. is supported by a training grant from the National Cancer Institute. H. M. T. holds a research career development award from the National Cancer Institute. We thank Diane Brown and Atsumi Kato for their assistance and Dr W. F. Dove and Dr M. Susman for help.

Received March 18; revised June 18, 1970.

[1] Temin, H. M., *Virology*, **23**, 486 (1964).
[2] Bader, J. P., *Virology*, **22**, 462 (1964).
[3] Temin, H. M., *Virology*, **20**, 577 (1963).
[4] Temin, H. M., *Proc. US Nat. Acad. Sci.*, **52**, 323 (1964).
[5] Baluda, M. B., and Nayak, D., in *Biology of Large RNA Viruses* (edit. by Barry, R., and Mahy, B.) (Academic Press, New York, 1970).
[6] Temin, H. M., in *Biology of Large RNA Viruses* (edit. by Barry, R., and Mahy, B.), 233 (Academic Press, New York, 1970).
[7] Stahl, F. W., Craseman, J. M., Okum, L., Fox, W., and Laird, C., *Virology*, **13**, 98 (1961).
[8] Szybalski, W., and Opara-Kubinska, Z., in *Proc. Eighteenth Symp. Ann. Fundamental Cancer Res.*, 241 (1964).
[9] Kao, F., and Puck, T. T., *Proc. US Nat. Acad. Sci.*, **60**, 1275 (1968).
[10] Temin, H. M., *J. Nat. Cancer Inst.*, **37**, 167 (1966).
[11] Temin, H. M., *J. Cell Physiol.*, **69**, 377 (1967).
[12] Temin, H. M., *Cancer Res.*, **28**, 1835 (1968).
[13] Temin, H. M., *Cold Spring Harbor Symp. Quant. Biol.*, **27**, 407 (1962).
[14] Hakala, M. T., *J. Biol. Chem.*, **234**, 3072 (1959).
[15] Wacker, A., Mennigmann, H. D., and Szybalski, W., *Nature*, **196**, 685 (1962).
[16] Bader, J. P., in *The Molecular Biology of Viruses* (edit. by Colter, J., and Parenchych, W.), 697 (Academic Press, New York, 1967).
[17] Kara, J., *Biochem. Biophys. Res. Commun.*, **32**, 817 (1968).
[18] Temin, H. M., *J. Cell. Physiol.*, **69**, 53 (1967).
[19] Duesberg, P. H., and Vogt, P. K., *Proc. US Nat. Acad. Sci.*, **64**, 939 (1969).
[20] Temin, H. M., and Rubin, H., *Virology*, **6**, 669 (1958).
[21] Altaner, C., and Temin, H. M., *Virology*, **40**, 118 (1970).

45

Chapter 4

Reprint of Temin and Mizutani's 1970 Paper Reporting the Discovery of Reverse Transcriptase

Although experiments with metabolic inhibitors continued to provide evidence that Rous sarcoma virus (RSV) replicated via a DNA provirus, it was the discovery of a viral enzyme capable of carrying out the synthesis of DNA from an RNA template that finally led to widespread acceptance of the DNA provirus hypothesis in 1970. This paper by Temin and Mizutani, published together with a similar paper by David Baltimore, unambiguously demonstrated a biochemical mechanism for synthesis of the DNA provirus. The existence of RNA-dependent DNA polymerase (now known as reverse transcriptase) clearly showed that the "central dogma" could be reversed and brought widespread acceptance of the new mode of information transfer that Howard Temin had predicted in 1964.

The key to detecting reverse transcriptase was looking for the enzyme in virus particles rather than in cells. It was known that other viruses carried a variety of enzymes, including RNA polymerases, in their virions. In addition, Temin and Mizutani had found that formation of the RSV provirus in infected cells did not require new protein synthesis following virus infection. These considerations suggested that an enzyme capable of synthesizing DNA from the viral RNA template might be present in virions of RSV. In striking confirmation of this prediction, incubation of detergent-treated RSV virions with deoxyribonucleoside triphosphates resulted in DNA synthesis. The RSV enzyme was further shown to specifically catalyze the synthesis of DNA, not RNA. Moreover, the virion DNA polymerase activity was dependent upon the presence of intact viral RNA, indicating that the virion enzyme catalyzed the synthesis of DNA from an RNA template.

As Temin and Mizutani concluded in their paper, these results provided "strong evidence that the DNA provirus hypothesis is correct." As they also pointed out, the discovery of reverse transcriptase not only had major implications for understanding viral carcinogenesis but would also change the way we think about information transfer in biological systems.

Geoffrey M. Cooper

(Reprinted from *Nature*, Vol. 226, No. 5252, pp. 1211–1213, June 27, 1970)

RNA-dependent DNA Polymerase in Virions of Rous Sarcoma Virus

INFECTION of sensitive cells by RNA sarcoma viruses requires the synthesis of new DNA different from that synthesized in the S-phase of the cell cycle (refs. 1, 2 and D. Boettiger and H. M. T. (*Nature*, in the press)); production of RNA tumour viruses is sensitive to actinomycin D[3,4]; and cells transformed by RNA tumour viruses have new DNA which hybridizes with viral RNA[5,6]. These are the basic observations essential to the DNA provirus hypothesis—replication of RNA tumour viruses takes place through a DNA intermediate, not through an RNA intermediate as does the replication of other RNA viruses[7].

Formation of the provirus is normal in stationary chicken cells exposed to Rous sarcoma virus (RSV), even in the presence of 0·5 μg/ml. cycloheximide (our unpublished results). This finding, together with the discovery of polymerases in virions of vaccinia virus and of reovirus[8-11], suggested that an enzyme that would synthesize DNA from an RNA template might be present in virions of RSV. We now report data supporting the existence of such an enzyme, and we learn that David Baltimore has independently discovered a similar enzyme in virions of Rauscher leukaemia virus[12].

The sources of virus and methods of concentration have been described[13]. All preparations were carried out in sterile conditions. Concentrated virus was placed on a layer of 15 per cent sucrose and centrifuged at 25,000 r.p.m. for 1 h in the 'SW 25.1' rotor of the Spinco ultracentrifuge on to a cushion of 60 per cent sucrose. The virus band was collected from the interphase and further purified by equilibrium sucrose density gradient centrifugation[14]. Virus further purified by sucrose velocity density gradient centrifugation gave the same results.

The polymerase assay consisted of 0·125 μmoles each of dATP, dCTP, and dGTP (Calbiochem) (in 0·02 M Tris-HCl buffer at pH 8·0, containing 0·33 Mm EDTA and 1·7 mM 2-mercaptoethanol); 1·25 μmoles of $MgCl_2$ and 2·5 μmoles of KCl; 2·5 μg phosphoenolpyruvate (Calbiochem); 10 μg pyruvate kinase (Calbiochem); 2·5 μCi of ^{3}H-TTP (Schwarz) (12 Ci/mmole); and 0·025 ml. of enzyme (10^8 focus forming units of disrupted Schmidt-Ruppin virus, $A_{280 nm} = 0·30$) in a total volume of 0·125 ml. Incubation was at 40° C for 1 h. 0·025 ml. of the reaction mixture was withdrawn and assayed for acid-insoluble counts by the method of Furlong[15].

Table 1. ACTIVATION OF ENZYME

System	³H-TTP incorporated (d.p.m.)
No virions	0
Non-disrupted virions	255
Virions disrupted with 'Nonidet'	
At 0° + DTT	6,730
At 0° − DTT	4,420
At 40° + DTT	5,000
At 40° − DTT	425

Purified virions untreated or incubated for 5 min at 0° C or 40° C with 0·25 per cent 'Nonidet P–40' (Shell Chemical Co.) with 0 or 1 per cent dithiothreitol (DTT) (Sigma) were assayed in the standard polymerase assay.

To observe full activity of the enzyme, it was necessary to treat the virions with a non-ionic detergent (Tables 1 and 4). If the treatment was at 40° C the presence of dithiothreitol (DTT) was necessary to recover activity. In most preparations of virions, however, there was some activity, 5–20 per cent of the disrupted virions, in the absence of detergent treatment, which probably represents disrupted virions in the preparation. It is known that virions of RNA tumour viruses are easily disrupted[16,17], so that the activity is probably present in the nucleoid of the virion.

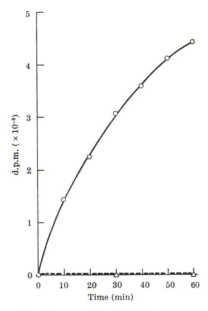

Fig. 1. Kinetics of incorporation. Virus treated with 'Nonidet' and dithiothreitol at 0° C and incubated at 37° C (○—○) or 80° C (△ - - - △) for 10 min was assayed in a standard polymerase assay.

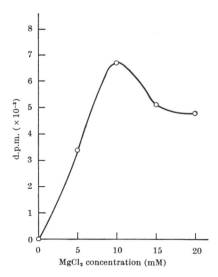

Fig. 2. MgCl$_2$ requirement. Virus treated with 'Nonidet' and dithiothreitol at 0° C was incubated in the standard polymerase assay with different concentrations of MgCl$_2$.

The kinetics of incorporation with disrupted virions are shown in Fig. 1. Incorporation is rapid for 1 h. Other experiments show that incorporation continues at about the same rate for the second hour. Preheating disrupted virus at 80° C prevents any incorporation, and so does pretreatment of disrupted virus with crystalline trypsin.

Fig. 2 demonstrates that there is an absolute requirement for MgCl$_2$, 10 mM being the optimum concentration. The data in Table 2 show that MnCl$_2$ can substitute for MgCl$_2$ in the polymerase assay, but CaCl$_2$ cannot. Other experiments show that a monovalent cation is not required for activity, although 20 mM KCl causes a 15 per cent stimulation. Higher concentrations of KCl are inhibitory: 60 per cent inhibition was observed at 80 mM.

Table 2. REQUIREMENTS FOR ENZYME ACTIVITY

System	^{3}H-TTP incorporated (d.p.m.)
Complete	5,675
Without MgCl$_2$	186
Without MgCl$_2$, with MnCl$_2$	5,570
Without MgCl$_2$, with CaCl$_2$	18
Without dATP	897
Without dCTP	1,780
Without dGTP	2,190

Virus treated with 'Nonidet' and dithiothreitol at 0° C was incubated in the standard polymerase assay with the substitutions listed.

Table 3. RNA DEPENDENCE OF POLYMERASE ACTIVITY

Treatment	^{3}H-TTP incorporated (d.p.m.)
Non-treated disrupted virions	9,110
Disrupted virions preincubated with ribonuclease A (50 µg/ml.) at 20° C for 1 h	2,650
Disrupted virions preincubated with ribonuclease A (1 mg/ml.) at 0° C for 1 h	137
Disrupted virions preincubated with lysozyme (50 µg/ml.) at 0° C for 1 h	9,650

Disrupted virions were incubated with ribonuclease A (Worthington) which was heated at 80° C for 10 min, or with lysozyme at the indicated concentration in the specified conditions, and a standard polymerase assay was performed.

When the amount of disrupted virions present in the polymerase assay was varied, the amount of incorporation varied with second-order kinetics. When incubation was carried out at different temperatures, a broad optimum between 40° C and 50° C was found. (The high temperature of this optimum may relate to the fact that the normal host of the virus is the chicken.) When incubation was carried out at different pHs, a broad optimum at pH 8–9·5 was found.

Table 2 demonstrates that all four deoxyribonucleoside triphosphates are required for full activity, but some activity was present when only three deoxyribonucleoside triphosphates were added and 10–20 per cent of full activity was still present with only two deoxyribonucleoside triphosphates. The activity in the presence of three deoxyribonucleoside triphosphates is probably the result of the presence of deoxyribonucleoside triphosphates in the virion. Other host components are known to be incorporated in the virion of RNA tumour viruses[18,19].

The data in Table 3 demonstrate that incorporation of thymidine triphosphate was more than 99 per cent abolished if the virions were pretreated at 0° with 1 mg ribonuclease per ml. Treatment with 50 µg/ml. ribo-

Table 4. SOURCE OF POLYMERASE

Source	^{3}H-TTP incorporated (d.p.m.)
Virions of SRV	1,410
Disrupted virions of SRV	5,675
Virions of AMV	1,875
Disrupted virions of AMV	12,850
Disrupted pellet from supernatant of uninfected cells	0

Virions of Schmidt-Ruppin virus (SRV) were prepared as before (experiment of Table 2). Virions of avian myeloblastosis virus (AMV) and a pellet from uninfected cells were prepared by differential centrifugation. All disrupted preparations were treated with 'Nonidet' and dithiothreitol at 0° C and assayed in a standard polymerase assay. The material used per tube was originally from 45 ml. of culture fluid for SRV, 20 ml. for AMV, and 20 ml. for uninfected cells.

Table 5. NATURE OF PRODUCT

Treatment	Residual acid-insoluble ^{3}H-TTP (d.p.m.)	
	Experiment A	Experiment B
Buffer	10,200	8,350
Deoxyribonuclease	697	1,520
Ribonuclease	10,900	7,200
KOH	—	8,250

A standard polymerase assay was performed with 'Nonidet' treated virions. The product was incubated in buffer or 0·3 M KOH at 37° C for 20 h or with (A) 1 mg/ml. or (B) 50 μg/ml. of deoxyribonuclease I (Worthington), or with 1 mg/ml. of ribonuclease A (Worthington) for 1 h at 37° C, and portions were removed and tested for acid-insoluble counts.

nuclease at 20° C did not prevent all incorporation of thymidine triphosphate, which suggests that the RNA of the virion may be masked by protein. (Lysozyme was added as a control for non-specific binding of ribonuclease to DNA.) Because the ribonuclease was heated for 10 min at 80° C or 100° C before use to destroy deoxyribonuclease it seems that intact RNA is necessary for incorporation of thymidine triphosphate.

To determine whether the enzyme is present in supernatants of normal cells or in RNA leukaemia viruses, the experiment of Table 4 was performed. Normal cell supernatant did not contain activity even after treatment with 'Nonidet'. Virions of avian myeloblastosis virus (AMV) contained activity that was increased ten-fold by treatment with 'Nonidet'.

The nature of the product of the polymerase assay was investigated by treating portions with deoxyribonuclease, ribonuclease or KOH. About 80 per cent of the product was made acid soluble by treatment with deoxyribonuclease, and the product was resistant to ribonuclease and KOH (Table 5).

To determine if the polymerase might also make RNA, disrupted virions were incubated with the four ribonucleoside triphosphates, including ^{3}H-UTP (Schwarz, 3·2 Ci/mmole). With either MgCl$_2$ or MnCl$_2$ in the incubation mixture, no incorporation was detected. In a parallel incubation with deoxyribonucleoside triphosphates, 12,200 d.p.m. of ^{3}H-TTP was incorporated.

These results demonstrate that there is a new polymerase inside the virions of RNA tumour viruses. It is not present in supernatants of normal cells but is present in virions of avian sarcoma and leukaemia RNA tumour viruses. The polymerase seems to catalyse the incorporation of deoxyribonucleoside triphosphates into DNA from an RNA template. Work is being performed to characterize further the reaction and the product. If the present results and Baltimore's results[12] with Rauscher leukaemia virus are upheld, they will constitute strong evidence that the DNA provirus hypothesis is correct and that RNA tumour viruses have a DNA genome when they are in cells and an RNA genome when they are in virions. This result would have strong implications for theories of viral carcinogenesis and, possibly, for theories of information transfer in other biological systems[20].

This work was supported by a US Public Health Service research grant from the National Cancer Institute. H. M. T. holds a research career development award from the National Cancer Institute.

HOWARD M. TEMIN
SATOSHI MIZUTANI

McArdle Laboratory for Cancer Research,
University of Wisconsin,
Madison,
Wisconsin 53706.

Received June 15, 1970.

[1] Temin, H. M., *Cancer Res.*, **28**, 1835 (1968).
[2] Murray, R. K., and Temin, H. M., *Intern. J. Cancer*, **5**, **320** (1970).
[3] Temin, H. M., *Virology*, **20**, 577 (1963).
[4] Baluda, M. B., and Nayak, D. P., *J. Virol.*, **4**, 554 (1969).
[5] Temin, H. M., *Proc. US Nat. Acad. Sci.*, **52**, 323 (1964).
[6] Baluda, M. B., and Nayak, D. P., in *Biology of Large RNA Viruses* (edit. by Barry, R., and Mahy, B.) (Academic Press, London, 1970).
[7] Temin, H. M., *Nat. Cancer Inst. Monog.*, **17**, 557 (1964).
[8] Kates, J. R., and McAuslan, B. R., *Proc. US Nat. Acad. Sci.*, **57**, 314 (1967).
[9] Munyon, W., Paoletti, E., and Grace, J. T., *Proc. US Nat. Acad. Sci.*, **58**, 2280 (1967).
[10] Borsa, J., and Graham, A. F., *Biochem. Biophys. Res. Commun.*, **33**, 895 (1968).
[11] Shatkin, A. J., and Sipe, J. D., *Proc. US Nat. Acad. Sci.*, **61**, 1462 (1968).
[12] Baltimore, D., *Nature*, **226**, 1209 (1970) (preceding article).
[13] Altaner, C., and Temin, H. M., *Virology*, **40**, 118 (1970).
[14] Robinson, W. S., Pitkanen, A., and Rubin, H., *Proc. US Nat. Acad. Sci.*, **54**, 137 (1965).
[15] Furlong, N. B., *Meth. Cancer Res.*, **3**, 27 (1967).
[16] Vogt, P. K., *Adv. Virus. Res.*, **11**, 293 (1965).
[17] Bauer, H., and Schafer, W., *Virology*, **29**, 494 (1966).
[18] Bauer, H., *Z. Naturforsch.*, **21**b, 453 (1966).
[19] Erikson, R. L., *Virology*, **37**, 124 (1969).
[20] Temin, H. M., *Persp. Biol. Med.* ,**5**, 320 (1970).

Chapter 5

Reprint of Temin's 1971 Paper Proposing the Protovirus Hypothesis

Following the discovery of reverse transcriptase and widespread acceptance of the provirus hypothesis, Howard Temin turned his attention to the possible roles of RNA-directed DNA synthesis (reverse transcription) in the cells of healthy organisms. He predicted that this unique mode of information transfer was not restricted to viruses but would also be found in healthy cells, where the transfer of information from RNA to DNA might play a role both in normal development and in generating the mutations responsible for non-virus-induced carcinogenesis. These considerations formed the basis of the protovirus hypothesis, as discussed in this 1971 editorial.

The central predictions of the protovirus hypothesis, the existence of reverse transcriptase and RNA $\rightarrow$ DNA information transfer in cells as well as in viruses, have been thoroughly substantiated. Reverse transcriptases not only are widely distributed in eukaryotes (including yeasts, plants, and animals) but also have been found in bacteria. In mammals, transposable elements that move via RNA $\rightarrow$ DNA information transfer (retrotransposons) constitute approximately 10% of genomic DNA. Also as predicted in the protovirus hypothesis, mutations induced by the movement of retrotransposons can contribute to the development of cancer both by activating oncogenes and by inactivating tumor suppressor genes.

The normal cellular role of reverse transcription, however, remains unclear. The retrotransposons that have been described to date appear to be parasitic self-replicating elements that transpose to random sites throughout the genome. The mutations they induce may play a role in evolution, but retrotransposons do not appear to be of direct benefit to the cell or organism in which they reside. However, much remains to be learned concerning the molecular mechanisms of development and differentiation, and a normal physiological role for RNA $\rightarrow$ DNA information transfer may still await discovery.

Geoffrey M. Cooper

Reprinted from *J. Natl. Cancer Inst.* **46**:III–VI, 1971

GUEST EDITORIAL━━━━━━━━━━━━━━**by Howard M. Temin**

McArdle Laboratory,
University of Wisconsin,
Madison, Wisconsin 53706

The Protovirus Hypothesis: Speculations on the Significance of RNA-Directed DNA Synthesis for Normal Development and for Carcinogenesis

THE NEOPLASTIC phenotype usually persists in progeny of cancer cells. An understanding of the mechanism of this persistence appears necessary to understanding the etiology of the cancer. Theories of persistence include genetic theories such as those involving somatic mutations or viruses, and epigenetic theories such as those involving embryonic arrests and activation. The oncogene theory (*1*) appears to combine viral and epigenetic theories by postulating that neoplasia results from the activation of a vertically transmitted provirus. However, this oncogene theory is basically an epigenetic theory, since it postulates a stable change in phenotype due to a switch in a regulatory system.

Recently, I have suggested a new genetic hypothesis which combines elements of somatic mutation and viral theories (*2, 3*). This protovirus hypothesis proposes apparent vertical transmission of the information for cancer, even though the germ line does not contain this information on its chromosomes (proviruses or oncogenes) or off its chromosomes (virions). The germ line is postulated to contain in its chromosomes the potential for genetic evolution by the somatic cells that may lead to the *de novo* formation of the information for cancer. Changes leading to cancer are not the normal roles of this genetic system. Normally the protoviruses from the germ line evolve in the somatic cells to differentiate the genomes of these cells by the creation of new DNA.

This apparent vertical transmission of the information for cancer differs from classical vertical transmission which involves the whole virus as an infectious particle in, for example, the milk or egg. It also differs from transfer of the DNA provirus of an RNA tumor virus, since in the protovirus hypothesis only the potential for genetic evolution of the information for cancer is transmitted in the DNA.

The earlier DNA provirus hypothesis, from which the protovirus hypothesis was derived, stated that replication of RNA tumor viruses involved a DNA intermediate and that permanent establishment in a normal cell of the genetic information for cancer required information transfer from viral RNA to DNA (*4*). Once this information transfer from RNA to DNA had occurred, further transfer of information from RNA to DNA was not required for the maintenance or the expression of the neoplasia. The DNA provirus hypothesis did not elucidate how the establishment in a cell of this new genetic information led to the neoplastic state.

Editor's note: Periodically the Journal will publish solicited guest editorials as a means of transmitting to investigators in cancer research the essence of current work in a special field of study. The Board of Editors will welcome suggestions for future editorials that succinctly summarize current work toward a clearly defined hypothesis regarding the causes or cure of cancer.

III

Normal Development

In speculating about the origin of viruses with this special mode of information transfer, RNA→DNA, I felt that RNA→DNA information transfer might play a role in normal organisms. Such reverse transcription could provide a mechanism for variability in the genome of somatic cells without disrupting the stability of the germ line. In the germ line only the usual modes of information transfer occur, that is DNA→DNA, DNA→RNA, and RNA→protein. Variation of expression occurs by control of the rate of these reactions, but variation of the DNA is supposed to occur only by random mutation followed by selection. The protovirus hypothesis states that somatic cells, in addition, use the RNA→DNA mode of information transfer, and new DNA sequences are formed by this process during the lifetime of a single organism.[1]

An organism needs to identify cells in a stable way, so that one cell is identified as a retinal cell at a particular position, and another cell is committed to make antibody to a particular antigen. The most stable storage place for such information appears to be the cellular DNA. RNA→DNA information transfer in somatic cells would provide a mechanism for stable differentiation of DNA. Other mechanisms are possible, that is stable cycles of the type discussed by Monod and Jacob (5) or classical DNA episomes. The protovirus mechanism might be preferred to cycles because it involves a structure rather than a dynamic process. The protovirus mechanism might be preferred to classical DNA episomes because it does not require excision of DNA for information transfer. However, it is possible that all of these mechanisms exist.

The process of protovirus transfer might work as follows (text-fig. 1). A region of DNA in cell A serves as template for synthesis of an RNA which is transferred to cell B. In cell B, a new DNA is made by an RNA-dependent DNA polymerase, using the transferred RNA as template. The new DNA then integrates into the DNA of cell B. This integration could be next to the homologous DNA or at a different place. In either case, cell B would differ from cell A, which remains unchanged. This process could be repeated. If the new DNA in cell B was integrated in some new place, it might act as a template for RNA from itself and from a

neighboring region. This RNA in turn could be transferred to cell C, transcribed into DNA, and integrated into the genome of cell C. Cell C would then differ from cell B and from cell A. These processes could continue and could involve a cell more than once. They would lead to the formation of duplications, tandem or distant, and possibly replacements.

In addition, there might be times when the RNA→DNA information transfer takes place only in cell A, with or without concomitant integration, thereby leading to gene amplification (6). However, if, as I believe (3), there is great genetic instability in this type of information transfer, variants would appear and wholly new DNA sequences could be formed.

There would be regions in the genome predetermined for this type of transfer. The specificities of the appropriate polymerase and integration systems would select these regions on the basis of their base sequences. The time and extent of protovirus transfer would be controlled by the availability of the polymerase and integration systems, and the absence of possible inhibitory systems, especially nucleases. Selection of randomly appearing variants could also be through the properties of the polymerase, integration, and, possibly, inhibitory systems. This selection also would involve selection of cells altered by the presence of new protovirus DNA. These selection processes would allow random variations to persist and to become amplified. Together these processes could lead to differentiation among the genomes of the somatic cells in an organism.[2] In the development of an organism, information transfer from DNA→RNA→DNA would allow variability and

---

[1] Information transfer from RNA to DNA could occur in the germ line without affecting progeny individuals if this DNA was not integrated. Strictly speaking, only integration of such DNA would usually be forbidden in the germ line.

[2] The successful experiments on nuclear transplantation from somatic cells to eggs (7) do not invalidate the protovirus hypothesis. There are two aspects of this hypothesis which can lead to reversibility. One is crossing-over to remove a duplication, as seen in the bar region of Drosophila (8). The other is epigenetic controls, which could inactivate regions of new genetic information in embryonic cells. These hypotheses might also encompass the cases of reversibility of the malignant phenotype (9).

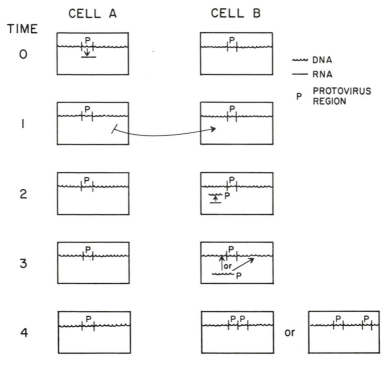

TEXT-FIGURE 1.—One step in protovirus evolution. Two separate cells, A and B, are shown at five successive times, 0–4. Cell A transfers a protovirus-derived RNA to cell B at time 1. DNA derived from this RNA is integrated into the DNA of cell B at time 3.

amplification; information transfer from DNA→ DNA would allow stability and storage.

The normal physiological evolution of the protovirus-derived DNA's would fall within a pattern predetermined by the rest of the cell genome and by the state of the cell and of the developing organism. Integration of protovirus-derived DNA, specifying some polymerases, for example, next to a region controlling membranes or other aspects of the cell surface could affect surface specificities of the cell. Integration next to a region controlling cell multiplication could affect multiplication control of the cell. Continued evolution along these lines could put together in a contiguous region of the chromosomes the information necessary for formation of an enveloped virion.

Abnormal Development (Carcinogenesis)

In the normal physiological evolution of the protovirus DNA, information from the environment would be effective only as it affected the general state and differentiation of the organism. However, protovirus evolution could by some random events give rise to a virus which could enter an organism from the outside. Because the virus has the ability to insert information into the cell, it could short-circuit the requirement for protovirus evolution. For example, once RNA sarcoma viruses were formed, they would upon infection make new DNA which would contain all of the information necessary to bring about neoplastic transformation. No multiple cycles of DNA→ RNA→DNA information transfer need occur

However, apart from laboratory conditions, such single-step transfer processes might never lead to cancer.

In normal development, this DNA→RNA→ DNA information transfer could be used to identify cells as being of a particular type and to recruit other cells into a related or identical form. A cell's particular differentiated state could be partially specified by the types of new protovirus-derived DNA it contained. RNA transcribed from this new protovirus-derived DNA could be transferred to other cells and thus induce an orderly conversion of the other cells into the same or a related differentiated state. Examples of such differentiation might be the secondary antibody response (10) and embryonic induction.

The usual process leading to cancer could be a variation in the normal physiological evolution of the protovirus DNA, so that variants which contain information for the cancer appeared either by mutation of the base sequences or by integration in incorrect places or both. Since DNA→RNA→ DNA information transfer involves DNA, RNA, and enzymes (protein), the formation of variants may result from the response of any of these molecules to irradiation or chemicals. This oncogenic process would not require the formation of an RNA tumor virus. It would require only alteration of the information in cells, so that there was the information which was necessary for cancer. This information for cancer may or may not be homologous to the information in RNA tumor viruses. The nature of this information is not considered here (see 2, 11).

In extreme cases, one could imagine that a product of protovirus evolution would infect the germ line, become integrated there, and thus also affect progeny organisms. Such a process could provide part of a mechanism for inheritance of some acquired characters. Selection at the level of the germ cells and at the level of the organism would also be involved.

This protovirus theory was derived *a priori* from consideration of the origin of RNA sarcoma viruses. It also explained the presence in uninfected cells of DNA homologous to a portion of the RNA of RNA tumor viruses (12, 13), and the apparent vertical transmission of the information for RNA tumor viruses (1, 14).

Since the protovirus theory was derived, several new experimental findings have been made which are relevant to it. An RNA-directed DNA polymerase has been found in virions of visna virus and simian foamy virus (15, 16). Neither of these viruses has so far been associated with neoplasia. Preliminary evidence suggests that RNA-directed DNA polymerase may also be involved in gene amplification in oocytes of Xenopus (5). Uninfected chicken cells may contain an element which is inherited as a single Mendelian dominant gene and can become infectious after phenotypic mixing with an RNA tumor virus (17). This element could be a protovirus-derived element that was not yet a complete virus genome. It also could be a defective provirus. Antigens related to RNA tumor viruses are found in normal mouse embryos (18). These antigens could represent the RNA-directed DNA polymerase and other proteins of the protoviruses.

Finding an RNA-directed DNA polymerase in uninfected tissues of normal organisms may be the first direct evidence bearing on this hypothesis. Knowledge of the specificities and time of occurrence of such a polymerase would aim in demonstrating the type of information transfer postulated here.

However, even if this protovirus theory is correct in its essentials, it raises the question of how carcinogens cause the misevolution of protoviruses to give rise to information for the neoplastic phenotype and how such information in protovirus-derived DNA actually causes the neoplastic phenotype.

ACKNOWLEDGMENTS

The work in my laboratory is supported by Public Health Service research grant CA 07175 from the National Cancer Institute. I hold Research Career Development Award 10K 3–CA 8182 from the National Cancer Institute. I thank my colleagues for useful suggestions about this manuscript.

REFERENCES

(1) HUEBNER R, TODARO GJ: Oncogenes of RNA tumor viruses as determinants of cancer. Proc Nat Acad Sci USA 64:1087–1094, 1969
(2) TEMIN HM: Malignant transformation of cells by viruses. Perspect Biol Med. In press

(*3*) Temin HM: The role of the DNA provirus in carcinogenesis by RNA tumor viruses. Proc 2d Lepetit Colloquium. North Holland Publ. Co., 1970

(*4*) ———: Nature of the provirus of Rous sarcoma. Nat Cancer Inst Monogr 17:557–570, 1964

(*5*) Monod J, Jacob F: Teleonomic mechanisms in cellular metabolism, growth, and differentiation. Sympos Quant Biol 26:389–401, 1961

(*6*) Tocchini-Valentini G: Gene amplification in Xenopus. Proc 2d Lepetit Colloquium. North Holland Publ. Co., 1970

(*7*) Gurdon JB, Laskey RA: The transplantation of nuclei from single cultured cells into enucleate frog's eggs. J Embryol Exp Morph 24:227–248, 1970

(*8*) Sturtevant AH, Morgan TH: Reverse mutation of the bar gene correlated with crossing-over. Science 57:746–747, 1923

(*9*) Braun AC: On the origin of the cancer cell. Amer Sci 58:307–320, 1970

(*10*) Dutton RW, Mishell RI: Cellular events in the immune response. The *in vitro* response of normal spleen cells to erythrocyte antigens. Sympos Quant Biol 32:407–414, 1967

(*11*) Temin HM: Mechanism of transformation by RNA tumor viruses. Ann Rev Microbiol. In press

(*12*) Harel J, Harel L, Goldé A, et al: Homologie entre génome du virus du sarcome de Rous (RSV) et génome cellulaire. C R Acad Sci [D] (Paris) 263:745–748, 1966

(*13*) Baluda MA, Nayak DP: DNA complementary to viral RNA in leukemia cells induced by avian myeloblastosis virus. Proc Nat Acad Sci USA 66:329–336, 1970

(*14*) Bentvelzen P, Doams JH, Hageman P, et al: Genetic transmission of viruses that incite mammary tumor in mice. Proc Nat Acad Sci USA 67:377–384, 1970

(*15*) Lin FH, Thormar H: Ribonucleic acid-dependent deoxyribonucleic acid polymerase in visna virus. J Virol 6:702–704, 1970

(*16*) Scolnik E, et al: Personal communication, 1970

(*17*) Hanafusa J, Miyamoto T, Hanafusa T: A cell-associated factor essential for formation of an infectious form of Rous sarcoma virus. Proc Nat Acad Sci USA 66:314–321, 1970

(*18*) Huebner RJ, Kelloff GJ, Sarma PS, et al: Group-specific (gs) antigen expression of the C-type RNA tumor virus genome during embryogenesis: Implications for ontogenesis and oncogenesis. Proc Nat Acad Sci USA 67:366–376, 1970

PART II

CANCER

The DNA Provirus: Howard Temin's Scientific Legacy
Edited by G. M. Cooper, R. Greenberg Temin, and B. Sugden
© 1995 American Society for Microbiology, Washington, DC 20005

Chapter 6

On the Origin of Oncogenes

Geoffrey M. Cooper

I first learned of Howard Temin's work in 1970, with the publication of his and David Baltimore's papers describing the discovery of reverse transcriptase (6, 88). At that time, I was a biochemistry graduate student working on the development of inhibitors of pyrimidine nucleoside metabolism as cancer chemotherapeutic drugs. It had become all too clear that this classic approach to the treatment of cancer depended on relatively minor biochemical differences between cancer cells and normal cells. I was therefore fascinated by the revelation that RNA tumor viruses employ a novel mechanism of replication that might permit the development of rational strategies for cancer treatment based on more fundamental differences between cancer cells and their normal counterparts.

The discovery of reverse transcriptase stimulated me to read Howard's earlier papers, and I realized that the work on which the provirus hypothesis was based was at least as fascinating as the enzyme itself. Howard had developed the provirus hypothesis through a logical series of experiments in which he had the intellectual strength and integrity to follow his data where it led, even though his results pointed to what was, at the time, an obviously heretical and unpopular conclusion. This period of Howard's career struck me as a model of scientific pursuit, and I decided at that time to try to join his laboratory as a postdoctoral fellow.

I first had the opportunity to meet Howard in 1972, when he visited my graduate student department as a seminar speaker. My advisor arranged for the two of us to have lunch with Howard and take him to the airport, so I was fortunate to have the chance to speak with Howard at some length on that occasion. This good fortune continued, and Howard eventually agreed to accept me as a postdoctoral fellow in his laboratory, which I joined in the summer of 1973. Thus began not only the most stimulating and important period of my professional training but also a relationship with Howard as mentor and friend that I was privileged to enjoy for the next two decades.

One of the questions that particularly intrigued Howard during the 1970s was the nature of the genetic changes responsible for non-virus-induced cancers. In this chapter, I first review his thoughts on the origin of cancer genes: he correctly

---

Geoffrey M. Cooper • Division of Molecular Genetics, Dana-Farber Cancer Institute, and Department of Pathology, Harvard Medical School, Boston, Massachusetts 02115.

predicted many aspects of our current understanding of the molecular alterations responsible for the development of malignant tumors. I then discuss some of the functions of proto-oncogenes in controlling normal cell growth and differentiation as well as the possibility that oncogene proteins provide novel targets for cancer chemotherapy—areas of current interest which represent direct outgrowths of some of Howard Temin's seminal contributions to cancer research.

GENES FOR NEOPLASIA

Following the discovery of reverse transcriptase and the widespread acceptance of the provirus hypothesis, Howard's attention turned to the possible roles of reverse transcription in cells. In the protovirus hypothesis (85, 86), he proposed that cells normally contain movable genetic elements that utilize reverse transcriptase to transpose via RNA intermediates. With respect to the genetic changes leading to cancer, he particularly suggested that cancer genes (which we now call oncogenes) arose by mutations of normal cell genes (proto-oncogenes) and that transposition of cellular sequences via RNA intermediates might play a role in this process.

In a 1974 paper entitled "On the Origin of the Genes for Neoplasia" (87), Howard drew a clear distinction between two kinds of hypotheses to account for the nature of the genetic changes leading to the development of cancer. In differentiation hypotheses, normal cells were assumed to contain oncogenes, but these genes were not expressed. Cancer then resulted not from mutations but from epigenetic changes that led to the expression of normally silent genes. The oncogene hypothesis of Huebner and Todaro (40) was considered a special case of a differentiation hypothesis in which the oncogenes were part of an inactive germ line provirus of a strongly transforming retrovirus. In contrast, mutation hypotheses assumed that oncogenes were not part of the genomes of normal cells. Instead, oncogenes were formed as a result of mutations that occurred during the process of carcinogenesis. The protovirus hypothesis was a special case of a mutation hypothesis in which genetic systems using reverse transcription were responsible for producing the mutations leading to oncogene formation. Importantly, Howard further suggested that the oncogenes of both strongly transforming retroviruses and non-virus-induced cancers were formed by mutations of similar normal cell genes.

Howard's hypothesis for the origin of cancer genes thus incorporated three key elements. First, cancer genes do not normally exist in cells but are formed by mutation. Second, strongly transforming viruses, such as Rous sarcoma virus, are formed by recombination between cellular cancer genes and retroviral genomes. Third, mutations giving rise to cancer genes in both retroviruses and cells can occur during the transposition of cellular genetic elements through RNA intermediates. These ideas have proven central to the development of cancer research, particularly to understanding the relationship between proto-oncogenes in normal cells and the oncogenes found in both retroviruses and non-virus-induced tumors.

One of the papers that Howard cited in support of the protovirus hypothesis was a 1973 paper from Ed Scolnick's laboratory (73) that reported that the strongly transforming Kirsten sarcoma virus was formed by recombination between viral and host cell sequences, a paper I vividly remember as being the subject of my first journal club presentation in Madison. However, the first definitive experimental evidence to elucidate the origin of retroviral oncogenes was reported by Harold Varmus, Mike Bishop, and their colleagues in 1976 (79). These studies demonstrated that the *src* oncogene of Rous sarcoma virus originated from normal-cell DNA sequences that had been incorporated into a retroviral genome. Subsequent experiments in a number of laboratories demonstrated that the *src* oncogene is a mutated version of its normal cell precursor (the *src* proto-oncogene) and that these mutations endow the oncogene with the ability to induce cell transformation (12). Further experiments have established that the oncogenes of other strongly transforming retroviruses likewise represent mutant forms of normal cell genes that have been incorporated into retroviral genomes as a result of recombination between viral and host sequences. Moreover, this recombination between viral and cell sequences has been shown to be a consequence of several features of the retrovirus life cycle, particularly reverse transcription (81). Studies of retroviral oncogenes have thus substantiated two aspects of Howard's hypothesis: (i) oncogenes are formed from proto-oncogenes by mutation and (ii) oncogenes of cellular origin are incorporated into retroviruses by reverse transcription and recombination.

As Howard further predicted, cancers that are not caused by strongly transforming viruses also result from mutations of normal cell genes, giving rise to oncogenes related to those of the retroviruses. In 1975, while I was still in Howard's laboratory as a postdoctoral fellow, we attempted to detect oncogenes in non-virus-induced chicken tumors by using transfection assays. These experiments were unsuccessful, but the same approach in the laboratories of both Bob Weinberg and myself revealed the first oncogenes in human tumors in 1981 (49, 74). By 1982, these human tumor oncogenes had been identified as mutated forms of the *ras* proto-oncogenes, which had previously been characterized by Scolnick and his colleagues as the cellular homologs of the oncogenes of Harvey and Kirsten sarcoma viruses (23, 67, 71, 72, 83, 84). Similar point mutations are responsible for converting *ras* proto-oncogenes to oncogenes in both retroviruses and human tumors, and studies of *ras* genes have continued to provide a key model for the role of mutations in the formation of both viral and cellular oncogenes.

Further studies have established that mutations of a variety of proto-oncogenes regularly occur in human tumors and contribute to tumor development (21) (Table 1). Like *ras,* many of the oncogenes active in human tumors have also been incorporated into chicken or mouse retroviruses. In addition, the development of human tumors frequently involves mutations that inactivate tumor suppressor genes (21, 53, 96) (Table 2). Indeed, it appears that most human cancers develop as a result of mutations affecting multiple proto-oncogenes and tumor suppressor genes, with the accumulation of such mutations being responsible for the stepwise progression of normal cells to malignancy. Human colorectal carcinomas are par-

TABLE 1
Examples of oncogenes in human cancers

Oncogene	Type(s) of cancer	Activating mutation
abl	Chronic myelogenous and acute lymphocytic leukemia	Translocation
bcl-2	Follicular B-cell lymphoma	Translocation
E2A/pbx1	Acute pre-B-cell leukemia	Translocation
erbB	Glioblastoma	Amplification
erbB-2	Breast and ovarian carcinomas	Amplification
gip	Adrenal cortical and ovarian carcinomas	Point mutation
gli	Glioblastoma	Amplification
gsp	Pituitary tumors	Point mutation
mdm-2	Sarcomas	Amplification
c-myc	B-cell lymphoma	Translocation
	Breast and lung carcinomas	Amplification
L-myc	Lung carcinoma	Amplification
N-myc	Neuroblastoma, lung carcinoma	Amplification
PML/RAR	Acute promyelocytic leukemia	Translocation
PRAD-1	B-cell leukemia, thyroid adenoma	Translocation
	Breast and squamous carcinomas	Amplification
rasH	Thyroid carcinoma	Point mutation
rasK	Colon, lung, pancreatic, and thyroid carcinomas	Point mutation
rasN	Acute leukemias and thyroid carcinomas	Point mutation
ret	Thyroid carcinoma	Rearrangement

ticularly well characterized at the molecular level, with studies by Bert Vogelstein and his colleagues having established that development of these tumors involves mutations that result in the formation of *ras* oncogenes and the inactivation of three distinct tumor suppressor genes (*APC, DCC,* and *p53*) (92). The key role of mutations in the development of non-virus-induced cancers, as Howard predicted in 1974, is thus clearly recognized.

TABLE 2
Tumor suppressor genes

Gene	Type(s) of cancer
Rb	Retinoblastoma; sarcomas; bladder, breast, esophageal, and lung carcinomas
p53	Bladder, breast, colorectal, esophageal, liver, lung, and ovarian carcinomas; brain tumors; sarcomas; lymphomas and leukemias
APC	Colorectal, stomach, and pancreatic carcinomas
DCC	Colorectal carcinomas
NF1	Neurofibromatosis type 1
NF2	Neurofibromatosis type 2
WT1	Wilms tumor
VHL	Renal cell carcinoma
MTS1	Melanoma; brain tumors; leukemias; sarcomas; bladder, breast, kidney, lung, ovarian, and pancreatic carcinomas
BRCA1	Breast carcinoma

Most of the genetic alterations affecting proto-oncogenes or tumor suppressor genes in human tumors result from point mutations, chromosome translocations, DNA amplification, or deletions, processes that do not involve reverse transcription. However, a number of mutations affecting both proto-oncogenes and tumor suppressor genes result from the transposition of genetic elements that move by reverse transcription of RNA intermediates, as predicted in the protovirus hypothesis. The genomes of eukaryotes contain several types of such elements (called retrotransposons), many of which are highly repeated sequences. The prototype of mutations induced by retrotransposons is the activation of the c-*myc* proto-oncogene by avian leukosis virus, a weakly oncogenic retrovirus that does not contain a viral oncogene. Studies in the laboratories of Bill Hayward and Susan Astrin demonstrated in 1981 that avian leukosis virus induces lymphomas in chickens as a result of the integration of proviral DNA adjacent to the c-*myc* proto-oncogene, leading to increased c-*myc* expression (37). Subsequent studies established that mutations induced by insertion of proviral DNA are similarly responsible for the activation of proto-oncogenes in many types of tumors. Moreover, insertional mutagenesis is not restricted to infectious retroviruses but can also result from the movement of nonviral retrotransposons. Examples include mutational activation of proto-oncogenes in mouse tumors by intracisternal A particles (18, 25, 101), inactivation of the *NF1* tumor suppressor gene by an *Alu* element in human neurofibromatosis (94), and inactivation of the *APC* tumor suppressor gene by a LINE sequence in human colon cancer (60). The movement of retrotransposons therefore represents at least one mechanism that leads to mutational alterations of both proto-oncogenes and tumor suppressor genes in human and animal neoplasms.

Thus, all three key elements of Howard's hypothesis for the origin of cancer genes have proven correct. First, mutations play a critical role in converting proto-oncogenes to oncogenes as well as in inactivating tumor suppressor genes. Second, similar oncogenes are found in both strongly transforming retroviruses and non-virus-induced tumors, including human cancers. Third, reverse transcription is involved in at least some mutations of proto-oncogenes and tumor suppressor genes in non-virus-induced tumors as well as being responsible for the incorporation of oncogenes into retroviral genomes. Most important, Howard realized that the same molecular mechanisms are ultimately responsible for both viral and non-viral carcinogenesis. The finding that related oncogenes are involved in both retroviruses and human cancers was particularly important in unifying these previously distinct areas of cancer research and contributed substantially to the major advances in understanding the molecular biology of human cancer that have taken place over the last decade.

FUNCTIONS OF PROTO-ONCOGENES IN CONTROL OF CELL PROLIFERATION

The discovery that oncogenes are mutated versions of normal cell genes (proto-oncogenes) focused attention on the roles of proto-oncogenes in normal cells and on the nature of the molecular alterations that convert proto-oncogenes to oncogenes. Howard frequently expressed the view that cancer results from mutations in genetic elements that normally function to control cell growth and differentiation (85, 86). This notion has been thoroughly substantiated by characterization of the proteins encoded by proto-oncogenes and by studies of the functions of both proto-oncogene and oncogene products. In this and the following section, I discuss some of the functions of proto-oncogene proteins (particularly the Ras proteins) in signal transduction pathways that regulate cell proliferation and differentiation in response to extracellular stimuli.

The proteins encoded by most proto-oncogenes can be divided into five functional groups: growth factors, protein-tyrosine kinases, guanine nucleotide binding proteins, protein-serine/threonine kinases, and transcription factors (1, 19, 21) (Table 3). Many of the protein-tyrosine kinases encoded by proto-oncogenes are growth factor receptors, which can be converted to oncogenes by mutations that result in their unregulated activity. The fact that a variety of proto-oncogenes encode either growth factors or growth factor receptors clearly indicates that cell transformation can result from aberrant functioning of proteins that regulate proliferation of normal cells. The Ras proteins also play a critical role in control of normal cell proliferation, serving to couple protein-tyrosine kinase receptors to the activation of a cascade of cytosolic protein-serine/threonine kinases (Fig. 1). Activation of these protein kinases, including the Raf proto-oncogene protein, then results in the phosphorylation of a variety of cytosolic and nuclear proteins, including transcription factors. Many of the proto-oncogenes thus encode proteins that normally function as elements in signal transduction pathways linking growth factor receptors to programmed changes in gene expression, ultimately leading to cell proliferation. The oncogenes encode altered forms of these proteins, which function abnormally to drive unregulated cell proliferation and transformation.

In addition to being the first oncogenes identified in human cancers, the *ras* genes have provided an important model for studies of the functions of proto-oncogene and oncogene proteins (55). Three cellular *ras* genes—*rasH*, *rasK*, and *rasN*—have been identified as oncogenes in human tumors. They encode closely

TABLE 3
Functions of representative proto-oncogenes

Cellular function	Proto-oncogene proteins[a]
Growth factors	PDGF, FGF, Hst, Int-2, EGF, IL-2
Protein-tyrosine kinases	
Receptors	ErbB, ErbB-2, Fms, Kit, Met, Ret, Trk
Nonreceptor kinases	Src, Yes, Fgr, Lck, Abl, Fes
GTP binding proteins	Ras, G_s, G_q, G_i, G_{12}
Protein-serine/threonine kinases	Raf, Mos, Akt, Cot, Tpl-2
Transcription factors	ErbA, PML/RAR, Fos, Jun, Myc, Myb, Ets, Rel, E2A/Pbx

[a] PDGF, platelet-derived growth factor; FGF, fibroblast growth factor; EGF, epidermal growth factor; IL-2, interleukin 2.

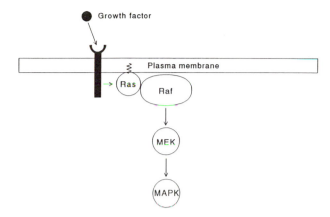

FIGURE 1. Ras proteins in mitogenic signal transduction. Ras proteins are activated downstream of receptor protein-tyrosine kinases. The Raf protein-serine/threonine kinase then binds to Ras, which serves to recruit Raf to the plasma membrane. Activated Raf then phosphorylates and activates MEK, which in turn phosphorylates and activates MAP kinase (MAPK).

related proteins of 188 or 189 amino acids which differ primarily in a region of 20 amino acids near the carboxy terminus. Downstream of this variable region, a conserved cysteine is the site of posttranslational addition of lipid (a farnesyl isoprenoid), which is required for both plasma membrane localization and biological activity of Ras proteins (20, 36). As discussed below, this processing step may provide a specific target for the action of chemotherapeutic drugs against tumors containing *ras* oncogenes.

The activity of Ras proteins is controlled by guanine nucleotide binding and hydrolysis, such that they alternate between inactive GDP-bound and active GTP-bound states (55). Ras proteins are coupled to receptor protein-tyrosine kinases by guanine nucleotide exchange factors that stimulate the exchange of Ras-GDP for GTP following growth factor stimulation (11). Activity of the Ras-GTP complex is then terminated by GTP hydrolysis. The mutations that convert *ras* proto-oncogenes to oncogenes alter the interaction of Ras proteins with guanine nucleotides either by decreasing Ras GTPase activity or by increasing the rate of exchange of bound GDP for free GTP. Both of these alterations increase the amount of GTP-bound active Ras protein, resulting in the unregulated constitutive activity of the Ras oncogene proteins that leads to cell transformation.

The initial insights into the role of Ras in signal transduction pathways in mammalian cells came from experiments in which microinjection of an anti-Ras monoclonal antibody was used to block Ras function (63, 76). Microinjection of cells with this antibody was found to inhibit cell proliferation and transformation by *src* or *ras* oncogenes but not by *raf*. Similar results were obtained using a dominant inhibitory Ras mutant (Ras N-17), which preferentially binds GDP rather than GTP and appears to interfere with normal Ras activity by blocking the actions of guanine nucleotide exchange factors (17, 27). These experiments thus suggested that these oncogene proteins function in a signal transduction cascade. Protein-tyrosine kinases, such as Src, require Ras to induce transformation and therefore act upstream of Ras. On the other hand, transformation by protein-serine/threonine kinases, such as Raf, does not require Ras activity, suggesting that Raf functions downstream of Ras. Consistent with this hypothesis, dominant inhibitory mutants of Raf also inhibit cell proliferation and block the action of both protein-tyrosine kinases and Ras (48).

The hypothesis that Raf functions downstream of Ras was then supported by biochemical analysis of Raf activation in response to growth factor treatment (90, 98). Growth factor stimulation of protein-tyrosine kinase receptors leads to activation of the Raf protein kinase. Importantly, such growth factor-stimulated activation of Raf is blocked by expression of the dominant inhibitory Ras N-17 mutant, indicating that Ras function is required for Raf activation. Taken together with the ability of a *raf* oncogene to induce transformation even when normal Ras activity is inhibited, these results indicate that Raf functions downstream of Ras in a signaling pathway leading to cell proliferation.

Recently, several research groups have demonstrated that the Ras and Raf proteins form a complex via direct protein-protein interactions (62, 91, 93, 95, 102). This association between Ras and Raf is specific for the active GTP-bound form of Ras and is dependent on the presence of a functional Ras effector domain. It thus appears that the conformational change induced by GTP binding allows Ras to bind to Raf, indicating that the Raf protein-serine/threonine kinase is a direct effector of Ras in mammalian cells. Importantly, however, interaction with Ras does not by itself appear to activate the Raf protein kinase. Rather, the Ras-Raf interaction is thought to recruit Raf to the plasma membrane, where it can then be activated by membrane lipids or other protein kinases. Consistent with this possibility, addition of a membrane-targeting signal to Raf results in localization of the protein to the plasma membrane and constitutive activation independent of Ras function (52, 80). The mechanism of Raf activation thus remains an open question in our understanding of mitogenic signal transduction, as is further discussed below.

Activation of Raf initiates a protein kinase cascade that ultimately results in phosphorylation of transcription factors and consequent alterations in gene expression. In particular, stimulation of Raf leads to activation of mitogen-activated protein (MAP) kinases, which are stimulated by a variety of mitogenic growth factors and other extracellular stimuli (10) (Fig. 1). Raf phosphorylates and activates another protein kinase, called MAP kinase kinase, or MEK, which in turn phosphorylates and activates MAP kinase (22, 39, 50). MAP kinase then phosphorylates a variety of proteins, including another protein kinase called RSK, which phosphorylates ribosomal protein S6. Importantly, both MAP kinase and RSK translocate between the nucleus and the cytoplasm. Furthermore, their substrates include a number of transcription factors (10, 29, 38, 59), so activation of the MAP kinase pathway provides a clear link between the cytosol and the nucleus, coupling the stimulation of growth factor receptors to changes in gene expression.

As noted above, the finding of protein-protein interactions between Ras and Raf indicates that Raf is a direct Ras effector. However, it appears that the Ras-

Raf interaction does not itself activate the Raf kinase. Instead, the major role of the Ras-Raf interaction is to recruit Raf to the plasma membrane, where it can be activated by other signals which remain to be determined. One possibility would be the direct activation of Raf by membrane lipids, such as diacylglycerol, but such lipid activators of Raf have not yet been identified. Alternatively, Raf activation may involve its phosphorylation by other membrane-associated protein kinases. Raf is primarily phosphorylated on serine residues in mitogen-stimulated cells, but tyrosine phosphorylation of Raf also occurs and may play a role in regulating Raf activity (26). In addition, Raf is phosphorylated by several isotypes of protein kinase C (PKC) (47, 77). Although the effect of PKC phosphorylation on Raf activity is controversial (57), both protein-tyrosine and serine/threonine kinases presently appear to be plausible candidates for Raf activators.

Additional experiments further implicate second messengers derived from phosphatidylcholine (PC) in Raf activation. A variety of growth factors stimulate hydrolysis of PC in addition to hydrolysis of phosphatidylinositol bisphosphate. PC hydrolysis yields diacylglycerol but not inositol triphosphate, so it is thought to activate nonclassic calcium-independent PKC isotypes (64). Treatment of cells with exogenous PC-phospholipase C (PC-PLC) is itself mitogenic (51), and overexpression of a cloned bacterial PC-PLC gene induces cell transformation (43), indicating that PC hydrolysis is sufficient to signal cell proliferation. Although the mechanism by which growth factors stimulate PC hydrolysis is not known, it appears to be activated downstream of Ras, since it is inhibited by expression of the dominant negative Ras N-17 mutant (15). Interestingly, however, PC hydrolysis appears to function upstream of Raf, since overexpression of PC-PLC overcomes the inhibitory effects of dominant negative Ras but not dominant negative Raf mutants (15, 16). Consistent with this, treatment of cells with exogenous PC-PLC activates the Raf protein kinase (16).

It thus appears that growth factor stimulation of PC hydrolysis provides at least one pathway for the generation of second messengers (e.g., diacylglycerol) that can induce Raf activation, possibly mediated by nonclassic isotypes of PKC. The nonclassic PKC isotype, PKCε, is known to be activated by PC-derived diacylglycerol (31), and overexpression of PKCε induces cell transformation (14, 61), suggesting that PKCε is a good candidate for participation in mitogenic signal transduction downstream of PC-PLC. In recent studies, we therefore investigated the possible role of PKCε in Raf activation (14a). A dominant inhibitory mutant in which the catalytic C-terminal domain of PKCε was deleted was found to inhibit cell proliferation, and this inhibitory effect was overcome by Raf but not by Src, which is consistent with the possible role of PKCε upstream of Raf. Moreover, in transient expression assays in COS cells, the dominant inhibitory PKCε mutant blocked Raf activation in response to either growth factors or PC-PLC. Conversely, overexpression of wild-type PKCε or a constitutively active mutant stimulated Raf activity in either untreated or growth factor-treated cells. These results thus suggest that stimulation of PKCε by PC-derived diacylglycerol is one mechanism of Raf activation. Whether this involves direct phosphorylation of Raf by PKCε or the intermediary activity of still other protein kinases requires further study.

PROTO-ONCOGENES IN DEVELOPMENT AND DIFFERENTIATION

In addition to their roles in controlling cell proliferation, many proto-onco-genes also function to regulate cell differentiation. Furthermore, since carefully regulated cell proliferation and differentiation are both critical to normal development, it is not surprising that a number of proto-oncogenes play critical roles in embryogenesis, a connection that Howard anticipated in 1970 (85, 86) and that has now provided important insights into the molecular mechanisms that regulate the fundamental processes of development and differentiation.

The action of Ras and Raf in signal transduction pathways downstream of receptor protein-tyrosine kinases provides an important model for the roles of proto-oncogenes in several developmental systems as well as in mitogenic signaling. For example, a number of growth factors and growth factor receptors, as well as Ras and Raf, are expressed in early mouse embryos, suggesting that auto-crine growth factor signaling may play a role in early embryo development (66, 69, 70). Consistent with this hypothesis, interference with the Ras-Raf signaling pathway by expression of the dominant negative Ras N-17 mutant blocks development of mouse embryos at the two-cell stage (99). Since this is the stage at which transcription of the embryonic genome initiates during mouse development, it is possible that autocrine signaling mediated by the Ras-Raf pathway plays a role in this period of transition between maternal and embryonic programs of gene expression.

In *Xenopus* embryos, the Ras-Raf pathway plays a critical role in formation of mesoderm, a classic example of embryonic induction. Mesoderm formation involves the differentiation of animal pole cells in response to an inductive factor produced by vegetal pole cells of the *Xenopus* embryo. The differentiation of animal pole cells to mesoderm can be induced experimentally by synergistic action of the mammalian growth factors fibroblast growth factor (FGF) and transforming growth factor β, suggesting the possibility that these growth factors function as natural inducers of mesoderm formation. This hypothesis is supported by the identification of *Xenopus* maternal RNAs encoding basic FGF and members of the transforming growth factor β family as well as FGF receptor. Moreover, dominant negative mutants of the FGF receptor, Ras, and Raf all block mesoderm formation, providing direct evidence that growth factor stimulation of the Ras-Raf signaling pathway plays a critical role in this process of embryonic induction (3, 58, 97).

The extensive genetic analysis that is possible in *Drosophila melanogaster* and *Caenorhabditis elegans* has also provided critical insights into the normal developmental functions of several proto-oncogene proteins, including Ras and Raf. In *D. melanogaster*, development of the terminal anterior and posterior regions of the embryo is governed by several genes expressed as maternal messages. One of these genes, *torso*, encodes a receptor tyrosine kinase which is similar to the mammalian platelet-derived growth factor (PDGF) receptor (78). Another member of this group of genes, *l(1)polehole*, is the *Drosophila* homolog of *raf*, which has been demonstrated genetically to function downstream of *torso* (4). Further experiments have shown that Ras is also required for Torso signaling and that it functions upstream of Raf (56). It thus appears that determination of cell fate at the termini of *Drosophila* embryos involves localized stimulation of the Torso receptor tyrosine kinase, leading to activation of the Ras-Raf pathway and transcriptional activation of genes that are specifically expressed in terminal regions of the embryo.

The Ras-Raf pathway is similarly involved in *Drosophila* eye development. The compound eye of *D. melanogaster* is composed of about 800 simple eyes, each of which contains eight photoreceptor neurons (called R1 through R8). These cells differentiated in a fixed sequence, starting with R8 and ending with R7. Differentiation of the R7 cell is induced by interaction with the neighboring R8 cell, and several genes required for R7 differentiation have been identified. One of these, *sevenless,* encodes a receptor protein-tyrosine kinase (32) which recognizes a ligand expressed on the surface of the R8 cell. Further genetic analysis has again established that signaling from Sevenless proceeds via the Ras-Raf pathway and activation of MAP kinase (9, 24, 75).

Vulval induction in *C. elegans* is a third well-characterized developmental pathway in which Ras and Raf are involved. In this system, the developmental fates of vulval precursor cells are determined by an inductive signal from a gonadal anchor cell. The genes required for vulval induction include *let-23,* which encodes a receptor tyrosine kinase related to the epidermal growth factor receptor (5). Downstream of Let-23, vulval differentiation requires the *let-60* and *lin-45* genes, which encode Ras and Raf proteins, respectively (8, 34, 35). Stimulation of the Ras-Raf pathway by tyrosine kinase receptors is thus a common mechanism of signaling cell differentiation in both *C. elegans* and *D. melanogaster.*

Signaling from tyrosine kinase receptors via Ras and Raf has also been implicated in the differentiation of some mammalian cell types, particularly neurons. Development of the mammalian nervous system is regulated by neurotropic factors, which signal differentiation and survival of neuronal cell types. The prototype neurotrophin is nerve growth factor (NGF), which is related to three other factors: brain-derived neurotrophic factor, neurotrophin-3 (NT-3), and NT-4. The receptors for these factors are protein-tyrosine kinases encoded by the *trk* family of proto-oncogenes (30, 44, 45). In particular, Trk is the receptor for NGF, TrkB is the receptor for brain-derived neurotrophic factor and NT-4, and TrkC is the receptor for NT-3. Thus, a family of protein-tyrosine kinases initially identified as oncogenes that induce transformation of fibroblasts normally functions to regulate neuronal differentiation.

The PC12 pheochromocytoma cell line has been widely used as a cell culture model for studying the induction of neuronal differentiation by NGF. Upon NGF treatment, PC12 cells differentiate to form cells resembling sympathetic neurons, as characterized by neurite extension and changes in gene expression. Importantly, introduction of activated *src, ras,* or *raf* oncogenes into PC12 cells similarly induces neuronal differentiation rather than cell proliferation (2, 7, 65, 90). Conversely, interference with Ras function by antibodies or the dominant negative Ras N-17 mutant blocks differentiation induced by NGF (33, 82) as well as activation of Raf and MAP kinase (89, 90, 98). Thus, activation of the same Ras-Raf-MAP kinase pathway that signals proliferation of most cells (e.g., fibroblasts) causes PC12 cells to cease proliferation and differentiate.

In addition to inducing differentiation, NGF prevents programmed cell death, thus playing a key role in development of the nervous system (68). In vertebrates, up to 50% of developing neurons are eliminated by apoptosis, with survival of those neurons that make the correct connections being signaled by NGF or other neurotrophins. The action of NGF in signaling cell survival is also reflected by its activity on PC12 cells in culture, where it prevents apoptosis in serum-free medium. NGF induces cell survival, as well as differentiation, by stimulating the Trk receptor (41, 54). In contrast to its action in differentiation, however, the action of NGF in inhibiting apoptosis is not blocked by expression of the dominant inhibitory Ras N-17 mutant and therefore appears to be mediated by a Ras-independent signaling pathway (100). Recent studies (100) further indicate that this pathway involves phosphatidylinositol 3-kinase (PI 3-kinase), which is activated downstream of a number of protein-tyrosine kinases, including Trk (19). A specific inhibitor of PI 3-kinase, wortmannin, induces apoptosis of PC12 cells, indicating that PI 3-kinase is required for cell survival (100). These results were further substantiated by studies of PC12 cells into which genes encoding either wild-type or mutant PDGF receptors had been introduced. PDGF acted similarly to NGF in inducing both differentiation and survival of PC12 cells expressing the wild-type receptor. On the other hand, PDGF was able to induce differentiation but not able to prevent apoptosis of PC12 cells expressing a mutant receptor that failed to activate PI 3-kinase.

It thus appears that distinct signaling pathways activated downstream of the Trk receptor protein-tyrosine kinases are responsible for induction of differentiation and prevention of apoptosis by NGF (Fig. 2). Differentiation is induced by activation of the Ras-Raf-MAP kinase pathway and does not require PI 3-kinase. In contrast, cell survival is dependent upon the activation of PI 3-kinase and does not require Ras. In terms of Howard's thoughts on the origin of cancer genes from genes normally involved in embryonic development, it is noteworthy that not only the Trk receptor but also many of the downstream signaling elements that mediate differentiation and survival of developing neurons were initially identified either as oncogenes (e.g., Ras and Raf) or as oncogene-associated proteins (e.g., PI 3-kinase) involved in cell transformation.

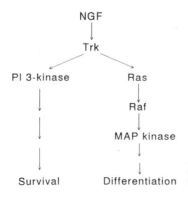

FIGURE 2. Distinct pathways signal differentiation and survival of PC12 cells in response to NGF.

ONCOGENES AS TARGETS FOR CANCER TREATMENT

In the 1970s, Howard took the position that human cancers were not caused by strongly transforming retroviruses; therefore, studies of these viruses were not directly applicable to human cancer prevention or treatment. Indeed, Howard's assertion, at the time he received the Nobel prize in 1975, that stopping smoking was more important to cancer prevention than retroviruses has been vividly remembered. Although Howard's view of the lack of importance of strongly transforming retroviruses in human cancer was not appreciated by many of his virologist colleagues in the 1970s, he once again proved to be correct. Thus, Howard felt that eventual inroads into the treatment of cancer might come from understanding the molecular basis of the conversion of normal cells to cancer cells rather than through attempts to isolate retroviruses that cause cancer in humans. In this context, it seems appropriate to close this chapter with a brief consideration of the possibility that oncogenes provide novel targets against which drugs with increased selectivity for cancer cells could be designed.

The development of drugs that target Ras proteins is being actively pursued by a number of research groups and provides an example of potential therapies that target an oncogene protein. The *ras* genes are the most commonly mutated proto-oncogenes in human cancers, with *ras* mutations playing a role in ~50% of colorectal cancers, ~25% of acute myeloid leukemias, ~25% of lung cancers, and ~90% of pancreatic cancers (13). Overall, mutations of *ras* proto-oncogenes are found in ~15% of all cancers in the United States. Thus, a successful therapy targeted against Ras oncogene proteins might be expected to make a significant contribution to cancer treatment. Since the Ras oncogene proteins differ from their normal homologs, the possibility of designing drugs that would specifically inhibit proteins encoded by mutated *ras* oncogenes has been widely considered. However, no such compounds have been identified, and the design of drugs that would selectively inhibit Ras oncogene proteins but not their normal proto-oncogene homologs is unquestionably a difficult undertaking.

On the other hand, substantial progress has been made in the development of drugs that inhibit Ras protein processing (28). Although Ras proto-oncogene proteins are clearly required for normal cell proliferation, these inhibitors of Ras processing display surprising selectivity in interfering with the proliferation of cells transformed by *ras* oncogenes. As noted earlier in this chapter, posttranslational modification of Ras proteins by addition of a farnesyl isoprenoid is required for their membrane association and biological activity. Although farnesylation is not unique to Ras, it is a relatively uncommon modification of cellular proteins. Consequently, the importance of farnesylation in Ras function prompted the development of inhibitors of farnesyl-protein transferase as potential Ras-targeted drugs. Several inhibitors of this enzyme have now been identified and found to selectively interfere with transformation by *ras* oncogenes under conditions in which they do not inhibit normal cell growth (28, 42, 46). It thus appears that *ras*-transformed cells may be more sensitive than normal cells to interference with Ras processing. Although the molecular basis for this selectivity is not yet understood, it appears to offer a therapeutic window that might allow specific treatment of tumors in which *ras* oncogenes are involved. Should this or other oncogene-targeted therapies ultimately prove to be clinically useful, Howard Temin's ideas on the origin of cancer genes will have been translated to a molecular basis for the rational treatment of human neoplastic disease.

REFERENCES

1. **Aaronson, S. A.** 1991. Growth factors and cancer. *Science* **254:**1146–1153.
2. **Alema, S., P. Casalbore, E. Agostini, and F. Tato.** 1985. Differentiation of PC12 phaeochromocytoma cells induced by v-*src* oncogene. *Nature* (London) **316:**557–559.
3. **Amaya, E., T. J. Musci, and M. W. Kirschner.** 1991. Expression of a dominant negative mutant of the FGF receptor disrupts mesoderm formation in *Xenopus* embryos. *Cell* **66:**257–270.
4. **Ambrosio, L., A. P. Mahowald, and N. Perrimon.** 1989. Requirement of the *Drosophila raf* homologue for *torso* function. *Nature* (London) **342:**288–291.
5. **Aroian, R. V., M. Koga, J. E. Mendel, Y. Ohshima, and P. W. Sternberg.** 1990. The *let-23* gene necessary for *Caenorhabditis elegans* vulval induction encodes a tyrosine kinase of the EGF receptor subfamily. *Nature* (London) **348:**693–698.
6. **Baltimore, D.** 1970. RNA-dependent DNA polymerase in virions of RNA tumour viruses. *Nature* (London) **226:**799–801.
7. **Bar-Sagi, D., and J. R. Feramisco.** 1985. Microinjection of the *ras* oncogene protein into PC12 cells induces morphological differentiation. *Cell* **42:**841–848.
8. **Beitel, G. J., S. G. Clark, and H. R. Horvitz.** 1990. *Caenorhabditis elegans ras* gene *let-60* acts as a switch in the pathway of vulval induction. *Nature* (London) **348:**503–509.
9. **Biggs, W. H., III, K. H. Zavitz, B. Dickson, A. van der Straten, D. Brunner, E. Hafen, and S. L. Zipursky.** 1994. The *Drosophila rolled* locus encodes a MAP kinase required in the sevenless signal transduction pathway. *EMBO J.* **13:**1628–1635.
10. **Blenis, J.** 1993. Signal transduction via the MAP kinases: proceed at your own RSK. *Proc. Natl. Acad. Sci. USA* **90:**5889–5892.
11. **Boguski, M. S., and F. McCormick.** 1993. Proteins that regulate Ras and its relatives. *Nature* (London) **366:**643–654.
12. **Bolen, J. B.** 1993. Nonreceptor tyrosine protein kinases. *Oncogene* **8:**2025–2031.
13. **Bos, J. L.** 1989. *Ras* oncogenes in human cancer: a review. *Cancer Res.* **49:**4682–4689.
14. **Cacace, A., S. N. Guadagno, R. S. Krauss, D. Fabbro, and I. B. Weinstein.** 1993. The epsilon isoform of protein kinase C is an oncogene when overexpressed in rat fibroblasts. *Oncogene* **8:**2094–2114.
14a. **Cai, H., and G. M. Cooper.** Unpublished results.
15. **Cai, H., P. Erhardt, J. Szeberenyi, M. T. Diaz-Meco, J. Moscat, and G. M. Cooper.** 1992. Hydrolysis of phosphatidylcholine is stimulated by Ras proteins during mitogenic signal transduction. *Mol. Cell. Biol.* **12:**5329–5335.
16. **Cai, H., P. Erhardt, J. Troppmair, M. T. Diaz-Meco, G. Sithanandam, U. R. Rapp, J. Moscat, and G. M. Cooper.** 1993. Hydrolysis of phosphatidylcholine couples Ras to activation of Raf protein kinase during mitogenic signal transduction. *Mol. Cell. Biol.* **13:**7645–7651.
17. **Cai, H., J. Szeberenyi, and G. M. Cooper.** 1990. Effect of a dominant inhibitory Ha-*ras* mutation on mitogenic signal transduction in NIH 3T3 cells. *Mol. Cell. Biol.* **10:**5314–5323.
18. **Canaani, E., O. Dreazen, A. Klar, G. Rechavi, D. Ram, J. B. Cohen, and D. Givol.** 1983. Activation of the c-*mos* oncogene in a mouse plasmacytoma by insertion of an endogenous intracisternal A-particle genome. *Proc. Natl. Acad. Sci. USA* **80:**7118–7122.
19. **Cantley, L. C., K. R. Auger, C. Carpenter, B. Duckworth, A. Graziani, R. Kapeller, and S. Soltoff.** 1991. Oncogenes and signal transduction. *Cell* **64:**281–302.
20. **Casey, P. J., P. A. Solski, C. J. Der, and J. E. Buss.** 1989. p21^{ras} is modified by a farnesyl isoprenoid. *Proc. Natl. Acad. Sci. USA* **86:**8323–8327.
21. **Cooper, G. M.** 1995. *Oncogenes,* 2d ed. Jones and Bartlett Publishers, Boston.
22. **Dent, P., W. Haser, T. A. J. Haystead, L. A. Vincent, T. M. Roberts, and T. W. Sturgill.** 1992.

Activation of mitogen-activated protein kinase kinase by v-Raf in NIH 3T3 cells and *in vitro*. *Science* **257**:1404–1407.

23. **Der, C. J., T. G. Krontiris, and G. M. Cooper.** 1982. Transforming genes of human bladder and lung carcinoma cell lines are homologous to the *ras* genes of Harvey and Kirsten sarcoma viruses. *Proc. Natl. Acad. Sci. USA* **79**:3637–3640.

24. **Dickson, B., F. Sprenger, D. Morrison, and E. Hafen.** 1992. Raf functions downstream of Ras1 in the Sevenless signal transduction pathway. *Nature* (London) **360**:600–603.

25. **Duhrsen, U., J. Stahl, and N. M. Gough.** 1990. *In vivo* transformation of factor-dependent hemo-poietic cells: role of intracisternal A-particle transposition for growth factor gene activation. *EMBO J.* **9**:1087–1096.

26. **Fabian, J. R., I. O. Daar, and D. K. Morrison.** 1993. Critical tyrosine residues regulate the enzymatic and biological activity of Raf-1 kinase. *Mol. Cell. Biol.* **13**:7170–7179.

27. **Feig, L. A., and G. M. Cooper.** 1988. Inhibition of NIH 3T3 cell proliferation by a mutant *ras* protein with preferential affinity for GDP. *Mol. Cell. Biol.* **8**:3235–3243.

28. **Gibbs, J. B., A. Oliff, and N. E. Kohl.** 1994. Farnesyltransferase inhibitors: Ras research yields a potential cancer therapeutic. *Cell* **77**:175–178.

29. **Gille, H., A. D. Sharrocks, and P. E. Shaw.** 1992. Phosphorylation of transcription factor p62^{TCF} by MAP kinase stimulates ternary complex formation at c-*fos* promoter. *Nature* (London) **358**:414–417.

30. **Glass, D. J., and G. D. Yancopoulos.** 1993. The neurotrophins and their receptors. *Trends Cell Biol.* **3**:262–268.

31. **Ha, K. S., and J. H. Exton.** 1993. Differential translocation of protein kinase C isozymes by thrombin and platelet-derived growth factor. A possible function for phosphatidylcholine-derived diacylglycerol. *J. Biol. Chem.* **268**:10534–10539.

32. **Hafen, E., K. Basler, J.-E. Edstroem, and G. M. Rubin.** 1987. *Sevenless*, a cell-specific homeotic gene of *Drosophila*, encodes a putative transmembrane receptor with a tyrosine kinase domain. *Science* **236**:55–63.

33. **Hagag, N., S. Halegoua, and M. Viola.** 1986. Inhibition of growth factor-induced differentiation of PC12 cells by microinjection of antibody to *ras* p21. *Nature* (London) **319**:680–682.

34. **Han, M., A. Golden, Y. Han, and P. W. Sternberg.** 1993. C. elegans lin-45 raf gene participates in *let-60 ras*-stimulated vulval differentiation. *Nature* (London) **363**:133–140.

35. **Han, M., and P. W. Sternberg.** 1990. *let-60*, a gene that specifies cell fates during *C. elegans* vulval induction, encodes a *ras* protein. *Cell* **63**:921–931.

36. **Hancock, J. F., A. I. Magee, J. E. Childs, and C. J. Marshall.** 1989. All *ras* proteins are polyiso-prenylated but only some are palmitoylated. *Cell* **57**:1167–1177.

37. **Hayward, W. S., B. G. Neel, and S. M. Astrin.** 1981. Activation of a cellular *onc* gene by promoter insertion in ALV-induced lymphoid leukosis. *Nature* (London) **290**:475–480.

38. **Hill, C. S., R. Marais, S. John, J. Wynne, S. Dalton, and R. Treisman.** 1993. Functional analysis of a growth factor-responsive transcription factor complex. *Cell* **73**:395–406.

39. **Howe, L. R., S. J. Leevers, N. Gomez, S. Nakielny, P. Cohen, and C. J. Marshall.** 1992. Activation of the MAP kinase pathway by the protein kinase raf. *Cell* **71**:335–342.

40. **Huebner, R. J., and G. J. Todaro.** 1969. Oncogenes of RNA tumor viruses as determinants of cancer. *Proc. Natl. Acad. Sci. USA* **64**:1087–1094.

41. **Ibanez, C. F., T. Ebendal, G. Barbany, J. Murray-Rust, T. L. Blundell, and H. Persson.** 1992. Disruption of the low affinity receptor-binding site in NGF allows neuronal survival and differen-tiation by binding to the *trk* gene product. *Cell* **69**:329–341.

42. **James, G. L., J. L. Goldstein, M. S. Brown, T. E. Rawson, T. C. Somers, R. S. McDowell, C. W. Crowley, B. K. Lucas, A. D. Levinson, and J. C. Marsters, Jr.** 1993. Benzodiazepine peptidomimetics: potent inhibitors of Ras farnesylation in animal cells. *Science* **260**:1937–1942.

43. **Johansen, T., G. Bjorkoy, A. Overvatn, M. T. Diaz-Meco, T. Traavik, and J. Moscat.** 1994. NIH 3T3 cells stably transfected with the gene encoding phosphatidylcholine-hydrolyzing phospholi-pase C from *Bacillus cereus* acquire a transformed phenotype. *Mol. Cell. Biol.* **14**:646–654.

44. **Kaplan, D. R., B. L. Hempstead, D. Martin-Zanca, M. V. Chao, and L. F. Parada.** 1991. The *trk* proto-oncogene product: a signal transducing receptor for nerve growth factor. *Science* **252**:554–558.

45. **Klein, R., S. Jing, V. Nanduri, E. O'Rourke, and M. Barbacid.** 1991. The *trk* proto-oncogene encodes a receptor for nerve growth factor. *Cell* **65:**189–197.

46. **Kohl, N. E., S. D. Mosser, S. J. deSolms, E. A. Giuliani, D. L. Pompliano, S. L. Graham, R. L. Smith, E. M. Scolnick, A. Oliff, and J. Gibbs.** 1993. Selective inhibition of *ras*-dependent transformation by a farnesyltransferase inhibitor. *Science* **260:**1934–1937.

47. **Kolch, W., G. Heidecker, G. Kochs, R. Hummel, H. Vahidi, H. Mischak, G. Finkenzeller, D. Marme, and U. R. Rapp.** 1993. PKC-α activates Raf-1 by direct phosphorylation. *Nature* (London) **364:**426–428.

48. **Kolch, W., G. Heidecker, P. Lloyd, and U. R. Rapp.** 1991. Raf-1 protein kinase is required for growth of induced NIH 3T3 cells. *Nature* (London) **349:**249–252.

49. **Krontiris, T. G., and G. M. Cooper.** 1981. Transforming activity of human tumor DNAs. *Proc. Natl. Acad. Sci. USA* **78:**1181–1184.

50. **Kyriakis, J. M., H. App, X. Zhang, P. Banerjee, D. L. Brautigan, U. R. Rapp, and J. Avruch.** 1992. Raf-1 activates MAP kinase-kinase. *Nature* (London) **358:**417–421.

51. **Larrodera, P., M. E. Cornet, M. T. Diaz-Meco, M. Lopez-Barrahona, I. Diaz-Laviada, P. H. Guddal, T. Johansen, and J. Moscat.** 1990. Phospholipase C-mediated hydrolysis of phosphatidyl-choline is an important step in PDGF-stimulated DNA synthesis. *Cell* **61:**1113–1120.

52. **Leevers, S. J., H. F. Paterson, and C. J. Marshall.** 1994. Requirement for Ras in Raf activation is overcome by targeting Raf to the plasma membrane. *Nature* (London) **369:**411–414.

53. **Levine, A. J.** 1993. The tumor suppressor genes. *Annu. Rev. Biochem.* **62:**623–651.

54. **Loeb, D. M., J. Maragos, D. Martin-Zanca, M. V. Chao, L. F. Parada, and L. A. Greene.** 1991. The *trk* proto-oncogene rescues NGF responsiveness in mutant NGF-nonresponsive PC12 cell lines. *Cell* **66:**961–966.

55. **Lowy, D. R., and B. M. Willumsen.** 1993. Function and regulation of Ras. *Annu. Rev. Biochem.* **62:**851–891.

56. **Lu, X., T.-B. Chou, N. G. Williams, T. M. Roberts, and N. Perrimon.** 1993. Control of cell fate determination by p21*^{ras}*/Ras1, an essential component of *torso* signaling in *Drosophila*. *Genes Dev.* **7:**621–632.

57. **Macdonald, S. G., C. M. Crews, L. Wu, J. Driller, R. Clark, R. L. Erikson, and F. McCormick.** 1993. Reconstitution of the Raf-1–MEK–ERK signal transduction pathway in vitro. *Mol. Cell. Biol.* **13:**6615–6620.

58. **MacNicol, A. M., A. J. Muslin, and L. T. Williams.** 1993. Raf-1 kinase is essential for early *Xenopus* development and mediates the induction of mesoderm by FGF. *Cell* **73:**571–583.

59. **Marais, R., J. Wynne, and R. Treisman.** 1993. The SRF accessory protein Elk-1 contains a growth factor-regulated transcriptional activation domain. *Cell* **73:**381–393.

60. **Miki, Y., I. Nishisho, A. Horii, Y. Miyoshi, J. Utsunomiya, K. W. Kinzler, B. Vogelstein, and Y. Nakamura.** 1992. Disruption of the *APC* gene by a retrotransposal insertion of L1 sequence in a colon cancer. *Cancer Res.* **52:**643–645.

61. **Mischak, H., J. Goodnight, W. Kolch, G. Martiny-Baron, C. Schaechtle, M. G. Kazanietz, P. M. Blumberg, J. H. Pierce, and J. F. Mushniski.** 1993. Overexpression of protein kinase C-δ and -ε in NIH 3T3 cells induces opposite effects on growth, morphology, anchorage dependence, and tumorigenicity. *J. Biol. Chem.* **268:**6090–6096.

62. **Moodie, S. A., B. M. Willumsen, M. J. Weber, and A. Wolfman.** 1993. Complexes of Ras-GTP with Raf-1 and mitogen-activated protein kinase kinase. *Science* **260:**1658–1661.

63. **Mulcahy, L. S., M. R. Smith, and D. W. Stacey.** 1985. Requirement for *ras* proto-oncogene function during serum-stimulated growth of NIH 3T3 cells. *Nature* (London) **313:**241–243.

64. **Nishizuka, Y.** 1992. Intracellular signaling by hydrolysis of phospholipids and activation of protein kinase C. *Science* **258:**607–614.

65. **Noda, M., M. Ko, A. Ogura, D. Liu, T. Amano, T. Takano, and Y. Ikawa.** 1985. Sarcoma viruses carrying *ras* oncogenes induce differentiation-associated properties in a neuronal cell line. *Nature* (London) **318:**73–75.

66. **Pal, S. K., R. Crowell, A. A. Kiessling, and G. M. Cooper.** 1993. Expression of proto-oncogenes in mouse eggs and preimplantation embryos. *Mol. Reprod. Dev.* **35:**8–15.

67. **Parada, L. F., C. J. Tabin, C. Shih, and R. A. Weinberg.** 1982. Human EJ bladder carcinoma oncogene is homologue of Harvey sarcoma virus *ras* gene. *Nature* (London) **297:**474–478.

68. **Raff, M. C., B. A. Barres, J. F. Burne, H. S. Coles, Y. Ishizaki, and M. D. Jacobson.** 1993. Programmed cell death and the control of cell survival: lessons from the nervous system. *Science* **262:**695–700.

69. **Rappolee, D. A., C. A. Brenner, R. Schultz, D. Mark, and Z. Werb.** 1988. Developmental expression of PDGF, TGF-α, and TGF-β genes in preimplantation mouse embryos. *Science* **241:** 708–712.

70. **Rappolee, D. A., K. S. Sturm, O. Behrendtsen, G. A. Schultz, R. A. Pedersen, and Z. Werb.** 1992. Insulin-like growth factor II acts through an endogenous growth pathway regulated by imprinting in early mouse embryos. *Genes Dev.* **6:**939–952.

71. **Reddy, E. P., R. K. Reynolds, E. Santos, and M. Barbacid.** 1982. A point mutation is responsible for the acquisition of transforming properties by the T24 human bladder carcinoma oncogene. *Nature* (London) **300:**149–152.

72. **Santos, E., S. R. Tronick, S. A. Aaronson, S. Pulciani, and M. Barbacid.** 1982. T24 human bladder carcinoma oncogene is an activated form of the normal human homologue of BALB- and Harvey-MSV transforming genes. *Nature* (London) **298:**343–347.

73. **Scolnick, E. M., E. Rands, D. Williams, and W. P. Parks.** 1973. Studies on the nucleic acid sequences of Kirsten sarcoma virus: a model for formation of a mammalian RNA-containing sarcoma virus. *J. Virol.* **12:**458–463.

74. **Shih, C., L. C. Padhy, M. Murray, and R. A. Weinberg.** 1981. Transforming genes of carcinomas and neuroblastomas introduced into mouse fibroblasts. *Nature* (London) **290:**261–264.

75. **Simon, M. A., D. D. L. Bowtell, G. S. Dodson, T. R. Laverty, and G. M. Rubin.** 1991. Ras1 and a putative guanine nucleotide exchange factor perform crucial steps in signaling by the sevenless protein tyrosine kinase. *Cell* **67:**701–715.

76. **Smith, M. R., S. J. DeGudicibus, and D. W. Stacey.** 1986. Requirement for c-*ras* proteins during viral oncogene transformation. *Nature* (London) **320:**540–543.

77. **Sozeri, O., K. Vollmer, M. Liyanage, D. Frith, G. Kour, G. E. Mark III, and S. Stabel.** 1992. Activation of the c-Raf protein kinase by protein kinase C phosphorylation. *Oncogene* **7:** 2259–2262.

78. **Sprenger, F., L. M. Stevens, and C. Nusslein-Volhard.** 1989. The *Drosophila* gene *torso* encodes a putative receptor tyrosine kinase. *Nature* (London) **338:**478–483.

79. **Stehelin, D., H. E. Varmus, J. M. Bishop, and P. K. Vogt.** 1976. DNA related to the transforming gene(s) of avian sarcoma viruses is present in normal avian DNA. *Nature* (London) **260:**170–173.

80. **Stokoe, D., S. G. Macdonald, K. Cadwallader, M. Symons, and J. F. Hancock.** 1994. Activation of Raf as a result of recruitment to the plasma membrane. *Science* **264:**1463–1467.

81. **Sugden, B.** 1993. How some retroviruses got their oncogenes. *Trends Biochem. Sci.* **18:**233–235.

82. **Szeberenyi, J., H. Cai, and G. M. Cooper.** 1990. Effect of a dominant inhibitory Ha-*ras* mutation on neuronal differentiation of PC12 cells. *Mol. Cell. Biol.* **10:**5324–5332.

83. **Tabin, C. J., S. M. Bradley, C. I. Bargmann, R. A. Weinberg, A. G. Papageorge, E. M. Scolnick, R. Dhar, D. R. Lowy, and E. H. Chang.** 1982. Mechanism of activation of a human oncogene. *Nature* (London) **300:**143–149.

84. **Taparowsky, E., Y. Suard, O. Fasano, K. Shimizu, M. Goldfarb, and M. Wigler.** 1982. Activation of the T24 bladder carcinoma transforming gene is linked to a single amino acid change. *Nature* (London) **300:**762–765.

85. **Temin, H. M.** 1970. Malignant transformation of cells by viruses. *Perspect. Biol. Med.* **14:**11–26.

86. **Temin, H. M.** 1971. The protovirus hypothesis: speculations on the significance of RNA-directed DNA synthesis for normal development and for carcinogenesis. *J. Natl. Cancer Inst.* **46:**III–VII.

87. **Temin, H. M.** 1974. On the origin of the genes for neoplasia. *Cancer Res.* **34:**2835–2841.

88. **Temin, H. M., and S. Mizutani.** 1970. RNA-dependent DNA polymerase in virions of Rous sarcoma virus. *Nature* (London) **226:**1211–1213.

89. **Thomas, S. M., M. DeMarco, G. D'Arcangelo, S. Halegoua, and J. S. Brugge.** 1992. Ras is essential for nerve growth factor- and phorbol ester-induced tyrosine phosphorylation of MAP kinases. *Cell* **68:**1031–1040.

90. **Troppmair, J., J. T. Bruder, H. App, H. Cai, L. Liptak, J. Szeberenyi, G. M. Cooper, and U. R. Rapp.** 1992. Ras controls coupling of growth factor receptors and protein kinase C in the membrane to Raf-1 and B-Raf protein serine kinases in the cytosol. *Oncogene* **7:**1867–1873.

91. **Van Aelst, L., M. Barr, S. Marcus, A. Polverino, and M. Wigler.** 1993. Complex formation between RAS and RAF and other protein kinases. *Proc. Natl. Acad. Sci. USA* **90**:6213–6217.
92. **Vogelstein, B., and K. W. Kinzler.** 1993. The multistep nature of cancer. *Trends Genet.* **9**:138–141.
93. **Vojtek, A. B., S. M. Hollenberg and J. A. Cooper.** 1993. Mammalian Ras interacts directly with the serine/threonine kinase Raf. *Cell* **74**:205–214.
94. **Wallace, M. R., L. B. Andersen, A. M. Saulino, P. E. Gregory, T. W. Glover, and F. S. Collins.** 1991. A *de novo Alu* insertion results in neurofibromatosis type 1. *Nature* (London) **353**:864–866.
95. **Warne, P. H., P. R. Viciana, and J. Downward.** 1993. Direct interaction of Ras and the amino-terminal region of Raf-1 *in vitro*. *Nature* (London) **364**:352–355.
96. **Weinberg, R. A.** 1991. Tumor suppressor genes. *Science* **254**:1138–1146.
97. **Whitman, M., and D. A. Melton.** 1992. Involvement of p21^{ras} in *Xenopus* mesoderm induction. *Nature* (London) **357**:252–254.
98. **Wood, K. W., C. Sarnecki, T. M. Roberts, and J. Blenis.** 1992. Ras mediates nerve growth factor receptor modulation of three signal-transducing protein kinases: MAP kinase, Raf-1, and RSK. *Cell* **68**:1041–1050.
99. **Yamauchi, N., A. A. Kiessling, and G. M. Cooper.** 1994. The Ras/Raf signaling pathway is required for progression of mouse embryos through the two-cell stage. *Mol. Cell. Biol.* **14**:6655–6662.
100. **Yao, R., and G. M. Cooper.** 1995. Requirement for phosphatidylinositol-3 kinase in the prevention of apoptosis by nerve growth factor. *Science* **267**:2003–2006.
101. **Ymer, S., W. Q. J. Tucker, C. J. Sanderson, A. J. Hapel, H. D. Campbell, and I. G. Young.** 1985. Constitutive synthesis of interleukin-3 by leukaemia cell line WEHI-3B is due to retroviral insertion near the gene. *Nature* (London) **317**:255–258.
102. **Zhang, X., J. Settleman, J. M. Kyriakis, E. Takeuchi-Suzuki, S. J. Elledge, M. S. Marshall, J. T. Bruder, U. R. Rapp, and J. Avruch.** 1993. Normal and oncogenic p21^{ras} proteins bind to the amino-terminal regulatory domain of c-Raf-1. *Nature* (London) **364**:308–313.

The DNA Provirus: Howard Temin's Scientific Legacy
Edited by G. M. Cooper, R. Greenberg Temin, and B. Sugden
© 1995 American Society for Microbiology, Washington, DC 20005

Chapter 7

Viruses, Genes, and Cancer: a Lineage of Discovery

J. Michael Bishop

. . . the brain of the lonely scientist, the 'adventurer' isolated in his study or laboratory, remains the preferred instrument, without which all that is done is routine.
True innovation does not come from the group, it is the fruit of reason, and reason
is individual.

—P. Levi, The Mirror Maker

I first encountered Howard Temin at a Gordon Conference in the summer of
1968. At the time, I was just becoming acquainted with RNA tumor viruses (as
retroviruses were then called), but I already knew Howard to be a renegade,
renowned and often belittled for his belief that the RNA genomes of retroviruses
were replicated by means of a DNA intermediate, which he called the provirus
(25).

The provirus hypothesis would eventually lead to the discovery of reverse
transcriptase, making Howard both a Nobel laureate and a household name in the
world of biology. As of 1968, however, the hypothesis had earned him little but
ridicule and grief. So that summer evening, I watched with interest (and from a
respectful distance) as Howard argued long into the night with skeptics and detrac-
tors. It was my first experience with a scientist who was essentially alone in his
beliefs. What I witnessed was a lesson for a lifetime.

The opposition to the provirus hypothesis that evening was strong, even vitri-
olic. In response, Howard was unfailingly patient and reasoned. He had no doubt
that his hypothesis was correct, but he was open to constructive criticism, and
he painstakingly tried to refute each opposing argument, even those that had no
force other than their animus. The strength of character that Howard displayed
convinced me then and there that I should take his ideas seriously. Whereas before
I had been only a desultory student of retroviruses, within the year that followed,
I redirected my research almost entirely to their study. The evening also gave me
a first glimpse of Howard's unflinching integrity and strength of conviction, the
qualities for which I most admired him.

J. Michael Bishop • The G. W. Hooper Research Foundation and Department of Microbiology and
Immunology, University of California, San Francisco, San Francisco, California 94143.

Howard spent years constructing his evidence for the provirus. He used virtually every means at hand, no matter how frail (24, 25): metabolic inhibitors to show that the synthesis and transcription of DNA were required for the replication of retroviruses, molecular hybridization in attempts to detect proviral DNA, and wonderfully subtle biological arguments that were particularly vexing to the more literal-minded among his opponents. Two of these subtle arguments were to resonate through my own research career.

First, Howard found that different strains of Rous sarcoma virus disfigured cells in different ways, as if the virus could exert a genetic control over the cells (22). In particular, he described how some strains of Rous sarcoma virus caused the infected and transformed cell to assume a fusiform shape, whereas other strains elicited a rounded morphology. In his characteristically single-minded manner, Howard used these findings to argue for a genetic interaction between viral and host genomes in the form of a provirus. There was, however, a plainer meaning that Howard also recognized and that proved far more important to me: transformation of the infected cell (and, thus, tumorigenesis in the bird or animal) might be attributable to the direct action of a viral gene—an "oncogene," a "tumor gene." In due course, such a gene was found in Rous sarcoma virus (14) and eventually christened src because of the tumors it induces: sarcomas.

Second, Howard had found what he thought was evidence for genetic exchange between retrovirus and host cell (1, 23). Indeed, this was the subject of his lecture at that Gordon Conference in 1968. Again, Howard blended the findings into his evidence for the provirus. We now know that Howard had in this instance been misled (as he eventually recognized) by intrinsic genetic variation of the virus. More to the point, however, Howard's imaginative interpretation of the data prefigured the eventual discovery that retroviruses can transduce cellular genes, most spectacularly, those capable of becoming oncogenes and, in the first instance, src itself (2, 21). Even in error, Howard had planted a potent seed that eventually grew into our first sighting of potential cancer genes in vertebrate cells and that came to dominate my own research.

A GENETIC PARADIGM FOR CANCER

In 1966, Peyton Rous received the Nobel Prize for discovering the virus with which Howard Temin would later do his own momentous work. Rous opened his Nobel lecture with the following lines (19):

> Tumors destroy man in a unique and appalling way, as flesh of his own flesh which has somehow been rendered proliferative, rampant, predatory and ungovernable. [Tumors] are the most concrete and formidable of human maladies, yet despite more than 70 years of experimental study they remain the least understood. . . . What can be the why for these happenings?

We now believe that we know the "why for these happenings," and it takes a remarkable form (Fig. 1). Our genetic dowries contain two sorts of genes that

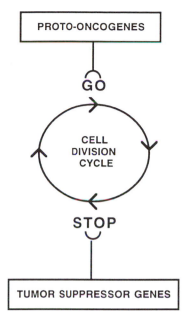

FIGURE 1. A genetic paradigm for cancer. The cell division cycle of vertebrate cells has an intrinsic pace and stride that can be moderated by both proto-oncogenes and tumor suppressor genes. Cancer arises from a combination of dominant gain-of-function mutations in proto-oncogenes and recessive loss-of-function mutations in tumor suppressor genes.

govern the proliferation of our cells: one sort provides accelerators to activate the engines of the cell (we call these proto-oncogenes), and the other sort provides brakes (we call these tumor suppressor genes). Jam an accelerator or remove a brake, and the cell may be unleashed to relentless proliferation, the beginnings of cancer.

I will briefly tell the story of how this imagery emerged. It is a story rich with illustrations of how science proceeds, of what it can and cannot do. It is a story that I will try to make accessible to even the most general reader. It is a story that does great credit to the legacy of Howard Temin.

CELLS AND CANCER

The story begins with cells, those microscopic, irreducible, living bricks from which our bodies are constructed. Each of us contains billions upon billions of cells arrayed into the complex tissues and organs that sustain our lives. In order to construct our bodies and keep them whole, our cells must proliferate.

The ability of cells to multiply lies at the root of life, but it also prefigures the baleful threat of cancer. Each second, 25 million cells divide in our bodies. Those divisions usually occur in the right place at the right time, governed by mechanisms we only dimly understand. When the governance fails, cancer may arise.

Why does the governance fail? How does it fail? What hope do we have of penetrating the complexities of cancer cells? Over the past several decades, we have taken giant strides toward answering those questions. The strides were made possible by two simplifications.

The first was to view cancer as a disease of individual cells and to study this disease not in an animal but in a petri dish. The second was to exploit viruses that rapidly and reproducibly convert cells to cancerous growth. Let us look at these two simplifications in a little more detail.

We can trace the use of cells in petri dishes to Alexis Carrel, who in 1912 received the first Nobel Prize for Medicine or Physiology awarded to a scientist in the United States (16). By birth and training, Carrel was French, but he had been run out of France by the academic community, which was annoyed by both his pungent personality and his interest in faith healing (he had proposed a scientific study of the miracles at Lourdes, which he regarded as authentic).

Carrel eventually settled at the Rockefeller Institute, faith healing no longer on his agenda. There he consolidated the pioneering work on vascular surgery and organ transplantation that eventually won him the Nobel Prize. His success in these fields had been made possible by the intricate stitching skills he learned from the great seamstresses of Lyon, one of whom was his mother.

But Carrel left another, more pertinent legacy to those of us who study cancer: his pathbreaking work growing vertebrate cells outside the living body, "cell culture," as we now call it. He claimed to have kept one culture of chicken cells alive and propagating for 32 years. We now know that he was probably fooled on that count, but there can be no denying the influence of his work.

How has the cell culture pioneered by Carrel figured in cancer research? Whole tumors are not easy objects for experimental study, so we resort to the belief that the properties of individual cancer cells probably explain the behavior of tumors. We can define these properties by growing the cancer cells outside of the animal, using an artificial mixture of nutrients to feed the cells and glass or plastic vessels to contain them.

In culture, cancer cells misbehave exactly as we might have expected from the properties of whole tumors: they continue to grow when they should not, i.e., when they have become crowded by their neighboring cells, and they develop a very different appearance from their normal counterparts. Normal cells are flat and extended, and they grow in orderly arrays; cancer cells are rounded, disordered, literally crawling all over one another—a convincing representation of the cells in an invasive cancer, a representation that is accessible to the experimentalist.

The use of cell culture for the study of cancer has often been faulted as a misleading artifice. The criticism is specious and misinformed. Much of what we now know of the mechanisms of tumorigenesis could not have been learned without the use of cells in culture. Indeed, Howard Temin's own pathbreaking work was dependent upon that technology.

VIRUSES AND CANCER

The second simplification came to us from Peyton Rous, who in 1911 discovered a virus that causes cancer in chickens, the Rous sarcoma virus (18). It must be said that neither Rous nor anyone else of his time really knew what a virus might be like; to them, it was merely an invisible and infectious poison (the meaning of the Latin word "virus"), a poison that seemed to have a life of its own.

Rous eventually gave up the study of his chicken virus. Tradition claims that he did this because of ridicule from other scientists. By his own report, however, Rous abandoned the work because neither he nor anyone else at the time could detect viruses in cancers of rodents, so even Rous himself (mistakenly) came to doubt the generality of his discovery (19). Fortunately, Peyton Rous had good genes. He lived long enough to see his original ideas vindicated and to receive the Nobel Prize at the age of 86 (although his son-in-law, Alan Hodgkin, beat him to Stockholm by 3 years).

Rous was honored for the discovery of his virus. Why was that discovery so important? First, because it established a precedent: some viruses can indeed cause cancer. By now, several have been implicated in the genesis of important human cancers (including carcinomas of the liver and of the uterine cervix). Second, and more important for my story here, it was important because cancer viruses provide a genetic simplification for the study of cancer.

Here is the nature of that simplification. Our DNA carries on the order of 50,000 genes. Each of these genes has its own specific chore, and among these chores, there must be many that are important to a cancer cell. In contrast, viruses generally have fewer than a dozen genes, and often only one of these genes is required to produce cancer. So, simply put, viruses can simplify the study of cancer by more than 1,000-fold.

The simplification provided by the virus of Peyton Rous could hardly have been more extreme. Rous sarcoma virus turned out to have only four genes (Fig. 2). Three of these genes are used to reproduce the virus, but the fourth is the oncogene *src,* whose only function in the life cycle of the virus is to transform infected cells to neoplastic growth (14).

The discovery that Rous sarcoma virus uses a gene to elicit cancer brought clarity to what had been a muddled business. There had been hints before that the elemental secrets of cancer might lie hidden in our genetic dowry, but here, in this virus, scientists found an explicit example of a gene that can switch a cell from normal to cancerous growth.

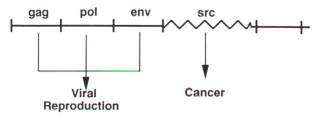

FIGURE 2. The genes of Rous sarcoma virus. The single-stranded RNA genome of Rous sarcoma virus carries four genes. Three are devoted to replication, encoding capsid proteins of the virus (*gag*), the enzymes for reverse transcription and integration of proviral DNA (*pol*), and the surface glycoprotein of the viral envelope (*env*). The fourth gene (*src*) plays no role in viral replication but elicits neoplastic transformation of cells and causes sarcomas in birds and mammals.

CANCER GENES IN CELLS

Now a further question arose. Might the cell itself have such cancer genes? Can the complexities of human cancer be reduced to the chemical vocabulary of DNA? The answer came from asking yet another question that is based on the logic of evolution. The *src* gene serves no obvious purpose for Rous sarcoma virus; it contributes nothing to the growth or survival of the virus. Why, then, is it there?

We have the answer. The oncogene *src* is present in the virus of Peyton Rous because of an accident of nature (2, 21). The gene was acquired from the cells in which the virus grows in an elaborate molecular ballet whose details we still do not fully understand. In other words, Rous sarcoma virus is an accidental pirate. The booty is a cellular gene with the potential to become a cancer gene, a cellular *src* gene that we now know to be more than a billion years old.

It quickly became apparent that *src* was but the tip of an iceberg. There are many vertebrate genes that can be pirated into viruses like that of Peyton Rous, there to become oncogenes. We now know of more than 20 (4). We have dubbed these cellular genes ''proto-oncogenes,'' because each has the potential to become an oncogene in a virus.

It seems most unlikely that evolution installed proto-oncogenes in our cells to cause cancer. These genes must normally have more noble purposes. Why then does their transfer into retroviruses give rise to oncogenes?

The answer lies in the elaborate molecular gymnastics by which proto-oncogenes are pirated into the genomes of retroviruses. During the pirating, proto-oncogenes suffer damage (mutations) that can convert them to oncogenes, from Dr. Jekyll to Mr. Hyde. It was the mechanism of this mutagenesis that most concerned Howard Temin during the last few years of his life (for his final appraisal of the problem, see reference 27).

PROTO-ONCOGENES AND CANCER

The existence of numerous proto-oncogenes also gave rise to a more expansive theory (Fig. 3). Perhaps proto-oncogenes exemplify a genetic keyboard on which all manner of carcinogens might play: any influence that can damage a proto-oncogene might give rise to an oncogene, even if the damage occurred without the gene ever leaving the cell, without the gene ever confronting a virus. In this view, proto-oncogenes become precursors to cancer genes within our cells, and damage to genes becomes the underpinning of all cancers, even those that are not caused by viruses.

In time, scientists learned to go beyond retroviruses by finding other ways to uncover proto-oncogenes (3). The tally of proto-oncogenes has now reached 100 or more, and we have learned that our hypothesis is at least in part correct. Once retroviruses had shown us where to look, we could examine the status of proto-oncogenes in human cancer cells. What we found was nothing short of

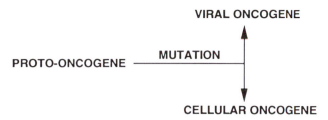

FIGURE 3. Proto-oncogenes as precursors of cancer genes. Proto-oncogenes can become oncogenes either by transduction into retroviruses or by disturbance at their sites of residence in chromosomes. In either event, the consequence is an abnormal gain of function in the form of either unleashed expression of the gene or deregulation of the protein product.

astonishing: most cancers that have been properly studied contain damage to one or more proto-oncogenes.

The damage takes three general forms (3): direct chemical or physical modification of the gene, known as mutation; scrambling of genes among different chromosomes by a process known as translocation; and overgrowth of genes, which we call amplification. In each instance, the damage somehow unleashes the gene or its protein product so that it runs amok, driving the cell to relentless proliferation. A genetic engine for the cancer cell has been uncovered.

A SECOND KIND OF CANCER GENE

The discovery and exploration of proto-oncogenes provided a new view of the cancer cell, rich in detail and prospect, but the story does not end there. As scientists continued to examine cancer cells, another sort of mischief was found: not damage that creates unwanted activity of genes but, instead, damage that eliminates the function of genes and often eliminates the genes themselves. Scientists found their way to this damage by two principal paths.

First, it is possible to induce the fusion of two different cells and thus to cause the intermingling of their genetic endowments in a single descendant. Experimental fusion of normal cells with cancer cells often suppresses the abnormalities of the cancer cells: the hybrid cells grow normally rather than as cancer cells (8, 9).

The cancer cells therefore appear to be defective in genetic functions that are required for the regulation of cellular proliferation and other behavior. Fusion with a normal cell restores the necessary genes and, as a result, suppresses cancerous growth. Thus, the responsible genes became known as "tumor suppressor genes," although at this point in the story, their existence remained hypothetical, and their possible nature remained obscure.

A second line of enquiry brought tumor suppressor genes into clear view. The cells of some human cancers contain chromosomes that have lost part of their DNAs, a deletion that is sometimes visible through a microscope. The loss of genes by deletion is akin to the deficiency imagined from the experiments with cell fusion.

TABLE 1
Tumor suppressor genes: a sampler[a]

Gene	Neoplasm(s)	Apparent function
APC	Colon carcinoma	Cell adhesion and signaling
DCC	Colon carcinoma	Cell adhesion and signaling
IRF1	Myeloid and lymphoid leukemias	Transcription factor
MCC	Colon carcinoma	?
MLH1	Colon carcinoma	DNA repair
MSH2	Colon carcinoma	DNA repair
MTS1	Diverse cancers	Inhibitor of cell cycle
NF1	Neurofibromatosis	Regulator of RAS protein (IRA/GAP)
NF2	Neurofibromatosis	Cytoskeleton-membrane link
p53	Diverse cancers	Transcription factor
RB1	Diverse cancers	Nuclear protein
VHL	Renal carcinoma, others	Cell adhesion and signaling
WT1	Wilms' tumor	Transcription factor

[a] The list represents the tumor suppressor genes isolated by molecular cloning as of this writing. The genes are designated by acronyms based on either the form of tumorigenesis in which they are involved or some inherent property. Also given are the forms of cancer commonly associated with loss of function of the genes and, where known, the biochemical functions of the gene products. Many of these latter remain poorly defined.

The first genetic deletions in cancer cells to be carefully studied involved a chromosome in human retinoblastoma, a malignant tumor of the retina that occurs only in young children (6). Retinoblastoma is a rare tumor, yet it was worried to death by scientists. Our critics might well have asked why this should be so. Would it not have been better to focus our attention and funds on more common ailments such as cancer of the breast, for example?

The answer lies in success. The peculiarities of retinoblastoma offered an accessible key that opened another large door to the inner sanctum of the cancer cell. Guided by their findings with retinoblastoma, scientists learned that probably all cancers suffer specific loss or inactivation of tumor suppressor genes (6, 13). So far, on the order of a dozen of these genes have been uncovered (Table 1), but there are probably many more. It is worth noting that even the earliest fruits of the often maligned human genome project will make these genes easier to find.

INHERITED CANCER

The discovery of cancer genes has also provided an explanation for inherited cancer. It was August Weissman who first gave impetus to the idea that the cells of our bodies are divided into two lineages (26): one lineage, the somatic lineage (from "soma," for body), assembles the body of each living creature and is a biological dead end; another lineage, the germinal lineage or germ line, the carrier of our genetic dowry, perpetuates the germ cells (sperm and egg) from one generation to the next.

Occasionally, the damage that creates a cancer gene occurs in the germinal lineage, i.e., in the sperm or the egg. Then, it is passed from generation to generation, creating an inherited predisposition to cancer, that is, cancer families of various sorts. We have now uncovered the nature of the inherited damage for several forms of cancer, and it has typically involved one or another of the tumor suppressor genes (11). However, inheritance of cancer is an exception, not the rule. Most cancer genes arise from damage in the somatic lineage and thus affect only a single individual. The mutant genes are not inherited but instead are extinguished with the individual in which they were created.

A UNIFYING VIEW OF CANCER

Thus, two sorts of genes figure in the genesis of human cancer: proto-oncogenes, which cause trouble only if they continue to produce an active protein, and tumor suppressor genes, which are pathogenic only when they are defective or lost. These are the jammed accelerators and missing brakes described at the outset of this essay, maladies that combine to maim and kill.

This is a powerful view of cancer. The seemingly countless causes of cancer—cigarette smoke, sunlight, asbestos, chemicals, viruses, and many others—all may work in a single way by playing on a genetic keyboard, by damaging a few of the genes in our DNA. An enemy has been found, and we are beginning to understand its lines of attack.

To see this point in miniature, consider cancer viruses once more (Fig. 4). There are many forms of cancer viruses, not only the sort discovered by Peyton Rous, but a fearful symmetry unites these viruses, because most if not all play on the keyboard of our genes. Some cancer viruses activate proto-oncogenes, as you learned for Rous sarcoma virus, but others inactivate tumor suppressor genes by one or another of several strategies (sometimes by direct assault on the genes; sometimes by attacking the proteins encoded by the genes). Thus, we have achieved a nearly universal and singular view of tumorigenesis, whatever its cause.

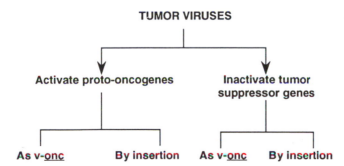

FIGURE 4. Mechanisms of viral tumorigenesis. A common strategy underlies tumorigenesis by diverse sorts of viruses. Retroviruses can activate proto-oncogenes by either transduction or insertional mutagenesis (see chapter 13 in this volume); alternatively, they can inactivate tumor suppressor genes by the latter mechanism. The oncogenes of DNA tumor viruses, such as the papovaviruses, adenoviruses, and papillomaviruses, generally inactivate the products of tumor suppressor genes by means of protein-protein interactions (13).

TUMOR PROGRESSION

Still, the story is not complete, for there is another great mystery to solve. We have known for a long time that the cancer cell does not emerge in one fell swoop. Multiple events, usually spread out over many years, are required (Fig. 5). We call this protracted but deadly sequence "tumor progression." It is as if each additional event adds insult to injury, the eventual sum being a malignant tumor (7, 17).

Take as an example metastasis, the ability of cancers to spread through the body. This spreading is not a trivial accomplishment: it is akin to playing the cello, throwing a curve ball, and writing a novel all in the same few moments. We cannot yet explain the mechanisms of metastasis in detail, but we do know that they are a composite of multiple abnormalities in the malignant cell.

How can we account for tumor progression? What are the individual steps in this deadly chain of events? The discovery of cancer genes provided an answer: each step may represent damage to an additional gene, either a proto-oncogene or a tumor suppressor gene (7). Here is one reason that aging is such an important factor in cancer: it takes time to accumulate all of this trouble. Here, too, is a reason that cancer is not more frequent among us: it is statistically unlikely that all of these events will combine in the same cell during a single lifetime.

Moreover, evolution was not content to rely on chance, so it constructed two other devices that help protect us from cancer (Fig. 5). First, we have genetic defenses that vastly diminish the frequency of the individual steps in tumorigenesis. We calculate that every gene in our DNA is damaged some 10 billion times in a lifetime, yet somehow, the machinery of our cells repairs virtually all of this damage. It will come as no surprise that deficiencies in these defenses can lead to cancer (20).

Second, the body can mount potent immune defenses against both foreign intruders (such as microbes and transplanted tissues) and errant natives (such as cancer cells). We know all too little of how these defenses act against cancer, but they are real, and deficiencies in them can also contribute to the genesis of cancer (5).

For the moment, we cannot assign distinct steps in tumorigenesis to individual genes, but everywhere we look, we find evidence that damage to multiple genes, both proto-oncogenes and tumor suppressor genes, figures in the genesis of vir-

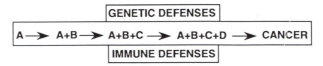

FIGURE 5. Tumorigenesis as a multistep progression. Most if not all tumors arise from a lengthy sequence of events, many of which represent damage to individual genes. Different combinations of events are apparently required to produce different types of tumors. The need for all of these events to occur in a single cellular lineage constitutes a formidable statistical barrier to the occurrence of cancer. In addition, repair mechanisms protect against the adverse effects of mutations in DNA, and immunological mechanisms defend against outlaw cells once they have emerged.

TABLE 2
Multiple genetic lesions in human cancer[a]

Type of cancer	Proto-oncogene(s)	Tumor suppressor genes
Carcinoma of colon	*KRAS*	*APC, DCC, p53*
Carcinoma of lung	*MYC, MYCL, MYCN*	*RB1, p53, 3p21*
Carcinoma of breast	*MYC, NEU*	*RB1, RB3, RB11, 17p, 17q21*
Neuroblastoma	*MYCN*	*1p36, 1p17*
Glioblastoma	*ERBB1, GL1, MYC, ROS, SIS*	*p53, p10, p13, p22*

[a] Most cancers typically display damage in a combination of proto-oncogenes and tumor suppressor genes, as represented here by several well-studied examples. The occurrence of each lesion is regarded as a separate step in tumor progression. The affected genes are represented by acronyms when their exact identity has been established or by chromosomal number or position when that is the only information presently available.

tually every cancer. Table 2 gives some prominent examples. Each of the tumors listed there features a relatively reproducible combination of lesions in proto-oncogenes and tumor suppressor genes, combinations that exemplify the fearful symmetry of the cancer cell.

A LARGER LESSON FROM CANCER RESEARCH

The catalogs of genetic lesions that we can now compile for human cancers are nothing short of astonishing. Less than 20 years ago, we knew nothing of these lesions and had no means by which to find them. No other field of modern biomedical science can boast of more rapid progress than the one dramatized here.

We owe most of this progress to what we scientists call "basic research." The laity often ask what this hallowed term implies. Modern cancer research provides an answer. Much of this research was not begun in search of an answer for cancer, much of it was not performed on human cancer or even on cancer of any sort, much of it employed mundane creatures (such as bacteria, yeasts, fruit flies, birds, and rodents) and was addressed to exotic problems (such as the cause of a chicken sarcoma), and all of it was fueled by the inquisitiveness of individual scientists, often without a particular medical mission.

The entire career of Peyton Rous dramatizes the point. Rous isolated his virus from chickens, beasts not renowned for glamor. Later, he turned his attention to tumor progression, which he studied in rabbits (19). Rous paid no heed to glamor. He chose a puzzle of nature that was close at hand and parlayed it into a discovery that would reach across more than half a century to remake cancer research in our time (Fig. 6).

The chicken virus discovered by Peyton Rous 70 years ago was the first cancer virus to be persuasively described. Decades later, the virus was recognized as a very special sort of creature whose replication reverses the usual flow of genetic information by means of an enzyme called reverse transcriptase (24). We owe this discovery to the imagination and courage of Howard Temin. It was a discovery that shook the foundations of biology and that proved vital to both the technology of recombinant DNA and the prompt isolation of the AIDS virus.

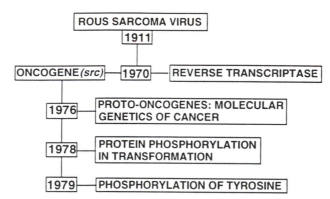

FIGURE 6. The retroviral lineage of discovery. The retrovirus discovered by Peyton Rous in 1911 eventually figured in four additional important advances: the discovery of viral oncogenes, the first genetic determinants implicated in tumorigenesis (7); the discovery of proto-oncogenes, the first sighting of potential cancer genes in cellular genomes (7); the discovery that protein phosphorylation can mediate neoplastic transformation of cells (10); and the discovery of protein-tyrosine kinases, a large and vital family of enzymes with central roles in cellular signaling (10).

In the same year that reverse transcriptase was discovered, the oncogene *src* became the first cancer gene to be decisively identified (14). Soon thereafter, the action of the *src* protein (a protein kinase that phosphorylates tyrosine) was uncovered, the first biochemical mechanism implicated in cancerous growth (10), and *src* was found in normal cells, the first sighting of proto-oncogenes and the first legitimation of the idea that our DNA can be the source of cancer (2, 21).

This litany offers a familiar but still neglected lesson: the proper conduct of science lies in the pursuit of nature's puzzles, wherever they may lead. We cannot prejudge the utility of any scholarship; we can only ask that it be sound. We rarely know where our explorations will take us; we virtually never know how quickly they will proceed.

The directions of science are ultimately determined by feasibility: science is the art of the possible, of the soluble, to recall a phrase from Peter Medawar (15). We usually cannot force nature's hand; she must tip it for us. We often cannot assault the great problems of biology at will; we must remain alert to nature's clues and seize on them whenever and wherever they appear, even if they appear in a chicken.

Bit by bit, the inner workings of the cancer cell are coming under our sway. With this knowledge, we seek devices for diagnosis, for prognosis, and for rational designs of therapy and prevention of cancer. However, a great intellectual adventure overshadows even these noble goals: the quest to understand the cell in all of its particulars, to know what keeps us whole and what rends us asunder. Howard Temin devoted his life to this quest, and we are all the better for it.

EPILOGUE

I never spoke to Howard Temin at that Gordon Conference in 1968. It was several years before we became personal friends while working together on the

Virology Study Section of the National Institutes of Health (I could have asked for no greater dividend from that labor).

It was a time when the study of reverse transcriptase had led to a surge of research on retroviruses, flooding the field with talented scientists. In our first conversation, I had the temerity to ask Howard how he felt about having so many Johnnies-come-lately (such as myself) cluttering the field in which he had once had the correct paradigm all to himself. "Oh, it doesn't bother me at all," he responded, "because I consider them all as my scientific children."

Later, in the published version of his Nobel lecture, he thanked some of those "children" (including me, I am proud to record) for their intellectual companionship (25). As I remember Howard now and mourn his loss, it is a comfort to know that his legacy lives on among several generations of scientists and that we have the example of his generous and robust spirit still before us to emulate.

ACKNOWLEDGMENTS. Work in my laboratory has been supported by the National Institutes of Health, the American Cancer Society, and the Markey Charitable Trust.

REFERENCES

1. **Altaner, C., and H. M. Temin.** 1970. Carcinogenesis by RNA sarcoma viruses. XII. A quantitative study of infection of rat cells in vitro by avian sarcoma viruses. *Virology* **40:**118–134.
2. **Bishop, J. M.** 1982. Oncogenes. *Sci. Am.* **246:**80–82.
3. **Bishop, J. M.** 1987. The molecular genetics of cancer. *Science* **235:**305–311.
4. **Bishop, J. M.** 1991. Molecular themes in oncogenesis. *Cell* **64:**235–248.
5. **Boon, T.** 1993. Teaching the immune system to fight cancer. *Sci. Am.* **268:**82–89.
6. **Cavenee, W. K.** 1986. The genetic basis of neoplasia: the retinoblastoma paradigm. *Trends Genet.* **2:**299–300.
7. **Fearon, E. R., and B. Vogelstein.** 1990. A genetic model for colorectal tumorigenesis. *Cell* **61:**759–767.
8. **Harris, H.** 1986. The genetic analysis of malignancy. *J. Cell Sci.* **4**(Suppl.):431–444.
9. **Harris, H., O. J. Miller, G. Klein, P. Worst, and T. Tachibana.** 1969. Suppression of malignancy by cell fusion. *Nature* (London) **223:**363–368.
10. **Hunter, T.** 1984. The proteins of oncogenes, p. 88–97. *In* E. Friedberg (ed.), *Cancer Biology: Readings from Scientific American.* W. H. Freeman & Co., New York.
11. **Knudson, A. G., Jr.** 1994. Genetics of cancer. *J. NIH Res.* **6:**63.
12. **Levi, P.** 1989. *The Mirror Maker: Stories and Essays,* p. 21. Schocken Books, Inc., New York. (Translated from Italian by R. Rosenthal.)
13. **Marshall, C. J.** 1991. Tumor suppressor genes. *Cell* **64:**313–326.
14. **Martin, G. S.** 1970. Rous sarcoma virus: a function required for the maintenance of the transformed state. *Nature* (London) **227:**1021–1023.
15. **Medawar, P.** 1967. *The Art of the Soluble.* Methuen Publishing Co., London.
16. **Moseley, J.** 1980. Alexis Carrel, the man unknown. *JAMA* **244:**1119–1121.
17. **Nowell, P. C.** 1986. Mechanisms of tumor progression. *Cancer Res.* **46:**2203–2207.
18. **Pitot, H. C.** 1983. The Rous sarcoma virus. *JAMA* **250:**1447–1448.
19. **Rous, P.** 1966. The challenge to man of the neoplastic cell, p. 162–171. *In Les Prix Nobel—the Nobel Prizes, 1966.* Norstedts Tryckeri AB, Stockholm.
20. **Service, R. W.** 1994. Stalking the start of colon cancer. *Science* **263:**1559–1560.
21. **Stehelin, D., H. E. Varmus, and J. M. Bishop.** 1976. DNA related to the transforming gene(s) of avian sarcoma viruses is present in normal avian DNA. *Nature* (London) **260:**170–173.
22. **Temin, H. M.** 1960. The control of cellular morphology on embryonic cells infected with Rous sarcoma virus in vitro. *Virology* **10:**182–197.
23. **Temin, H. M.** 1970. Formation and activation of the provirus of RNA sarcoma viruses, p. 233–249.

In R. D. Barry and B. W. J. Mahy (ed.), *The Biology of Large RNA Viruses*. Academic Press, Inc., New York.

24. **Temin, H. M.** 1972. RNA-directed DNA synthesis, p. 56–65. *In* E. Friedberg (ed.), *Cancer Biology—Readings from Scientific American*. W. H. Freeman & Co., New York.

25. **Temin, H. M.** 1976. The DNA provirus hypothesis: the establishment and implications of RNA-directed DNA synthesis. *Science* **192:**1075–1080.

26. **Weismann, A.** 1892. *Das Keimplasma. Eine Theorie der Verebung.* Fischer, Jena, Germany.

27. **Zhang, J. Y., and H. M. Temin.** 1993. 3′ junctions of oncogene-virus sequences and the mechanisms for formation of highly oncogenic retroviruses. *J. Virol.* **67:**1747–1751.

The DNA Provirus: Howard Temin's Scientific Legacy
Edited by G. M. Cooper, R. Greenberg Temin, and B. Sugden
© 1995 American Society for Microbiology, Washington, DC 20005

Chapter 8

Mammalian Development and Human Cancer: from the Phage Group to the Genetics of Intestinal Cancer

William F. Dove

PROLOGUE

Together in this volume, we construct a mosaic. I shall bring into focus one stone of that mosaic: the abiding interest in the intersection of mammalian development and human cancer that I have shared with Howard Temin.

The experimental format of my own research is the laboratory mouse. It is no coincidence that Howard's first serious venture into biology was taken in company with the laboratory mouse. As a high school student, he had the good fortune to participate in a summer research program at C. C. Little's mouse mecca, The Jackson Laboratory in Maine. It is also no coincidence that Alexandra Shedlovsky, Amy Moser, and I, along with our McArdle Laboratory colleagues, continue to relate to the resources of The Jackson Laboratory in developing mouse models with which to study human health and disease. As many know from personal scientific experience, one must remain vigilant not to presume identity between mouse and human. In this essay, I discuss a mouse model for intestinal cancer, Min (for multiple intestinal neoplasia). We shall see homology to the human condition but not identity.

Howard's interest in development was fostered during his graduate career at the California Institute of Technology (Caltech) by a novel fusion to the emergent discipline of virology. When asked how he got into cancer research, Temin replied,

> As a graduate student I was interested in *cell development*, and the Rous sarcoma virus provided a system to change normal cells into other kinds of cells.
> —McArdle Laboratory, Nobel Prize press conference, October 1975
> (emphasis added)

Howard's initial doctoral research home, the laboratory of embryologist Albert Tyler, did not permit the speed and quantitative character of experimental analysis

William F. Dove • McArdle Laboratory for Cancer Research, University of Wisconsin, 1400 University Avenue, Madison, Wisconsin 53706.

that was being fostered at Caltech by the Phage Group, founded by Max Delbrück. Delbrück served on Howard's thesis committee in 1960, just before I, a doctoral student in physical chemistry at Caltech, began to work in the Phage Group. Work from my laboratory at Wisconsin on regulation in bacteriophage lambda (7) and on the cell cycle and cytoskeleton of *Physarum polycephalum* (2) has continued the tradition of the Phage Group. Now I am studying mammalian biology, and again, it is no coincidence that the research on development and neoplasia which I discuss here draws strongly from the tradition of the Phage Group. Howard and I have marveled at the connectivity of science that generates such apparent coincidences.

Let me finish this prologue by drawing your attention to ways in which Howard Temin continued to nourish his early interest in development while his career as a virologist expanded:

> In speculating about the origin of viruses with this special mode of information transfer, RNA → DNA, I felt that [this] information transfer *might play a role in normal organisms.*
> —H. M. Temin, Guest editorial: The Protovirus Hypothesis (20) (emphasis added)

In this intentionally speculative essay, Howard considered the possibility that reverse transcription generates somatic diversity for metazoan development. In subsequent decades, Howard remained interested in this possibility. The fascination has been fueled by the continuing discovery of retrotransposons all the way back to prokaryotic genomes and by the observation of apparent cDNA integrations in complex genomes. Today, the question continues to burn, one of Howard's "eternal flames": What biological function(s) drives the evolution of retrotransposition?

INTRODUCTION

To focus on my stone in our mosaic: What are the molecular and biological intersections of mammalian development and human cancer? Specific evidence on these intersections comes from the analysis of mutants, mouse or human. In the simplest case, one determines what neoplastic processes occur spontaneously in a mutant that has lost the function of a single developmental gene. In human cancer genetics, this case is approximated by families in which a mutant allele inherited in heterozygous form predisposes the individual to a particular cancer syndrome. Often, but not always, it is found that the neoplastic lineage has somatically lost the remaining normal allele that was present in the zygote.

Familial retinoblastoma is the paradigm: complete deletions of *RB* predispose heterozygotes to the disease, which is manifested by expansion of the population of a cell type that shows markers characteristic of early retinoblasts. The tumor lineage has somatically lost the remaining normal allele by one of several possible mitotic mechanisms. A number of other cases have been found in the human: neurofibromatosis types I and II, Wilms' tumor, multiple endocrine neoplasia, familial adenomatous polyposis, and breast neoplasia (BRCA1 and BRCA2).

A full exploration of the intersection with normal mammalian development of these loss-of-function neoplasms requires analysis of embryos homozygous for the mutant allele. Generally speaking, material for this analysis is not available for the human. As the gene responsible for the human predisposition is identified, it has become possible in general to derive by gene targeting a mutant mouse line carrying a mutated allele of the mouse homolog. One intercrosses mice heterozygous for such a targeted allele and analyzes embryos that are homozygous for the mutant defect.

This exploration has been energetically pursued with mixed results. Mice homozygous for targeted loss-of-function alleles of the mouse homologs for *Rb* and *Nf-1* show defects in embryonic development. The defects are manifested after midgestation, affecting only certain organs.

However, the neoplastic phenotype of heterozygotes for targeted alleles of *Rb* and *Nf-1* does not encourage use of the mouse model for understanding the specific intersection of development and human cancer in which each of these genes is involved. The tissues that are predisposed to neoplasia differ in the mouse and the human. On this score, perhaps it will be informative to study development in mouse embryos homozygous for the *Min* mutant allele of the *Apc* gene. Here, the heterozygous mouse line shows a neoplastic phenotype that closely resembles that of corresponding human familial adenomatous polyposis families. This neoplastic process can best be appreciated in the context of the biology of the intestinal epithelium of the mammal. The intestinal epithelium is a self-renewing tissue, even in the adult mammal. Proliferation is limited to crypts. As cells exit the crypt and terminally differentiate, they move up the fingerlike villi that line the small intestine. These villi are replaced by a cuff that surrounds the crypt of the large intestine. The total mitotic risk of the intestinal epithelium is enormous. Clearly, this risk is attenuated by the strategy of shedding one daughter lineage from each stem cell mitosis within the crypt. But whether and how the retained daughter lineage reduces its risk of mitotic error remain deep mysteries (3).

The Min mouse carries a germ line mutation in one allele of the *Apc* gene and develops scores of adenomas in both large and small intestines during its lifetime. Homozygotes for the *Min* mutation cannot be found among the live-born progeny of a cross between two *Min*/+ heterozygotes (12). Thus, the *Apc* gene is an essential developmental gene in the mouse with an appropriate heterozygous phenotype of intestinal neoplasia. Indeed, the essential starting point of the experimental strategy used in our laboratory is finding the appropriate mutant phenotype at the level of the intact organism. The development of efficient point mutagenesis of the mouse germ line with ethylnitrosourea has made this strategy feasible (5, 15–17, 22). Among the more than 14 recessive mutations in essential developmental genes that we have isolated and mapped, only *Min* shows a pronounced heterozygous neoplastic phenotype (15a). I focus in this essay on two genes: *Apc* on chromosome 18, in which the *Min* nonsense allele has been induced by ethylni-

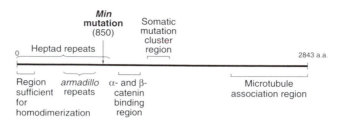

FIGURE 1. The APC/Apc polypeptide. a.a., amino acids.

trosourea (12, 18), and the first modifier-of-Min locus, *Mom-1*, on chromosome 4 (4).

The *APC/Apc* gene of mammals encodes a giant 300-kDa cytoplasmic polypeptide of unknown function (Fig. 1). Several regions involved in macromolecular interactions have been identified for *APC/Apc:* a homodimerization domain near the amino terminus; heptad repeats characteristic of polypeptides that can form coiled-coil oligomers throughout the amino-terminal third of the molecule; seven 42-amino-acid repeats homologous to those found in the plakoglobin homolog of *Drosophila melanogaster* (in *armadillo*) and in the *smg* family of 21-kDa G proteins of mammals (14); and a region that binds α- and β-catenin, polypeptides that also bind the calcium-dependent cell adhesion molecule E-cadherin (8). In the carboxy-terminal region of the *APC/Apc* polypeptide is a region that associates with the microtubular cytoskeletons of mammalian cells.

Using the Min mouse, I address five questions. (i) What is the developmental potential of the *Min/Min* zygote? (ii) What somatic genetic events occur when tumors form in *Min/+* heterozygotes? (iii) What is the developmental potential of *Min/0* adenomas? (iv) What other loci in the mouse genome can modify the severity of *Min*-induced neoplasia? (v) For each gene that is implicated in the developmental or neoplastic action of *Apc*, how is gene action partitioned between the tumor lineage, the microenvironment, and the systemic environment of the host?

DEVELOPMENT OF THE *Min/Min* EMBRYO (13)

When mice heterozygous for the *Min* nonsense allele of *Apc* are intercrossed, no live-born *Min/Min* homozygous progeny are found (12). Abnormal embryos can be found in intercross litters as early as 5.5 days of gestation, soon after implantation. Control crosses show abnormal embryos at much lower frequencies. It is, however, one thing to report abnormal embryos in the experimental crosses but quite another to demonstrate that the abnormal early embryos are the mutant homozygotes. We have therefore invested the effort needed to determine the genotypes of the individual pregastrulation embryos from these matings.

The most powerful method for determining genotypes of the mutant and wild-type sites of the *Apc* gene involves using mismatched PCR primers that amplify the region of the gene surrounding the *Min* site and create an allele-specific *Hin*dIII site in the wild-type product and control *Hin*dIII sites in both the *Min* and the wild-type products (9). Using this method, we have determined the genotypes of embryos at 7.5 days of gestation that have been dissected away from much of the maternal decidual tissue. The mutant homozygotes are poorly developed at this stage.

What tissues are specifically affected in the *Min/Min* embryo? Embryos isolated within their decidual swellings and stained by hematoxylin and eosin include abnormal embryos that are severely affected in their primitive ectoderm at the distal end of the egg cylinder at 6.5 days of gestation, before the onset of gastrulation is heralded by the formation of the primitive streak. Embryos of this stage have been dissected away from the decidua, prepared for histological analysis, and scored morphologically, and finally, their genotypes have been determined from DNA isolated from histological sections.

Determining the genotypes of embryos that have been sectioned for histology requires a demanding protocol of DNA isolation from single sections and the use of informative PCR markers. The presence of hematoxylin and eosin compromises genotyping by the site-specific protocol discussed above. The markers of choice, therefore, have been polymorphic, simple-sequence repeat markers that closely flank the *Apc* locus. The *Min/Min* embryos have been generated by intercrossing (AKR × C57BL/6-*Min/+*)F₁ parents. Embryos shown to be homozygous for the wild-type region surrounding *Apc* show normal primitive ectoderm. In contrast, embryos homozygously mutant for the *Apc* region show the same typically abnormal phenotype seen in the embryos that were fixed within the decidua, i.e., underdeveloped primitive ectoderm but relatively normal extraembryonic ectoderm and endoderm.

We have investigated earlier stages of development, i.e., those immediately following implantation. All 38 embryos from heterozygote intercrosses developed primitive ectoderm, though a subset, perhaps the mutant homozygotes, seemed to be retarded. This stage does not yet permit explicit genotyping by which we can rigorously assign the genotype, but we can conclude that the *Min/Min* embryo forms primitive ectoderm but is blocked in the maintenance or expansion of this totipotent early embryonic tissue.

What cell types can develop in the absence of the normal *Apc* product? Looking late in development, we find that the invasive trophoblast giant-cell population is abundant and apparently normal, even at 10.5 days of gestation.

In summary, an embryo homozygous for the *Min* allele of *Apc* is affected in the development or persistence of the primitive ectoderm soon after implantation. The primitive ectoderm has been shown to be the embryonic precursor of all three germ layers of the mouse embryo. However, the mutant condition is not one of general cell lethality; the invasive trophoblast giant cells, at least, can persist in the absence of wild-type *Apc*. One rule that we derive from this study is that the developmental genes that predispose to neoplastic processes by loss of function must be fundamentally regulatory rather than necessary for cell viability.

SOMATIC GENETIC EVENTS DURING TUMORIGENESIS IN *Min*/+ (9)

If the *Apc* gene controls neoplasia by a simple loss-of-function process, one would expect tumors arising in *Min*/+ carriers to lose all activity of the normal *Apc* gene. In the corresponding case in the human, however, at most only 70% of tumors show detectable loss of the normal allele.

Loss of the wild-type *Apc* site has been analyzed by sectioning tumors, staining 1 of every 10 sections, locating the region of neoplastic growth, and isolating DNA from the unstained slides. The site-specific PCR assay described above was used to quantitate the ratio of wild-type to *Min* mutant DNA in the tumor (Fig. 2). For every one of 47 adenomas, extensive loss of the normal allele was observed in tumor samples but not in adjacent normal tissue. Thus, even for adenomas smaller than 1 mm in diameter, most cells in the tumor have lost the function of *Apc*.

These experiments, performed on an inbred background, have been extended to an (AKR × B6)F₁ background so that the fate of the entire wild-type chromosome can be followed (Fig. 3). At loci everywhere along chromosome 18, the allele linked to the wild-type *Apc* gene is lost in the 50 tumors analyzed. Control loci on other chromosomes have not shown such losses in *Min*-induced intestinal tumors.

Titration of the copy number shows that the remaining B6-*Min* chromosome is haploid, not reduplicated. Therefore, in the Min mouse, it seems that all adenomas are hemizygous for the mutated copy of chromosome 18. However, although all tumors show extensive loss, each tumor retains 10 to 20% of cells with the wild-type allele. Although great care was taken to analyze only the neoplastic region of the tumor, it is conceivable that DNA from normal cells contaminated the samples. Alternatively, it is conceivable that the adenoma is not clonal.

Three scenarios for understanding the somatic events underlying adenoma formation in the mouse are now under consideration (Fig. 4). The first assumes that the *Min* nonsense allele has simply lost *Apc* function and that adenoma formation requires only a mitotic error in which the wild-type homolog is lost. A second suggests that the nonsense allele of *Min* has a heterozygous cellular phenotype, for instance, the enhancement of mitotic errors to produce the hemizygous adenoma. Regarding this second scenario, it is interesting to recall that the APC polypeptide

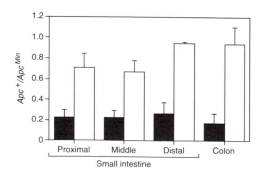

FIGURE 2. *Apc* allelic ratios for intestinal samples from B6-*Min* mice. ■, Tumor; □, control.

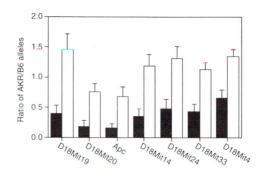

FIGURE 3. Allelic ratios for intestinal regions in (AKR × B6-*Min*)F₁ *Apc^Min^/Apc^+^* mice. ■, Tumor; □, control.

interacts with the microtubular cytoskeleton and so perhaps interacts with the mitotic apparatus as well. A third plausible scenario underlines the fact that we do not yet know that total loss of Apc function is sufficient for adenoma formation, even if it is necessary. There may be yet other somatic mutations, symbolized by *M*, that are also necessary. At least three genetic events are necessary for adenoma formation in this case.

THE DEVELOPMENTAL POTENTIAL OF *Min*/0 TUMORS (10)

In sections of *Min*-induced tumors, we have detected by immunohistochemistry the expression of cellular markers for three differentiated cell types of the intestinal epithelium: the Paneth cell (lysozyme), the enteroendocrine cell (serotonin), and the enterocyte (fatty acid-binding protein). Even in invasive tumors in older Min animals, we continue to find these markers.

It seems that the loss of Apc function does not compromise the expression of these differentiation markers of the intestinal epithelium, but we must remain cautious on at least two counts. First, we do not yet know whether the differentiation marker is expressed by the cells that have lost the Apc function rather than by the minority of cells in the tumor that may still retain it. Second, one differentiation marker does not constitute full demonstration of a cell type.

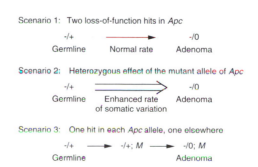

FIGURE 4. Clonal evolution of the adenoma.

WHAT OTHER LOCI INTERACT WITH Apc^{Min} IN TUMORIGENESIS?

Genetic changes accumulate as cancer progresses. Indeed, we are becoming acutely aware of situations in which cancer is driven by genetic instability syndromes.

If both *Apc* and *p53* are mutated in the germ line, is the probability of tumor formation enhanced? Note that the *Min*-induced adenoma has lost a copy of chromosome 18, and *p53* lesions are commonly associated with aneuploidy. We have intercrossed heterozygotes for a targeted loss-of-function allele of *p53* in the presence of the *Min* mutation and determined the multiplicity, size distribution, and histology of the resultant tumors. No significant differences have been detected for any of these three parameters at 80 days in the development of intestinal tumors (6a).

We have carried out an explicit search for loci that do modify the action of the *Min* mutation. Min mice on the sensitive C57BL/6 genetic background have been crossed with other inbred mouse strains, and the tumor multiplicity of the F_1 hybrids has been determined (Fig. 5). On the sensitive B6 background, shown as white bars in the histogram in Fig. 5, the mean tumor multiplicity is 29. In contrast, for Min animals in the (AKR × B6)F_1 generation (black bars, Fig. 5), the mean tumor multiplicity is only 5. The genetic basis of this reduction in tumor multiplicity has been determined by analyzing a population of animals in which the F_1 generation has been backcrossed to the sensitive B6 parent (gray bars, Fig. 5). Animals with low tumor multiplicity tend to inherit the AKR allele at a locus on mouse chromosome 4 that we call modifier-of-Min 1, or *Mom-1*. Those with a high tumor multiplicity tend to be homozygous for B6 alleles in the region of *Mom-1*. The location of *Mom-1* has been estimated to a resolution of about 15 centimorgans (cM).

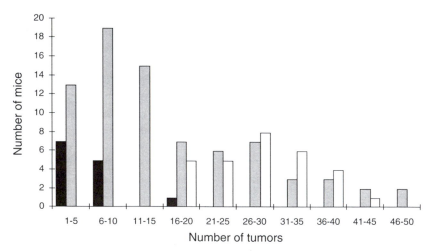

FIGURE 5. Tumor multiplicity in *Min/* + mice from the AKR backcross. ■, AKR × B6 mice; ▨, B6 × (AKR × B6) mice; □, B6 mice.

Mom-1 accounts for only a portion of the dominant effects of the AKR background on *Min*-induced intestinal neoplasia. However, no other dominant resistance allele can be mapped from these backcross data. Even further suppression of tumor multiplicity can be seen when the AKR background is homozygous (9a).

The effects of the AKR allele of *Mom-1* have been detected in heterozygotes; they are at least partially dominant. With congenic *Mom-1* derivatives, we are actively investigating whether two copies of the resistant AKR allele confer additional suppression of *Min*-induced neoplasia. Further, we want to know whether the *Mom-1* locus affects the other neoplastic processes to which Apc^{Min} predisposes an organism.

What molecule is encoded by the *Mom-1* locus? A region as large as 15 cM has many interesting candidates (4). To move toward a more restricted set of candidates, a resolution for the map position of *Mom-1* in the range of 1 to 2 cM is now being achieved by the analysis of recombinant chromosomes that dissect this region (7a).

The *APC*/*Apc* gene seems to be broadly expressed, as judged by RNA analysis, yet its mutations are normally first detected by their effects on the intestinal epithelium. This apparent tissue specificity might reflect an enhanced risk of somatic mutations or mitotic errors in the intestinal epithelium, with its very high proliferative index. However, further studies of human kindreds affected in *APC* and of Min mice have shown that other tissues are predisposed to developmental malformation or neoplasia by mutations in *APC*/*Apc* (Table 1).

One neoplastic pathway found in Min mice but not yet associated with germ line *APC* mutations in humans is that of the mammary gland (11). The penetrance of this phenotype is normally 10% or less in Min mice. A similarly low penetrance in the human might have escaped detection owing to a high background of other causes. Indeed, Thompson and his colleagues (21) have reported somatic mutations in *APC* in human mammary tumors.

Overall, we believe that the Min strain permits a general scan of the mammalian gene pool for genes that can modify the risk of particular neoplasms. Over the past decade, we all have been impressed by the steady progress in the identification of disease-predisposing genes in the human population, from familial hyper-

TABLE 1
Heterozygous phenotypes associated with germ line mutations in *APC*/*Apc*

Human	Mouse
Multiple colon adenomas	Multiple colon adenomas
Small intestine adenomas	Small intestine adenomas
Desmoid tumors	Desmoid tumors
Epidermoid cysts	Epidermoid cysts
Gastric polyps	Mammary adenocarcinomas
Osteomas of skull and mandible	Mammary keratoacanthomas
Hypertrophy of retinal pigment epithelium	
Abnormal dentition	

cholesterolemia to retinoblastoma to cystic fibrosis, but most of these genetic predispositions show variable expressivity ranging all the way to incomplete penetrance. Perhaps a part of this variation is due to modifier alleles at unlinked loci in the human. However, quantitative modifiers of disease severity are difficult to map within the available human disease families. By contrast, loci such as *Mom-1* can be mapped in appropriate mouse crosses. Thus, my colleagues and I foresee decades in which mouse strains that closely simulate human disease predispositions will be used to detect protective genes (6). Candidate genes detected by mouse genetics can then be tested within the corresponding human kindreds to determine whether corresponding resistant alleles exist in the human. Even if such mutant alleles cannot be found in the human population, the molecular identification of loci such as *Mom-1* by using the molecular genetics of the mouse may present opportunities for intervention in human disease.

It will be interesting to see what comes of this.

GENE ACTION INTRINSIC OR EXTRINSIC TO THE TUMOR LINEAGE

To carry forward the analysis, we need a way to distinguish between genes whose action is intrinsic or autonomous to the tumor lineage and those that act from surrounding tissue. Indeed, for several years in the 1960s, Howard Temin was involved in studying both the intrinsic and the extrinsic sides of cell conversion by Rous sarcoma virus (19). In this system, the transforming provirus is an element intrinsic to the neoplasm, and insulin-like growth factors to which converted cells respond differentially are extrinsic. The properties of the tumor can in principle be altered by changes either in the provirus or in the source of any differential growth factors.

We have recently made an observation consistent with the hypothesis that *Mom-1* may act extrinsically to the tumor lineage in *Min*-induced adenomas (8a). When tumors form in (AKR $\times$ B6-*Min*/+)F_1 hosts, the region surrounding the *Mom-1* locus remains heterozygous in the tumor lineage. Unlike the *Apc* locus, where the protective wild-type allele is always lost, the protective allele of *Mom-1* seems to be retained. One interpretation of this result is that *Mom-1* controls tumor formation from surrounding tissue.

This suggestion is far from proven. We want to be able to label one of the genetic components of the chimera in such a way that cells derived from it can be recognized unambiguously no matter what the state of differentiation or neoplasia. We are seeking a histochemical cell lineage marker to permit such an analysis of *Mom-1* action. Over the coming years, we may be able in these ways to tease apart the networks of genes that impinge upon the pathways of neoplasia predisposed by mutation of the developmental gene, *Apc*.

EPILOGUE

Howard and I talked more than once about the special opportunity for doing science away from the mainstream bustle of the urban centers of the United States. The metaphor he used was Backwater Science.

It is in the shadows that momentous encounters occur.

—Alfred Fabre-Luce

Every one of us in research knows the brave solitude of striking out into unmapped territory. Howard and I have enjoyed the special opportunity to do this in the Wisconsin community. In particular, McArdle Laboratory has been built by solid citizens of the cancer research community: H. P. Rusch, V. R. Potter, J. A. Miller, R. K. Boutwell, E. C. Miller, C. Heidelberger, and G. C. Mueller, the founders who hired Henry Pitot, Howard Temin, and Waclaw Szybalski in 1960 and Charles Kasper and me a few years later. These are the early investigators responsible for the well-built cabin and research discipline from which Backwater Science could be pursued.

In commemoration of Howard, we marvel at the series of successes in his research and that of his protegées. Let me add some texture from the early days of the mid-1960s. From my very first week in Madison in 1965, five of us gathered for Monday lunch. I thought of this lunch gathering as The Caltech Group. (In fact, only Howard, Millard Susman, and I were direct emigrés from Caltech; Masayasu Nomura and Bernie Weisblum were indirect descendants of the Caltech Phage Group through their experiences with Seymour Benzer at Purdue.)

In 1965, the notion of retrotranscription was far from accepted, much less the notion of retrotransposition. There were other interpretations of the effects of inhibition of DNA synthesis on the life cycle of Rous sarcoma virus, and Howard's hybridization evidence for a DNA provirus, offered a decade in advance of Southern blotting, was necessarily marginal.

> Hybridization experiments have been carried out, and there is an increase in . . . the DNA from infected cells, as compared with DNA from uninfected cells (Temin, 1964b). *These results, however, are as yet unconfirmed.*
> —H. M. Temin, Studies on Carcinogenesis by Avian Sarcoma Viruses. IV. The Molecular Biology of Viruses (1967) (emphasis added)

During those days, stretching into years, we lunching colleagues related to the DNA provirus hypothesis not as established fact, not as a loyalty contest between believers and nonbelievers, but as a plausible idea looking for stronger tests of its viability. The bromodeoxyuridine sensitization experiments of David Boettiger (1) evolved from Monday discussions of The Caltech Group. It is in the highest sense of Karl Popper that Howard would refer to his own claims as "unconfirmed." More important in science than to be right is to do it right.

There are many visionaries. The visions that move successfully into our base of knowledge are the ones that come into direct contact with feasible experiments. Howard's DNA provirus vision, based on strong biological inference, came into contact first with the experimental recognition of virus-borne enzymes by Kates and MacAusland, among others. Later, the vision contacted vast improvements in DNA analysis with restriction enzymes, mammalian cell DNA transformation, and low-noise Southern blotting.

All of these were matters of science, carried on quietly both in the backwaters and the mainstream over the 1960s. But science also has an essential public face, which can be ugly.

> Dr. H. M. Temin's *tantalizingly incomplete* work . . . suggests that avian sarcoma virions contain an enzyme capable of making DNA using a single stranded RNA template. . . . Temin *claims* that avian sarcoma virions, after treatment *with an unspecified detergent,* will incorporate deoxynucleoside triphosphates . . .
> —*Nature* (London) **226**:1003 (June 13, 1970) (emphasis added)

> The discovery of this unprecedented enzyme . . . is an *extraordinary personal vindication* for Dr. Howard Temin.
> —*Nature* (London) **226**:1198 (June 27, 1970) (emphasis added)

In a span of 2 weeks, one could read descriptions of Howard Temin's research that ranged from anonymous condescension to public idolatry. I have always believed that Howard had a sufficient store of skepticism to survive any such waves of public notoriety and fame. He recognized clearly the competitive urge that drives so much creative effort, both scientific and journalistic.

The vital partner to competition is a sense of community. When Howard was awarded the Nobel Prize along with Renato Dulbecco and David Baltimore, he gave a prominent nod of gratitude to the virology community. In recent years, Howard invested significant personal energy as a de facto ombudsman for certain ethical disputes within that community.

Here in the backwaters of Wisconsin, especially during Howard's brave confrontation with cancer, his colleagues have been paramount, with his family at the center.

I have given here a strictly personal view of current issues in the biology of cancer and in the outlook of Howard Temin. There should be other views. For example, had Sewall Wright lived to 105, as expected, right now he would be arguing with Howard that retrotranscription and retrotransposition have been selected not for a role in development but for their role in the genetic variation of populations.

ACKNOWLEDGMENTS. Studies of the Min strain, following its discovery by Amy Moser in my laboratory, have grown to involve many constructive collaborations. Biological and genetic studies of Min at McArdle Laboratory have involved Amy Moser; doctoral students Cindy Luongo, Karen Gould, and Al Shoemaker; our specialists, Melanie McNeley, Darren Katzung, Linda Clipson, and Natalie Borenstein; and our consultants, Norman Drinkwater and Henry Pitot. Our studies of general embryology and neoplasia in the mouse have depended on Camille Connelly, Paraj Mandrekar, and Alexandra Shedlovsky. To study mammary neoplasia, we have worked with Jill Haag and Michael Gould of the Clinical Cancer Center at Wisconsin. The intestinal cell type analysis was carried out with Kevin Roth and Jeff Gordon of Washington University, St. Louis. The molecular genetics of *Apc* has benefited from close collaboration with Li-Kuo Su, Dan Levy, Stan Hamilton, Ken Kinzler, and Bert Vogelstein of Johns Hopkins University. The knockout genetics of *p53* has involved Larry Donehower of Baylor University. To map the *Mom-1* locus, we have joined with Bill Dietrich and Eric Lander of The Whitehead Institute, Massachusetts Institute of Technology. Finally, the analysis of *Min/Min* embryos has enjoyed the expertise of Richard Gardner of the Imperial Cancer Research Fund, Oxford.

Studies of the Min strain will grow over the coming years now that it is genetically inbred and provided by The Jackson Laboratory to all investigators worldwide.

REFERENCES

1. **Boettiger, D., and H. M. Temin.** 1970. Light inactivation of focus formation by chicken embryo fibroblasts infected with avian sarcoma virus in the presence of 5-bromodeoxyuridine. *Nature* (London) **228**:622–624.

2. **Burland, T. G., L. Solnica-Krezel, J. Bailey, D. B. Cunningham, and W. F. Dove.** 1993. Patterns of inheritance, development and the mitotic cycle in the protist *Physarum polycephalum. Adv. Microb. Physiol.* **35:**1–69.

3. **Cairns, J.** 1975. Mutation selection and the natural history of cancer. *Nature* (London) **255:**197.

4. **Dietrich, W., E. Lander, A. Moser, K. Gould, C. Luongo, N. Borenstein, and W. F. Dove.** 1993. Genetic identification of *Mom-1,* a major modifier locus affecting *Min*-induced intestinal neoplasia in the mouse. *Cell* **75:**631–639.

5. **Dove, W. F.** 1987. Molecular genetics of *Mus musculus:* point mutagenesis and milliMorgans. *Genetics* **116:**5–8.

6. **Dove, W. F.** 1993. The gene, the polygene, and the genome. *Genetics* **134:**999–1002.

6a. **Dove, W. F., D. Katzung, and L. Donehower.** Unpublished data.

7. **Furth, M. E., J. L. Yates, and W. F. Dove.** 1979. Positive and negative control of bacteriophage lambda DNA replication. *Cold Spring Harbor Symp. Quant. Biol.* **43:**147–153.

7a. **Gould, K. A.** Unpublished data.

8. **Kemler, R.** 1993. From cadherins to catenins: cytoplasmic protein interactions and regulation of cell adhesion. *Trends Genet.* **9:**317–321.

8a. **Luongo, C.** Unpublished data.

9. **Luongo, C., A. R. Moser, S. Gledhill, and W. F. Dove.** 1994. Loss of *Apc*$^+$ in intestinal adenomas from Min mice. *Cancer Res.* **54:**5947–5952.

9a. **Moser, A. R.** Unpublished data.

10. **Moser, A. R., W. F. Dove, K. A. Roth, and J. I. Gordon.** 1992. The *Min* (multiple intestinal neoplasia) mutation: its effect on gut epithelial cell differentiation and interaction with a modifier system. *J. Cell Biol.* **116:**1517–1526.

11. **Moser, A. R., E. M. Mattes, W. F. Dove, M. J. Lindstrom, J. D. Haag, and M. N. Gould.** 1993. *Apc*^{Min}, a mutation in the murine *Apc* gene, predisposes to mammary carcinomas and focal alveolar hyperplasias. *Proc. Natl. Acad. Sci. USA* **90:**8977–8981.

12. **Moser, A. R., H. C. Pitot, and W. F. Dove.** 1990. A dominant mutation that predisposes to multiple intestinal neoplasia in the mouse. *Science* **247:**322.

13. **Moser, A. R., A. R. Shoemaker, C. S. Connelly, L. Clipson, K. A. Gould, C. Luongo, W. F. Dove, P. H. Siggers, and R. L. Gardner.** Homozygosity for the *Min* allele of *Apc* results in disruption of mouse development prior to gastrulation. *Dev. Dynamics*, in press.

14. **Peifer, M., S. Berg, and A. B. Reynolds.** 1994. A repeating amino acid motif shared by proteins with diverse cellular roles. *Cell* **76:**789–791.

15. **Russell, W. L., E. M. Kelly, P. R. Hunsicker, J. W. Bangham, S. C. Maddux, and E. L. Phipps.** 1979. Specific-locus test shows ethylnitrosourea to be the most potent mutagen in the mouse. *Proc. Natl. Acad. Sci. USA* **76:**5818–5819.

15a. **Shedlovsky, A., C. Connelly, and P. Mandrekar.** Unpublished data.

16. **Shedlovsky, A., J.-L. Guénet, L. L. Johnson, and W. F. Dove.** 1986. Induction of recessive mutations in the *T/t-H-2* region of the mouse genome by a point mutagen. *Genet. Res.* **47:**135–142.

17. **Shedlovsky, A., J. D. McDonald, D. Symula, and W. F. Dove.** 1993. Mouse models of human phenylketonuria. *Genetics* **134:**1205–1210.

18. **Su, L.-K., K. W. Kinzler, B. Vogelstein, A. C. Preisinger, A. R. Moser, C. Luongo, K. A. Gould, and W. F. Dove.** 1992. A germline mutation of the murine homolog of the APC gene causes multiple intestinal neoplasia. *Science* **256:**668–670.

19. **Temin, H. M.** 1967. Studies on carcinogenesis by avian sarcoma viruses. VI. Differential multiplication of uninfected and of converted cells in response to insulin. *J. Cell. Physiol.* **69:**377–384.

20. **Temin, H. M.** 1971. The protovirus hypothesis: speculations on the significance of RNA-directed DNA synthesis for normal development and for carcinogenesis. *J. Natl. Cancer Inst.* **46:**III–VII.

21. **Thompson, A. M., R. G. Morris, M. Wallace, A. H. Wyllie, C. M. Steel, and D. C. Carter.** 1993. Allele loss from 5q21 (APC/MCC) and 18q21 (DCC) and DCC mRNA expression in breast cancer. *Br. J. Cancer* **68:**64–68.

22. **Vitaterna, M. H., D. P. King, A.-M. Chang, J. M. Kornhauser, P. L. Lowrey, J. D. McDonald, W. F. Dove, L. H. Pinto, F. W. Turek, and J. Takahashi.** 1994. Mutagenesis and mapping of a mouse gene, *Clock,* essential for circadian behavior. *Science* **264:**719–725.

Chapter 9

Malignant Transformation of Cells by the v-Rel Oncoprotein

Thomas D. Gilmore, David W. White, Sugata Sarkar, and Saïd Sif

PROLOGUE

I (T.D.G.) joined Howard Temin's group at McArdle Laboratory for Cancer Research as a postdoctoral fellow in the summer of 1984, after completing my doctoral work with G. Steven Martin at the University of California. Having been trained in cell biology, biological chemistry, and retrovirology, I was eager to learn and apply the emerging tools of molecular biology to the study of cancer. (Even in 1984, many laboratories did not use what are now standard molecular biology techniques.)

Shortly after I had joined the Temin laboratory, it became very clear to me that Howard's primary interest in retrovirus-induced cancer was from a genetic point of view. For example, what changes have occurred in the viral genome that make a given retrovirus highly transforming? Eventually, we settled on a project that was a follow-up to one initiated by a former postdoctoral fellow, Gary Tarpley. Simply put, the project involved constructing avian retroviral vectors for the expression of the v-Ha-*ras* gene and the rat c-Ha-*ras* proto-oncogene to determine whether these murine genes were transforming in the avian system in tissue culture and (always important to Howard) in vivo in chickens.

Although I eventually did make the *ras* plasmids and was able to show that the *ras* genes were transforming in the avian system, my interests gradually drifted toward using molecular and biochemical approaches to understanding the mechanism by which the then-obscure v-*rel* oncogene transforms chicken spleen cells. (To my chagrin and Howard's consternation, I never did submit the work on the *ras*-containing retroviral vectors, even though I have used them in work since that time.) My interest in v-Rel was stimulated in part by my background at Berkeley and in part by many conversations with Bakary Sylla, who was also a postdoctoral fellow in the Temin laboratory at that time. Shortly after I had made anti-v-Rel antibodies and had done the initial localization studies (24), which suggested

---

Thomas D. Gilmore, David W. White, and Sugata Sarkar • Biology Department, Boston University, 5 Cummington Street, Boston, Massachusetts 02215. *Saïd Sif* • Department of Molecular Biology, Massachusetts General Hospital, Boston, Massachusetts 02114.

many of the properties of Rel as a transcription factor controlled by subcellular localization, Céline Gélinas joined the Temin laboratory. Bakary, Céline, and I had many conversations that drove the research toward a more protein-oriented approach to understanding transformation by v-Rel.

Although understanding the pathway by which v-Rel transforms cells was not Howard's primary interest, he nevertheless allowed Céline, Bakary, and me to pursue this avenue of experimentation. Howard later visited me in Boston, shortly after I had set up my laboratory at Boston University and at about the time the news about the relationship between Rel proteins and NF-κB was breaking. He told me with a degree of humility, "Now that you, Céline, and Bakary have left, I don't really have much interest in continuing with the work on the v-Rel protein, but I worry that people will think that I am afraid of the competition if I stop working on it!"

Howard Temin's laboratory had studied the v-*rel* oncogene for some time before I joined the laboratory. Although Howard had himself developed the focus assay for studying transformation by Rous sarcoma virus (73), he had switched his interest to a class of avian retroviruses, the reticuloendotheliosis viruses, that were (and to some extent still are) relatively obscure. These viruses include turkey, duck, and chicken retroviruses that cause a variety of diseases, including runting, immunosuppressive diseases, and malignancies, in young birds (55). At the sequence level, the reticuloendotheliosis viruses are more related to mammalian retroviruses than to other avian retroviruses, and they can infect a wide variety of cell types from many avian and mammalian species, including humans. Howard was interested in the reticuloendotheliosis viruses in two general ways: first, he was interested in their life cycles (see also chapters 2, 15, 16, 19, and 21) in terms of developing these viruses as vectors for gene therapy, and second, he was interested in the mechanism by which one of these viruses, Rev-T, had evolved so as to cause lymphomas in young poults.

Rev-T causes a rapidly fatal lymphoma when injected into young chickens: chickens die within 7 to 10 days postinfection and have multiple tumors in their spleens and livers (reviewed in references 20 and 58). In vitro, Rev-T primarily transforms chicken cells of early B- and T-cell lineages, but it can also transform cells of erythroid, myeloid, and fibroblast lineages. Research from Howard's laboratory had played a part in defining the structure and sequence of much of v-*rel* and chicken and turkey c-*rel* (11, 77, 78) and had helped establish a clear understanding of how Rev-T had evolved as a recombination between Rev-A and turkey c-*rel* (12, 13, 44, 45). These early studies set the stage for a more detailed description of the v-Rel and c-Rel proteins, in many cases by people who were in Howard's laboratory or who had been trained by Howard (see below). Interestingly, it also led to a second convergence of the research of Howard Temin and David Baltimore, when the homology between Rel proteins and the subunits of the NF-κB transcription factor was discovered.

In a fatherly sort of way, Howard had a great deal of interest in his former students and postdoctoral fellows. I am deeply indebted to Howard's continued interest in my career and work. My last conversation with Howard was typical of his interest in his former students: obviously in a weakened state, he called me about 2 months before his death to inquire about Eric Humphries. Eric Humphries, director of the Mary Babb Randolph Cancer Center at the University of West Virginia and one of Howard's first graduate students, had passed away unexpectedly in early December 1993. Howard was deeply concerned about Eric's wife and family, but in a surprisingly light tone considering his own condition, he also remarked, "In my state, I hardly imagined that one of my students would be dying before me."

CLASSES OF REL PROTEINS

Rel family transcription factors have been identified in species from *Drosophila melanogaster* to humans. In vertebrates, Rel transcription factors include p50, p52, RelA, RelB, c-Rel, and the retroviral oncoprotein v-Rel; in insects, Rel transcription factors include Dorsal and Dif (reviewed in references 19 and 22). With the exception of Dorsal (which is involved in the development of embryonic polarity), all Rel transcription factors appear to be primarily involved in immune and inflammatory responses (reviewed in references 2 and 21). That is, Rel proteins are generally expressed at the highest levels in immune cells, and in many cases, they control the expression of genes involved in the immune response. The classic Rel complex, studied extensively in David Baltimore's laboratory, is the NF-κB transcription complex, which is a heterodimer of p50 and RelA (formerly called p65) and controls the expression of the κ light-chain immunoglobulin gene (reviewed in reference 1). In addition, Rel complexes are involved in regulating the expression of several viral promoters, the most notorious being that of human immunodeficiency virus type 1 (reviewed in reference 21).

Rel proteins are related through a conserved domain of approximately 300 amino acids called the Rel homology (RH) domain (see Fig. 1 and reference 19). The RH domain contains sequences important for DNA binding, nuclear localization, formation of dimers, and interaction with IκB inhibitor proteins and heterologous transcription factors. Rel proteins can be subdivided into two classes that are based on sequences C terminal to the RH domain. One class of Rel proteins (class I) includes p50 and p52: these Rel proteins are synthesized as precursor proteins (p105 and p100) that are cleaved just C terminal to the RH domain to generate mature proteins (p50 and p52) that have higher affinities for DNA than the precursor proteins do. The cleaved sequences C terminal to the RH domain in class I Rel proteins contain multiple copies of a 33-amino-acid repeat, called the ANK repeat, which has been found in many proteins and is known to mediate protein-protein interactions (reviewed in reference 43). In addition, the RH domain of class I Rel proteins contains an insert that is not present in class II Rel proteins. Class II Rel proteins include c-Rel, RelA, RelB, Dorsal, and Dif. Class II Rel proteins are not cleaved and do not contain ANK repeats but do contain transcriptional activation domains in their C-terminal sequences.

Amost all vertebrate Rel proteins can form homodimers and heterodimers. This combinatorial diversity results in distinct DNA-binding specificities for the various Rel complexes; that is, different Rel complexes have different affinities for related DNA sequences, commonly called κB sites.

REGULATION OF REL PROTEINS

Although specific Rel proteins can be regulated in several ways (reviewed in reference 22), all Rel proteins are likely to be regulated by subcellular location. That is, Rel complexes are retained in the cytoplasm in an inactive form and can be induced to enter the nucleus and affect gene expression when a cell receives any one of various stimuli (reviewed in reference 2). In general, Rel complexes are retained in the cytoplasm and prohibited from binding DNA by direct interaction with one of a family of inhibitor proteins termed the IκB family of proteins.

The IκB family includes IκB-α, IκB-β, IκB-γ, and Bcl-3; moreover, the NF-κB precursor proteins p100 and p105 can also function as Iκ proteins by retaining other Rel proteins in the cytoplasm and inhibiting their DNA-binding activities (reviewed in reference 23). Different IκBs interact with various affinities with different Rel complexes to provide another layer of complexity to the regulation of Rel proteins. The IκB proteins are structurally related in that all contain five to eight copies of the ANK repeat, which is important for their interaction with Rel proteins. Although IκB proteins are usually thought to inhibit nuclear localization and DNA binding by Rel proteins, these may not be the only consequences of interaction of a given IκB protein with a Rel complex (see reference 23).

Induction of a Rel complex (to enter the nucleus) arises by degradation of IκB. The best-studied system is the regulation of NF-κB (p50-RelA) by IκB-α. The most commonly accepted model indicates that induction of NF-κB is initiated by modification of IκB-α (e.g., by phosphorylation) and that this modification induces the degradation of IκB-α, probably by the cellular proteasome (53). Free NF-κB can then enter the nucleus, bind to κB sites, and affect the expression of target genes.

STRUCTURE OF THE V-REL ONCOPROTEIN

The v-Rel oncoprotein is the transforming protein of the replication-defective avian Rev-T retrovirus, which is derived from the Rev-A replication-competent helper virus (reviewed in references 20, 29, and 58). As part of the mutational process that generated the highly oncogenic virus, Rev-T has lost part of its *gag* and *pol* genes and has had cellular *rel* sequences substituted for most of the *env* gene. As described by Kirk Wilhelmsen while a graduate student with Howard, the v-*rel* oncogene is an *env-rel-env* fusion gene that encodes a protein of 503 amino acids that has 11 N-terminal Env amino acids (with three mutations compared with the Rev-A Env protein) followed by 474 c-Rel-derived amino acids

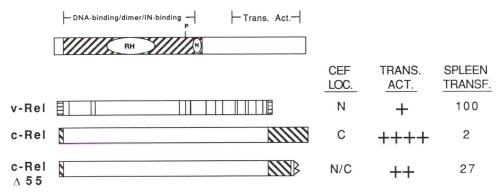

FIGURE 1. Comparison of v-Rel and c-Rel. (Top) Generalized structure of class II Rel proteins: the hatched area is the RH domain that contains sequences important for DNA binding, dimerization, inhibitor (IN) (IκB) binding, nuclear localization (N), and phosphorylation by protein kinase A (P); the C-terminal half (open box) contains multiple transcription activation domains (Trans. Act.). (Bottom) Comparative structures of v-Rel and c-Rel. The boxes at the ends of v-Rel indicate Env-derived amino acids (2 at the N terminus and 19 at the C terminus). The lines in v-Rel indicate small amino acid differences between v-Rel and turkey c-Rel. A truncated form of c-Rel (c-Rel Δ55 [36]), missing 55 C-terminal amino acids, is shown at the bottom. In c-Rel and c-Rel Δ55, the hatched boxes at the ends indicate sequences not present in v-Rel. To the right are indicated the subcellular locations of these Rel proteins in chicken fibroblasts (N, nuclear; C, cytoplasmic; N/C nuclear and cytoplasmic), the relative strength of transcriptional activation (TRANS. ACT.) of the C-terminal sequences in Rel proteins, and the relative in vitro chicken spleen cell transforming efficiencies (SPLEEN TRANSF.) of the Rel proteins.

and 19 out-of-frame Env-derived amino acids at its C terminus (78). When cDNAs encoding chicken c-Rel were identified (independently by Mark Hannink in Howard's laboratory and Tony Capobianco in our laboratory), it became clear that v-Rel had lost 2 N-terminal amino acids and 118 C-terminal amino acids compared with c-Rel and that there were about 18 internal changes (point mutations, deletions) between v-Rel and c-Rel (Fig. 1; references 11 and 27).

FUNCTIONAL DIFFERENCES BETWEEN v-REL AND c-REL

v-Rel and c-Rel differ in several ways. The deletion of C-terminal sequences in v-Rel has three functional consequences (Fig. 1). First, v-Rel and C-terminally truncated forms of c-Rel are located in the nuclei of chicken fibroblasts, whereas full-length c-Rel is located in the cytoplasm (11), probably because these truncated Rel proteins interact less strongly with IκB-α (p40) than does full-length c-Rel. Second, the C-terminal truncation in v-Rel has removed c-Rel sequences that can act as strong transcriptional activation domains (60). Third, v-Rel and C-terminally truncated forms of c-Rel are highly transforming (Fig. 1). Moreover, the C-terminal sequences in many spontaneously arising transforming c-Rel proteins have been deleted (Fig. 2; see also reference 32). However, some cell lines transformed by viruses designed to overexpress chicken c-Rel synthesize apparently full-length

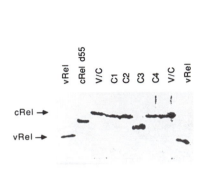

FIGURE 2. Expression of Rel proteins in transformed spleen cell lines. A Western blot (immunoblot) using anti-Rel antibody was performed on chicken spleen cell lines transformed by viruses overexpressing the following Rel proteins: v-Rel, a c-Rel protein with 55 C-terminal amino acids deleted (cRel d55 [Fig. 1; 36]), and a v-Rel–c-Rel recombinant virus (V/C) containing the RH domain of v-Rel and C-terminal sequences from c-Rel (11). Lanes C1, C2, C3, and C4 contain independent spleen cell lines derived from spleen cell colonies transformed by a retroviral vector (JDc-Rel [11]) for the overexpression of chicken c-Rel. Cell lines C1, C2, and C4 express c-Rel proteins that comigrate with full-length c-Rel; C3 expresses a truncated c-Rel protein. The positions of full-length c-Rel and v-Rel are indicated by arrows.

c-Rel proteins (Fig. 2; clones C1, C2, and C4), but it is not clear whether these transforming full-length c-Rel proteins have small changes compared with wild-type c-Rel.

The C-terminal deletion in v-Rel appears to be the most important mutation for activating the oncogenicity of c-Rel. It is likely that the original v-*rel* gene had a 3′ truncation of c-*rel* and thus encoded a truncated c-Rel protein; the internal differences between v-Rel and c-Rel that we now recognize probably came about during propagation of the virus and selection in vitro (or in vivo) for highly transforming variants (discussed in references 28 and 58). For example, early isolates of Rev-T appear to have caused death in young chickens with decreased frequency and increased latency compared with later Rev-T viruses (58). However, it is unlikely that any single internal mutation (between v-Rel and c-Rel) is essential for the transforming function of v-Rel. For example, Girish Bhat, while a graduate student with Howard, converted every v-Rel difference back to the amino acid found in c-Rel, and no single change completely inactivated the transforming function of v-Rel (7). Nevertheless, this and other studies (32, 49, 72) indicate that the internal changes in v-Rel contribute to its increased oncogenicity, probably by modulating the activity of v-Rel. One interesting example from Eric Humphries' group is the recent identification of a v-*rel* variant that encodes an amino acid change at position 40 (an Ala-to-Ser mutation), which increases the DNA-binding activity of v-Rel and changes its cell type-transforming spectrum to one that is more specific for B cells (62). Again, this suggests that v-Rel has acquired additional mutations through selective pressure in vitro.

FUNCTIONS NECESSARY FOR TRANSFORMATION BY v-REL

There are two general models to explain the mechanism by which v-Rel effects transformation of chicken spleen cells (reviewed in references 20 and 29). v-Rel may be binding to DNA and directly affecting the expression of cellular genes, leading to the transformed state. Alternatively, v-Rel may be disrupting a normal equilibrium between Rel complexes and IκB proteins, also leading to the transformed state; this could occur either because v-Rel sequesters IκB and therefore activates cellular Rel complexes or because v-Rel, by forming heterocomplexes with cellular Rel proteins, changes the normal pattern of gene activation mediated by cellular Rel complexes. In addition, v-Rel could be acting in both ways.

An analysis of many mutants shows a rigorous correlation between the ability of v-Rel to form homodimers and bind DNA and its ability to transform chicken spleen cells. Namely, there are no transforming v-Rel mutants that cannot also bind to DNA as homodimers (reviewed in reference 20). However, DNA binding by v-Rel is not sufficient for transformation, and sequences C terminal to the RH domain are also needed for transformation. In our laboratory, Sugata Sarkar performed a deletion analysis of sequences encoding the C-terminal half of v-Rel and found that the C-terminal v-Rel sequences that are necessary for transformation are likely to contain transcriptional activation domains (Fig. 3; 63). Specifically, there appear to be two weak transcriptional activation domains (domains B and C in Fig. 3) in the C-terminal half of v-Rel, and either of these domains is needed along with an intact RH domain to form a highly transforming v-Rel protein (see also reference 63).

This result is consistent with a model wherein v-Rel causes transformation by increasing the expression of certain cellular genes. Therefore, Sugata wished to determine whether a heterologous transcriptional activation domain could substitute for the C-terminal activation domain in v-Rel. She created a plasmid (v-Rel/Act) encoding a fusion protein containing the RH domain of v-Rel fused to an artificial activation domain derived from Bluescript plasmid sequences. These random Bluescript plasmid sequences activated transcription to approximately the same extent as the C-terminal activation domain of v-Rel when these sequences were fused to the DNA-binding domain of the yeast protein GAL4 and assayed for their ability to stimulate expression from a reporter chloramphenicol acetyltransferase (CAT) gene containing upstream binding sites for GAL4 (Fig. 4). However, the intact v-Rel–heterologous activation domain fusion protein (v-Rel–Act) was not transforming for chicken spleen cells (Fig. 4). Therefore, it appears that an artificial transcriptional activation domain cannot substitute for the C-terminal v-Rel activation domain. This indicates either that a specific type of C-terminal activation domain is needed for v-Rel to effect transformation or that there are other C-terminal functions in v-Rel necessary for transformation.

The model described above for transformation by v-Rel dictates that v-Rel must be in the nucleus, directly regulating the expression of cellular genes; however, certain experiments are hard to reconcile with this model, and many of these experiments come directly from work originating in Howard Temin's laboratory. First, most v-Rel is found in the cytoplasm of transformed chicken spleen cells (24, 69) in high-molecular-weight complexes with several cellular proteins (68); these proteins include c-Rel (11), IκB-α (or p40 [15]), and NF-κB p105 (10) and p100 (65). Second, the subcellular location of v-Rel does not appear to affect its transforming function: that is, v-Rel proteins that have been engineered to localize exclusively to the nucleus or to the cytoplasm are equally transforming (25). This remains a surprising result in light of other experiments and can be most easily

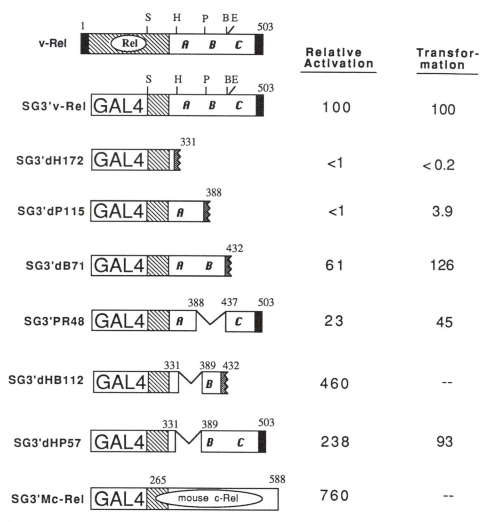

FIGURE 3. Correlation between the presence of a C-terminal transcriptional activation domain in v-Rel and transformation of chicken spleen cells. The general structure of v-Rel is shown at the top: ■, Env-derived amino acids; ▨, RH domain. Three C-terminal domains (A, B, and C) are delimited by restriction enzyme sites (H, *Hinc*II; P, *Pvu*II; B, *Bst*XI; E, *Eco*RI), and the numbers above the structure indicate the amino acids in v-Rel. Other structures represent GAL4 (1–147) fusion proteins that were created at a unique *Stu*I site (S) and have been assayed for their relative abilities to activate transcription from a GAL4-site-containing reporter plasmid in chicken fibroblasts. Each protein contains the v-Rel amino acids indicated above the structure (SG3'Mc-Rel is a GAL4 fusion protein containing C-terminal sequences from mouse c-Rel). Spleen cell transformation data are not for GAL4 fusion proteins but for v-Rel proteins that contain intact N-terminal RH domains and the indicated C-terminal deletions. --, Not determined. See Sarkar and Gilmore (63) for further details.

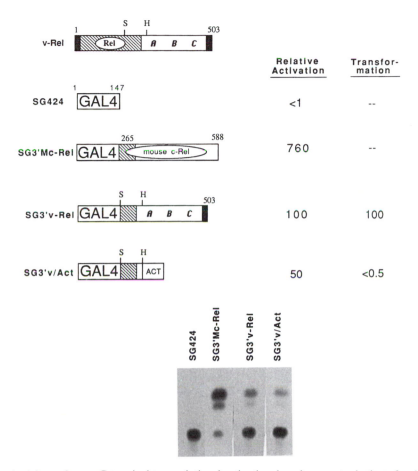

	Relative Activation	Transfor- mation
SG424	<1	--
SG3'Mc-Rel	760	--
SG3'v-Rel	100	100
SG3'v/Act	50	<0.5

FIGURE 4. A heterologous C-terminal transcriptional activation domain cannot substitute for the transcriptional activation domain of v-Rel. As in Fig. 3, the structures of v-Rel (top) and relevant GAL4 fusion proteins (middle) are shown. SG3'v/Act contains vector-encoded sequences (ACT) from pBluescript SK+ (Trp-Gln-Asp-Arg-Phe-Pro-Asp-Trp-Leu-Ala-Gly-Ser-Glu-Arg-Asn-Ala-Ile-Asn-Val-Ser) fused to v-Rel sequences at the unique *Hinc*II (H) site. The relative transcriptional activation data were determined in duplicate cotransfection assays in chicken cells as described for Fig. 3 (see also reference 63), and the relative spleen cell transformation data are for intact v-Rel proteins containing the RH domain and the indicated C-terminal sequences. A representative CAT transcriptional activation assay for the indicated GAL4 fusion proteins is shown at the bottom. In this assay, the extent of transcriptional activation is determined by quantitating the amount of radioactivity in the upper spot.

explained by the requirement for only a small amount of v-Rel in its important compartment (e.g., the nucleus) in order to transform. This experiment came directly out of a Friday discussion with Howard after I (T.D.G.) had told him that there appeared to be different locations for v-Rel in different cell types (24); he suggested that I try to alter the subcellular location of v-Rel and left the details to me to work out.

ISOLATION AND CHARACTERIZATION OF TEMPERATURE-SENSITIVE MUTANTS OF v-REL

As a first step toward identifying the cellular changes that occur during v-Rel-mediated transformation, David White in our laboratory created two temperature-sensitive (*ts*) v-Rel mutants: v-G37E and v-R273H (75). These mutants have single amino acid changes in the RH domain (Gly to Glu at amino acid 37 [mutant v-G37E] and Arg to His at amino acid 273 [v-R273H]) that are analogous to changes that have occurred in two *ts* Dorsal mutants (35). These *ts* v-Rel mutants induce the formation of transformed spleen cell colonies at 36.5°C but not at 41.5°C, whereas wild-type v-Rel efficiently transforms cells at both temperatures. In addition, both *ts* mutant v-Rel proteins, whether isolated from cells or synthesized in vitro, are *ts* for binding a κB-site-containing probe in electrophoretic mobility shift assays. It is likely that mutant v-G37E is *ts* for DNA binding per se, since the mutation is near sequences in v-Rel that have been shown to contact DNA (39); mutant v-R273H is likely to be *ts* for homodimer formation (and hence DNA binding), since other mutations near this site render v-Rel defective for the formation of homodimers (49). Several characteristics of these *ts* v-Rel mutant proteins are summarized in Table 1.

v-REL BLOCKS PROGRAMMED CELL DEATH

Chicken spleen cells transformed in vitro by wild-type v-Rel usually grow as large multicellular clumps. When *ts* v-Rel-transformed cells growing at 36.5°C are shifted to 41.5°C, the cells disaggregate and stop growing (75). Moreover, *ts* v-Rel-infected cells at 41.5°C appear by several criteria to be going through a form of programmed cell death, termed apoptosis: they have condensed chromatin, as judged by 4′,6-diamidino-2-phenylindole (DAPI) staining; DNA from these cells is cleaved into oligonucleosome-length fragments ("DNA ladders"); within 72 h at 41.5°C, most cells no longer stain with DAPI and yet still appear as whole cell bodies; and the appearance of apoptosis in these cells is delayed in the presence

TABLE 1
Characteristics of *ts* v-Rel mutants[a]

Function	Response at assay temp of:		
	30°C	36°C	41°C
Transformation	ND[b]	+	−
Immortalization (block to apoptosis)	ND	+	−
DNA binding	+ +	+	−
Binding to IκB-α (in vitro)	+ +	+	−
Subcellular location in chicken embryo fibroblasts	ND	Nuclear	Nuclear

[a] Data are summarized from White and Gilmore (75) and White et al. (76).
[b] ND, not determined.

of the antioxidant *N*-acetylcysteine (76). Similarly, Hockenbery et al. showed that *N*-acetylcysteine can block the onset of apoptosis in murine lymphoid cells (30). Our results suggest that v-Rel blocks a normal process of programmed cell death as part of its transforming and immortalizing process and are consistent with the ability of v-Rel to block the apoptosis that occurs in chicken bursal B cells (50).

To identify proteins that might be important for transformation by v-Rel or that are associated with apoptosis, we next sought to determine whether there are changes in specific proteins in *ts* v-Rel-infected cells shifted from 36.5 to 41.5°C. The overall profiles of the proteins, as judged by Coomassie blue staining, in *ts* transformed cells shifted to 41.5°C for up to 72 h are similar to those of proteins in cells maintained at 36.5°C (76). Moreover, the cellular levels of v-Rel, tumor suppressor proteins Rb and p53, oncoprotein c-Myc, and cell death-associated protein Bcl-2 remain quite stable over this period (76).

In contrast, p40 (chicken IκB-α), which is in a complex with v-Rel in transformed cells, undergoes proteolysis when *ts* v-Rel-infected cells are shifted to 41.5°C (76). In v-R273H-infected cells at 41.5°C, p40 is completely degraded within approximately 12 h after a shift to 41.5°C. The complete disappearance of p40 is similar to what is seen in normal cells when NF-κB is induced. In v-G37E-infected cells, p40 is processed to a form that is approximately 3 to 4 kDa smaller. This truncated form of p40 is missing sequences from the N terminus and is located in a detergent-insoluble fraction in cells, whereas most full-length p40 is in a detergent-soluble fraction. It is possible that this smaller form of p40 represents a normal processing intermediate. These results are consistent with a model wherein association with v-Rel stabilizes IκB-α in transformed cells (see also reference 33), and they suggest that IκB-α may undergo different types of proteolytic processing. Moreover, because a structural change is induced in *ts* v-Rel when cells are shifted to 41.5°C, these results suggest that structural changes in a Rel protein can also initiate degradation of IκB-α.

STRATEGY FOR ISOLATING TARGET GENES IMPORTANT FOR TRANSFORMATION BY v-REL

It is fairly clear that most nuclear oncoproteins, including Myc (46), Ets, Fos (79), Jun (3), Myb (40), and v-Rel (4, 63), need to bind to DNA and activate transcription to effect malignant transformation. However, without exception, the target genes relevant to the transformation process are not known. There are several brute-force approaches that can be or have been taken to identify such target genes; these include PCR-based differential display, differential or subtractive cDNA cloning, binding assays with active chromatin, and educated guessing (e.g., looking at the influence of the given oncogene on expression of other oncogenes or tumor suppressor genes). Such approaches are time-consuming and often result in the identification of target genes that do not appear to be relevant to the transformation process (52, 54). Therefore, it appears that the identification of transformation-relevant target genes is a field that is begging for new approaches.

For these reasons, we are considering a genetic approach to identify target genes important for transformation by v-Rel. The general scheme is outlined in Fig. 5 and has, in essence, been successfully used in yeast cells (37) and mammalian cells (56) to clone genes by function. The overall aim of our proposal is to use *ts* v-Rel mutants to isolate a limiting effector gene for transformation by v-Rel. That is, we will first determine a temperature (e.g., 39°C) at which *ts* v-Rel-infected cells form small colonies in agar, probably owing to a decreased proliferative or immortalization signal caused by the lack of a limiting gene product, i.e., a gene that cannot be fully activated by the weakened v-Rel transcription factor. Once such a temperature has been determined, we will introduce a constitutively expressed vertebrate cDNA library (either by infection or electroporation)

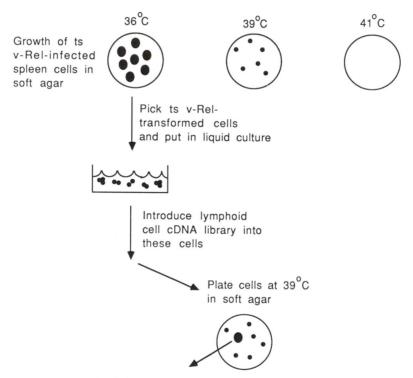

FIGURE 5. Genetic screen to identify activated target genes important for transformation by v-Rel. This screen employs *ts* v-Rel mutants that transform chicken spleen cells at 36°C but not at 41°C (75) and that can induce the formation of small, abortively transformed colonies at an intermediate temperature (e.g., 39°C). cDNAs that can restore full transformation to the *ts* v-Rel-transformed cells at the intermediate temperature will be selected, their sequences will be determined, and genomic clones for these cDNAs will be isolated. With the genomic clones corresponding to these cDNAs, it can be determined whether complementing genes are direct targets for activation by v-Rel. See text for further details.

into *ts* v-Rel-transformed cells growing at the permissive temperature (36.5°C). After enough time has passed for the introduced gene to be incorporated into the DNA of these cells, the cells will be plated at the intermediate temperature (i.e., 39°C), and large colonies will be selected. cDNAs whose expression complements the *ts* v-Rel mutants will be rescued from these cells, and their rescuing abilities will be confirmed in a fresh assay. In initial experiments, we have determined that v-R273H does indeed induce the formation of microscopic, abortive spleen cell colonies at 39°C, whereas v-G37E still induces the formation of large colonies at 39°C. Therefore, we are initiating such a screen with v-R273H-infected cells at 39°C.

This type of genetic approach has recently been used in yeast cells to identify a downstream target gene for the *Schizosaccharomyces pombe* Cdc10 transcription factor (37). As with all genetic screens, there are several artifact genes that one could imagine isolating. However, this type of approach is warranted by the knowledge that one is directly selecting for genes that complement transformation induced by the nuclear oncoprotein and by the relative lack of success of the brute-force approaches mentioned above. In other words, I (T.D.G.) have been converted to Howard's belief in a molecular genetic approach to the study of cancer (see Prologue above).

RELEVANCE OF TRANSFORMATION OF CHICKEN SPLEEN CELLS BY v-REL TO HUMAN CANCER

There have been several reports of rearranged *rel* family genes associated with human lymphoid cancers. Through a recombination event, human c-*rel* is fused to an unknown gene termed *nrg* (for non-*rel* gene) in a cell line derived from a pre-T diffuse large-cell lymphoma, and c-*rel* has been amplified in two follicular large-cell lymphoma cell lines (42). Furthermore, several deletions of the human *nfkb2* gene, which would delete C-terminal sequences of p100, have been identified in cell lines derived from B-cell lymphomas (16, 51, 74, 80). However, none of these rearranged c-*rel* or *nfkb2* genes has been shown to be transforming by itself, suggesting either that these genes are not causally related to the cancers or that these alterations are only one type of change that has contributed to the malignancy.

Because we have suggested that v-Rel transforms cells by activating genes and because the sequences encoding amino acids important for transcriptional activation by c-Rel have been deleted in the c-*rel*–*nrg* fusion gene (42, 60), we hypothesized that the Nrg sequences in the c-Rel–Nrg fusion protein might also contain a transcriptional activation domain. Therefore, as with v-Rel (Fig. 3), we fused c-Rel–Nrg sequences to the DNA-binding domain of GAL4 and tested this fusion protein for its ability to activate expression of a reporter plasmid in chicken fibroblasts (Fig. 6A). The GAL4-Nrg fusion protein did not activate transcription, even under conditions where other Rel activation domains were highly active. Therefore, it is unlikely that the c-Rel–Nrg fusion protein would function as an activator of transcription from genes containing κB binding sites. Thus, either

122 GILMORE ET AL.

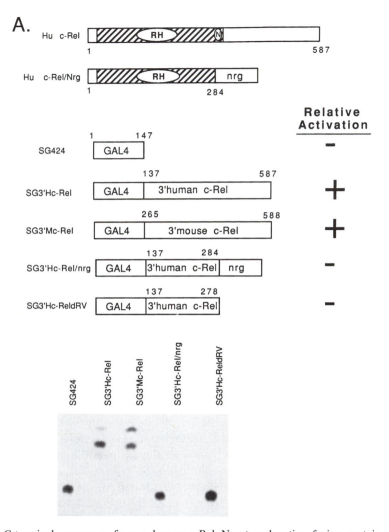

FIGURE 6. C-terminal sequences from a human c-Rel–Nrg translocation fusion protein and from chicken p100 do not activate transcription when fused to GAL4. Plasmids expressing the indicated GAL4 (1–147) fusion proteins, whose structures are shown, were cotransfected with a GAL4-site-containing reporter plasmid into chicken fibroblasts. A CAT assay was performed on cell lysates 48 h later as described previously (47, 48). (A) At the top are shown the general structures of human c-Rel and a human c-Rel protein (c-Rel–Nrg) that has arisen because of a chromosomal translocation in a pre-T diffuse large-cell lymphoma that results in the fusion of 166 amino acids encoded by an unrelated gene (nrg) onto c-Rel (42). In the middle are structures of five GAL4 fusion proteins: SG424 (GAL4 amino acids 1 through 147), SG3'Hc-Rel (GAL4–human c-Rel amino acids 137 through 587 [60]), SG3'Mc-Rel (GAL4–mouse c-Rel amino acids 265 through 588 [8]), SG3'Hc-Rel/nrg (GAL4–human c-Rel–Nrg fusion protein [42]), and SG3'Hc-ReldRV (GAL4–human c-Rel fusion protein deleted at an EcoRV site). The relative abilities of these proteins to activate transcription are indicated at the right, and a representative CAT assay is shown at the bottom (see legend to Fig. 4).

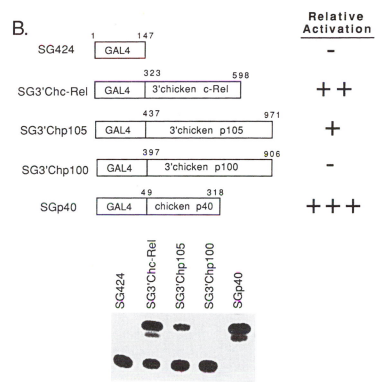

FIGURE 6. *Continued* (B) At the top are the structures of the five GAL4 fusion proteins: SG424 (GAL4 amino acids 1 through 147), SG-3′Chc (GAL4–chicken c-Rel amino acids 323 through 598 [47]), SG-3′Chp105 (GAL4-chicken p105 amino acids 437 through 971 [10]), SG-3′Chp100 (GAL4-chicken p100 amino acids 397 through 906), and SGChp40 (GAL4-chicken p40 [chicken IκB-α] amino acids 49 through 318 [47]). A representative CAT assay is shown at the bottom.

c-Rel–Nrg contributes to the oncogenic state in a way that is different from that of v-Rel, or the c-*rel*–*nrg* rearrangement is not contributing to the oncogenic state in these human lymphoid cancer cells. Alternatively, the model proposed here for transformation by v-Rel is incorrect.

In addition, the C-terminal sequences of NF-κB p100 do not appear to contain a transcriptional activation domain (Fig. 6B); therefore, the C-terminal truncation of p100 found in many human lymphomas (16, 51, 74, 80) is unlikely to be exposing a normally regulated transcription activation domain. Moreover, when inserted into an avian retroviral vector, neither the c-*rel*–*nrg* fusion gene nor truncated versions of chicken p100 can transform chicken spleen cells in vitro (65a).

How might these different results be reconciled with those of v-Rel in chicken spleen cells? It is possible that the truncations of c-Rel and p100 that occur in human lymphoid cancers do not by themselves produce unregulated transcription-activating proteins but that the truncated c-Rel and p100 proteins can constitutively enter the nucleus and bring with them heterologous Rel factors (e.g., RelA) that can activate transcription. Alternatively, the human Rel proteins might con-

TABLE 2

Important steps in understanding malignant transformation directed by v-Rel oncoprotein[a]

Step	Researchers and reference
1. Identification and isolation of Rev-T retrovirus as an oncogenic agent	Robinson and Twiehaus, 1974 (61)
2. Development of in vitro assay for malignant transformation of chicken spleen and bone marrow cells	Hoelzer et al., 1980 (31)
3. Identification of target cells for transformation by v-Rel	Beug et al., 1981 (6); Lewis et al., 1981 (41); Shibuya et al., 1982[a] (64); Barth and Humphries, 1988[a] (5)
4. Identification of v-rel non-helper virus-related sequences in Rev-T	Brietman et al., 1980 (8); Gonda et al., 1980 (26); Simek and Rice, 1980 (66); Cohen et al., 1981 (14); Hu et al., 1981 (34); Rice et al., 1982 (59)
5. Identification and characterization of avian c-rel genomic sequences	Chen et al., 1983[a] (13); Wilhelmsen et al., 1984[a] (78)
6. Cloning and sequencing of v-rel gene	Stephens et al., 1983 (70); Wilhelmsen et al., 1984[a] (78)
7. Identification of v-Rel protein and determination of its nuclear and cytoplasmic subcellular locations	Gilmore and Temin, 1986[a] (24); Rice et al., 1986 (57)
8. Demonstration that v-Rel can enhance expression from promoters and might therefore act as a transcription factor	Gélinas and Temin, 1988[a] (17)
Direct identification of sequences in v-Rel and c-Rel that can act as transcription activators	Bull et al., 1990 (9); Kamens et al., 1990[a] (36)
9. Identification of p68[c-rel]	Simek and Rice, 1988 (67)
Cloning of full-length chicken c-rel cDNA	Capobianco et al., 1990[a] (10)
Demonstration that truncation of c-Rel C terminus can activate its oncogenicity	Sylla and Temin, 1986[a] (72); Kamens et al., 1990 (36)
10. Cloning of D. melanogaster dorsal gene and sequence similarity to v-Rel	Steward, 1987 (71)
11. Cloning of cDNAs encoding a subunit of NF-κB transcription factor and identification of functional (DNA binding) and sequence homology to v-Rel	Ghosh et al., 1990 (18); Kieran et al., 1990 (38)
12. Identification of cellular proteins that interact with v-Rel in transformed cells:	Simek and Rice, 1988 (68)
as cellular Rel family proteins	Capobianco et al., 1992[a] [10]; Sif and Gilmore, 1993 (65)
and an IκB protein	Davis et al., 1991 (15)
13. Correlation between transformation by v-Rel and its ability to bind DNA	Ballard et al., 1990 (4)
14. Identification of translocation involving c-rel gene in a human cancer cell line	Lu et al., 1991 (42)
15. Correlation between transformation by v-Rel and its ability to activate transcription	Sarkar and Gilmore, 1993[a] (63)

[a] Research done in Howard Temin's laboratory or by people trained in his laboratory.

tribute in some other way to the malignant state, or, as discussed above, v-Rel might have other C-terminal functions (besides transcriptional activation) that are important for transformation in the avian system.

EPILOGUE

   Although Howard Temin is often remembered for his contributions to the study of the replication of retroviruses, the study of transformation of cells by the v-*rel* oncogene of Rev-T was another area in which he sponsored and mentored the research of a number of young scientists. Table 2 is a list of several key findings that have contributed to our understanding of the mechanism by which v-Rel transforms cells; many of these findings originated with people trained by Howard Temin.

   ACKNOWLEDGMENTS. This chapter is dedicated to the memory of Howard M. Temin for his seminal contributions to the understanding of the cellular and molecular mechanisms of cancer. His support of principal-investigator-driven research, including that in my own laboratory on transformation by v-Rel, was inspirational and greatly appreciated.

   We thank I. Verma for cDNA subclones containing the c-*rel*–*nrg* fusion genes used here. We also acknowledge the work of Paul Richardson, George Mosialos, and Tony Capobianco on transformation by v-Rel while they were graduate students in the laboratory of T.D.G. We apologize to any researchers whose contributions to our understanding of transformation by v-Rel we have overlooked.

   Recent research in the laboratory of T.D.G. on the v-Rel oncoprotein has been supported by National Institutes of Health grant CA47763 from the National Cancer Institute, and T.D.G. has been partially supported by a Faculty Research Award from the American Cancer Society.

REFERENCES

1. **Baeuerle, P. A.** 1991. The inducible transcription activator NF-κB: regulation by distinct protein subunits. *Biochim. Biophys. Acta* **1072**:63–80.
2. **Baeuerle, P. A., and T. Henkel.** 1994. Function and activation of NF-κB in the immune system. *Annu. Rev. Immunol.* **12**:141–179.
3. **Baichwal, V. R., and R. Tjian.** 1990. Control of c-Jun activity by interaction of a cell-specific inhibitor with regulatory domain delta: differences between v- and c-Jun. *Cell* **63**:815–825.
4. **Ballard, D. W., W. H. Walker, S. Doerre, P. Sista, J. A. Molitor, E. P. Dixon, N. J. Peffer, M. Hannink, and W. C. Greene.** 1990. The v-*rel* oncogene encodes a κB enhancer binding protein that inhibits NF-κB function. *Cell* **63**:803–814.
5. **Barth, C. F., and E. H. Humphries.** 1988. A non-immunosuppressive helper virus allows high efficiency induction of B-cell lymphomas by reticuloendotheliosis virus strain T. *J. Exp. Med.* **167**:89–108.
6. **Beug, H., H. Müller, S. Grieser, G. Doederlein, and T. Graf.** 1981. Hematopoietic cells transformed *in vitro* by REV-T avian reticuloendotheliosis virus express characteristics of very immature lymphoid cells. *Virology* **115**:295–309.
7. **Bhat, G. V., and H. M. Temin.** 1990. Mutational analysis of v-*rel,* the oncogene of reticuloendotheliosis virus strain T. *Oncogene* **5**:625–634.
8. **Brietman, M. L., M. M. C. Lai, and P. K. Vogt.** 1980. The genomic RNA of avian reticuloendotheiosis virus REV. *Virology* **100**:450–461.
9. **Bull, P., K. L. Morley, M. F. Hoekstra, T. Hunter, and I. M. Verma.** 1990. The mouse c-rel protein has an N-terminal regulatory domain and a C-terminal transcriptional transactivation domain. *Mol. Cell. Biol.* **10**:5473–5485.

10. **Capobianco, A. J., D. Chang, G. Mosialos, and T. D. Gilmore.** 1992. p105, the NF-κB p50 precursor, is one of the cellular proteins complexed with the v-Rel oncoprotein in transformed chicken cells. *J. Virol.* **66:**3758–3767.

11. **Capobianco, A. J., D. L. Simmons, and T. D. Gilmore.** 1990. Cloning and expression of a chicken c-*rel* cDNA: unlike p59[v-rel], p68[c-rel] is a cytoplasmic protein in chicken embryo fibroblasts. *Oncogene* **5:**257–265.

12. **Chen, I., and H. M. Temin.** 1982. Substitution of 5′ helper sequences into non-*rel* portion of reticuloendotheliosis virus strain T suppresses transformation of chicken spleen cells. *Cell* **31:**111–120.

13. **Chen, I., K. Wilhelmsen, and H. M. Temin.** 1983. Structure and expression of c-*rel*, the cellular homolog to the oncogene of reticuloendotheliosis virus strain T. *J. Virol.* **45:**104–113.

14. **Cohen, R. S., T. C. Wong, and M. M. C. Lai.** 1981. Characterization of transformation- and replication-specific sequences of reticuloendotheliosis virus. *Virology* **113:**672–685.

15. **Davis, N., S. Ghosh, D. L. Simmons, P. Tempst, H.-C. Liou, D. Baltimore, and H. R. Bose, Jr.** 1991. Rel-associated pp40: an inhibitor of the Rel family of transcription factors. *Science* **253:**1268–1271.

16. **Fracchiolla, N. S., L. Lombardi, M. Salina, A. Migliazza, L. Baldini, E. Berti, L. Cro, E. Polli, A. T. Maiolo, and A. Neri.** 1993. Structural alterations of the NF-κB transcription factor *lyt*-10 in lymphoid malignancies. *Oncogene* **8:**2839–2845.

17. **Gélinas, C., and H. M. Temin.** 1988. The v-*rel* oncogene encodes a cell-specific transcriptional activator of certain promoters. *Oncogene* **3:**349–355.

18. **Ghosh, S., A. M. Gifford, L. R. Riviere, P. Tempst, G. P. Nolan, and D. Baltimore.** 1990. Cloning of the p50 DNA binding subunit of NF-κB: homology to *rel* and *dorsal*. *Cell* **62:**1019–1029.

19. **Gilmore, T. D.** 1990. NF-κB, KBF1, *dorsal,* and *rel*ated matters. *Cell* **62:**841–843.

20. **Gilmore, T. D.** 1991. Malignant transformation by mutant Rel proteins. *Trends Genet.* **7:**318–322.

21. **Gilmore, T. D.** 1992. Role of *rel* family genes in normal and malignant lymphoid cell growth. *Cancer Surv.* **15:**69–87.

22. **Gilmore, T. D.** Regulation of Rel transcription complexes. *In* S. Goodbourn (ed.), *Frontiers in Molecular Biology: Eukaryotic Gene Transcription,* in press. Oxford University Press, Oxford, England.

23. **Gilmore, T. D., and P. J. Morin.** 1993. The IκB proteins: members of a multifunctional family. *Trends Genet.* **9:**427–433.

24. **Gilmore, T. D., and H. M. Temin.** 1986. Different localization of the product of the v-*rel* oncogene in chicken fibroblasts and spleen cells correlates with transformation by REV-T. *Cell* **44:**791–800.

25. **Gilmore, T. D., and H. M. Temin.** 1988. v-*rel* oncoproteins in the nucleus and in the cytoplasm transform chicken spleen cells. *J. Virol.* **62:**703–714.

26. **Gonda, M. A., N. R. Rice, and R. V. Gilden.** 1980. Avian reticuloendotheliosis virus: characterization of the high molecular weight viral RNA in transforming and helper virus populations. *J. Virol.* **34:**743–751.

27. **Hannink, M., and H. M. Temin.** 1989. Transactivation of gene expression by nuclear and cytoplasmic *rel* proteins. *Mol. Cell. Biol.* **9:**4323–4336.

28. **Hannink, M., and H. M. Temin.** 1991. Mutation-driven evolution of Rev-T, p. 99–109. *In* W. R. Paukovits (ed.), *Growth Regulation and Carcinogenesis.* CRC Press, Inc., New York.

29. **Hannink, M., and H. M. Temin.** 1991. Molecular mechanisms of transformation by the v-*rel* oncogene. *Crit. Rev. Oncog.* **2:**293–309.

30. **Hockenbery, D. M., Z. N. Oltvai, X.-M. Yin, C. L. Milliman, and S. J. Korsmeyer.** 1993. Bcl-2 functions in an antioxidant pathway to prevent apoptosis. *Cell* **75:**241–251.

31. **Hoelzer, J. D., R. B. Lewis, C. R. Wasmuth, and H. R. Bose, Jr.** 1980. Hematopoietic cell transformation by reticuloendotheliosis virus: characterization of the genetic defect. *Virology* **100:**462–474.

32. **Hrdlicková, R., J. Nehyba, and E. H. Humphries.** 1994. In vivo evolution of c-*rel* oncogenic potential. *J. Virol.* **68:**2371–2382.

33. **Hrdlicková, R., J. Nehyba, A. Roy, E. H. Humphries, and H. R. Bose, Jr.** 1995. The relocalization of v-Rel from the nucleus to the cytoplasm coincides with induction of expression of *Ikba* and *nfkb1* and stabilization of IκB-α. *J. Virol.* **69:**403–413.

34. **Hu, S. S. F., M. M. C. Lai, T. C. Wong, R. S. Cohen, and M. Sevoian.** 1981. Avian reticuloendothel-

iosis virus: characterization of the genome structure by heteroduplex mapping. *J. Virol.* **37:**899–907.

35. **Isoda, K., S. Roth, and C. Nüsslein-Volhard.** 1992. The functional domains of the *Drosophila* morphogen *dorsal:* evidence from the analysis of mutants. *Genes Dev.* **6:**619–630.

36. **Kamens, J., P. Richardson, G. Mosialos, R. Brent, and T. Gilmore.** 1990. Oncogenic transformation by v-Rel requires an amino-terminal activation domain. *Mol. Cell. Biol.* **10:**2840–2847.

37. **Kelly, T. J., G. S. Martin, S. L. Forsburg, R. J. Stephen, A. Russo, and P. Nurse.** 1993. The fission yeast *cdc18*$^+$ gene product couples S phase to START and mitosis. *Cell* **74:**371–382.

38. **Kieran, M., V. Blank, F. Logeat, J. Vandekerckhove, F. Lottspeich, O. Le Bail, M. B. Urban, P. Kourilsky, P. A. Baeuerle, and A. Israël.** 1990. The DNA binding subunit of NF-κB is identical to factor KBF1 and homologous to the *rel* oncogene product. *Cell* **62:**1007–1018.

39. **Kumar, S., A. B. Rabson, and C. Gélinas.** 1992. The RxxRxRxxC motif conserved in all rel/κB proteins is essential for the DNA-binding activity and redox regulation of the v-Rel oncoprotein. *Mol. Cell. Biol.* **12:**3094–3106.

40. **Lane, T. N., C. E. Ibanez, A. Garcia, T. Graf, and J. S. Lipsick.** 1990. Transformation by v-*myb* correlates with transactivation of gene expression. *Mol. Cell. Biol.* **10:**2591–2598.

41. **Lewis, R., J. McClure, B. Rup, D. W. Niesel, R. F. Garry, J. D. Hoelzer, K. Nazerian, and H. R. Bose, Jr.** 1981. Avian reticuloendotheliosis virus: identification of the hematopoietic target cell for transformation. *Cell* **25:**421–431.

42. **Lu, D., J. D. Thompson, G. K. Gorski, N. R. Rice, M. G. Mayer, and J. J. Yunis.** 1991. Alterations at the *rel* locus in human lymphoma. *Oncogene* **6:**1235–1241.

43. **Michaely, P., and V. Bennett.** 1992. The ANK repeat: a ubiquitous motif involved in macromolecular recognition. *Trends Cell Biol.* **2:**127–129.

44. **Miller, C. K., J. E. Embretson, and H. M. Temin.** 1988. Transforming viruses spontaneously arise from nontransforming reticuloendotheliosis virus strain T-derived viruses as a result of increased accumulation of spliced viral RNA. *J. Virol.* **62:**1219–1226.

45. **Miller, C. K., and H. M. Temin.** 1986. Insertion of several different DNAs in reticulendotheliosis virus strain T suppresses transformation by reducing the amount of subgenomic mRNA. *J. Virol.* **58:**75–80.

46. **Min, S., S. J. Crider-Miller, and E. J. Taparowsky.** 1994. The transcription activation domains of v-Myc and VP16 interact with common factors required for cellular transformation and proliferation. *Cell Growth Diff.* **5:**563–573.

47. **Morin, P., and T. D. Gilmore.** 1992. The C terminus of the NF-κB p50 precursor protein and an IκB isoform contain transcription activation domains. *Nucleic Acids Res.* **20:**2453–2458.

48. **Morin, P. J., G. S. Subramanian, and T. D. Gilmore.** 1993. GAL4-IκBα and GAL4-IκBγ activate transcription by different mechanisms. *Nucleic Acids Res.* **21:**2157–2163.

49. **Mosialos, G., and T. D. Gilmore.** 1993. v-Rel and c-Rel are differentially affected by mutations at a consensus protein kinase recognition sequence. *Oncogene* **8:**721–730.

50. **Neiman, P. E., S. J. Thomas, and G. Loring.** 1991. Induction of apoptosis during normal and neoplastic B-cell development in the bursa of Fabricius. *Proc. Natl. Acad. Sci. USA* **88:**5857–5861.

51. **Neri, A., C.-C. Chang, L. Lombardi, M. Salina, P. Corradini, A. T. Maiolo, R. S. K. Chaganti, and R. Dalla-Favera.** 1991. B cell lymphoma-associated chromosomal translocation involves candidate oncogene *lyt*-10, homologous to NF-κB p50. *Cell* **67:**1075–1087.

52. **Ness, S. A., A. Marknell, and T. Graf.** 1989. The v-*myb* oncogene product binds to and activates the promyelocyte-specific *mim-1* gene. *Cell* **59:**1115–1125.

53. **Palombella, V. J., O. J. Rando, A. L. Goldberg, and T. Maniatis.** 1994. The ubiquitin-proteasome pathway is required for processing the NF-κB1 precursor protein and the activation of NF-κB. *Cell* **78:**773–785.

54. **Prendergast, G. C., and M. D. Cole.** 1989. Post-transcriptional regulation of cellular gene expression by the c-*myc* oncogene. *Mol. Cell. Biol.* **9:**124–134.

55. **Purchase H. G., C. Ludford, K. Nazerian, and H. W. Cox.** 1973. A new group of oncogenic viruses: reticuloendotheliosis, chick syncytial, duck infectious anemia and spleen necrosis viruses. *J. Natl. Cancer Inst.* **51:**489–497.

56. **Raynor, J. R., and T. J. Gonda.** 1994. A simple and efficient procedure for generating stable expression libraries by cDNA cloning in a retroviral vector. *Mol. Cell. Biol.* **14:**880–887.

57. **Rice, N. R., T. D. Copeland, S. Simek, S. Oroszlan, and R. V. Gilden.** 1986. Detection and characterization of the protein encoded by the v-*rel* oncogene. *Virology* 149:217–229.

58. **Rice, N. R., and R. V. Gilden.** 1988. The rel oncogene, p. 495–512. *In* E. P. Reddy, A. M. Skalka, and T. Curran (ed.), *The Oncogene Handbook*. Elsevier Science Publishers, New York.

59. **Rice, N. R., R. R. Hiebsch, M. A. Gonda, H. R. Bose, Jr., and R. V. Gilden.** 1982. Genome of reticuloendotheliosis virus: characterization by use of cloned proviral DNA. *J. Virol.* 42:237–252.

60. **Richardson, P. M., and T. D. Gilmore.** 1991. v-Rel is an inactive member of the Rel family of transcriptional activating proteins. *J. Virol.* 65:3122–3130.

61. **Robinson, F. R., and M. J. Twiehaus.** 1974. Isolation of the avian reticuloendotheliosis virus (strain T). *Avian Dis.* 18:278–288.

62. **Romero, P. H., and E. H. Humphries.** 1995. A mutant v-*rel* with increased ability to transform B lymphocytes. *J. Virol.* 69:301–307.

63. **Sarkar, S., and T. D. Gilmore.** 1993. Transformation by the vRel oncoprotein requires sequences carboxy-terminal to the Rel homology domain. *Oncogene* 8:2245–2252.

64. **Shibuya, T., I. Chen, A. Mowstan, and T. Mak.** 1982. Morphological, immunological, and biochemical analysis of chicken spleen cells transformed by reticuloendotheliosis virus strain T (REV-T). *Cancer Res.* 42:2722–2728.

65. **Sif, S., and T. D. Gilmore.** 1993. NF-κB p100 is one of the high-molecular-weight proteins associated with the v-Rel oncoprotein in transformed chicken spleen cells. *J. Virol.* 67:7612–7617.

65a.**Sif, S., and T. D. Gilmore.** Unpublished results.

66. **Simek, S., and N. R. Rice.** 1980. Analysis of the nucleic acid components in reticuloendotheliosis virus. *J. Virol.* 33:320–329.

67. **Simek, S., and N. R. Rice.** 1988. Detection and characterization of the protein encoded by the chicken c-*rel* protooncogene. *Oncogene Res.* 2:103–119.

68. **Simek, S., and N. R. Rice.** 1988. p59^{v-rel}, the transforming protein of reticuloendotheliosis virus, is complexed with at least four other proteins in transformed chicken lymphoid cells. *J. Virol.* 62:4730–4736.

69. **Simek, S. L., R. M. Stephens, and N. R. Rice.** 1986. Localization of the v-*rel* protein in reticuloendotheliosis virus strain T-transformed lymphoid cells. *J. Virol.* 59:120–126.

70. **Stephens, R. M., N. R. Rice, R. R. Hiebsch, H. R. Bose, Jr., and R. V. Gilden.** 1983. Nucleotide sequence of v-*rel:* the oncogene of reticuloendotheliosis virus. *Proc. Natl. Acad. Sci. USA* 80:6229–6233.

71. **Steward, R.** 1987. *dorsal,* an embryonic polarity gene in *Drosophila,* is homologous to the vertebrate proto-oncogene, c-*rel. Science* 238:692–694.

72. **Sylla, B. S., and H. M. Temin.** 1986. Activation of the oncogenicity of the c-*rel* proto-oncogene. *Mol. Cell. Biol.* 6:4709–4716.

73. **Temin, H. M., and H. Rubin.** 1958. Characteristics of an assay for Rous sarcoma virus and Rous sarcoma cells in tissue culture. *Virology* 6:669–688.

74. **Thakur, S., H.-C. Lin, W.-T. Tseng, S. Kumar, R. Bravo, F. Foss, C. Gélinas, and A. B. Rabson.** 1994. Rearrangement and altered expression of the *NFKB-2* gene in human cutaneous T-lymphoma cells. *Oncogene* 9:2335–2344.

75. **White, D. W., and T. D. Gilmore.** 1993. Temperature-sensitive transforming mutants of the v-*rel* oncogene. *J. Virol.* 67:6876–6881.

76. **White, D. W., A. Roy, and T. D. Gilmore.** 1995. The v-Rel oncoprotein blocks apoptosis and proteolysis of IκB-α in transformed chicken spleen cells. *Oncogene* 10:857–868.

77. **Wilhelmsen, K., and H. M. Temin.** 1984. Structure and dimorphism of c-*rel* (turkey), the cellular homolog to the oncogene of reticuloendotheliosis virus strain T. *J. Virol.* 49:521–529.

78. **Wilhelmsen, K. C., K. Eggleton, and H. M. Temin.** 1984. Nucleic acid sequences of the oncogene v-*rel* in reticuloendotheliosis virus strain T and its cellular homolog, the proto-oncogene c-*rel. J. Virol.* 52:172–182.

79. **Wisdom, R., J. Yen, D. Rashid, and I. M. Verma.** 1992. Transformation by FosB requires a *trans*-activation domain missing in FosB2 that can be substituted by heterologous activation domains. *Genes Dev.* 6:667–675.

80. **Zhang, J., C.-C. Chang, L. Lombardi, and R. Dalla-Favera.** 1994. Rearranged NF*K*B2 gene in the HUT78 T-lymphoma cell line codes for a constitutively nuclear factor lacking transcriptional repressor functions. *Oncogene* 9:1931–1937.

The DNA Provirus: Howard Temin's Scientific Legacy
Edited by G. M. Cooper, R. Greenberg Temin, and B. Sugden
© 1995 American Society for Microbiology, Washington, DC 20005

Chapter 10

Progress in the Molecular Medicine of Cancer

Samuel Broder

Thus, we see how apparently inefficient science is, how it is impossible to predict the consequence of science, and how good science almost always has fruitful consequences.

—Howard Temin, 16 December 1991

Cancer is a formidable challenge, and we should have no illusions that major breakthroughs in the prevention, diagnosis, and treatment of this disease will come easily. Nevertheless, several converging lines of research are providing exciting new opportunities for clinical application. Some of these areas will be summarized in this chapter. The goal is to identify a few areas in which the molecular foundations of oncology are supporting ideas that are likely to have clinical ramifications in the near future.

FAMILIAL BREAST CANCER GENES

A collaboration involving scientists from several different institutions has recently led to the isolation of *BRCA-1,* the 17q-linked breast and ovarian cancer susceptibility gene (29, 64). (*BRCA-2,* a gene linked to certain forms of breast cancer and located on chromosome 13, has also been identified.) *BRCA-1* is thought to function as a tumor suppressor gene, encoding a protein with the capacity to negatively regulate tumor growth. As a rule, cancer-associated alleles carry mutations that impair or bring about a loss of normal function. In this context, an increased risk of cancer is inherited as a dominant genetic trait, while the predisposing allele acts recessively in somatic cells. In other words, for this category of cancer gene, a single inherited copy of the mutant allele predisposes to cancer, while the loss of inactivation of the wild-type allele (e.g., following exposure to an environmental carcinogen) completes one of the steps necessary to bring about a clinical malignancy. Breast (and ovarian) cancers from patients who carry *BRCA-1* lose the wild-type copy while retaining the mutant allele. It is also

Samuel Broder • National Cancer Institute, Bethesda, Maryland 20892.

theoretically possible that certain mutations of *BRCA-1* bring about a gain of function that plays a role in maintaining the malignant state. By means of a series of elegant strategies based on positional cloning, *BRCA-1* has recently been identified following a prodigious effort in several different laboratories around the world. The successful effort was led by Mark Skolnick, and this project depended heavily on the earlier work of Mary Claire King and her coworkers, who identified the probable site on the long arm of chromosome 17.

The *BRCA-1* gene encodes a predicted product of 1,863 amino acids. Of note, the amino-terminal region of this protein contains a zinc finger domain, a motif noted in many nucleic acid-binding proteins. While the mutations in *BRCA-1* account for only a small percentage of typical cases of breast and ovarian cancer, the isolation of this gene will have substantial implications for the clinical management of women from kindreds with familial breast cancer and will also provide important insights into the molecular pathogenesis of sporadic tumors.

MICROSATELLITE ALTERATIONS: A REVOLUTIONARY CONCEPT IN THE INHERITANCE AND DETECTION OF CANCER

The molecular dissection of colon cancer continues to provide new insights into the cascade of genetic events that lead to full-blown malignant transformation. It is now well established that there is a gene cluster on the long arm of chromosome 5 (5q21) that is made up of two neighboring and partially homologous tumor suppressor genes: the adenomatous polyposis coli (*APC*) gene and the "mutated in colon cancer" (*MCC*) gene, which appear to be causally related to the development of both familial and presumably nonfamilial (sporadic) colorectal cancers (35, 54, 72). This part of the story is comparatively straightforward and parallels traditional cancer genetics. Thus, *APC* is mutated in germ line cells from individuals with autosomal, dominantly inherited familial adenomatous polyposis (FAP), a disorder characterized by the early and extensive formation of colorectal polyps that undergo malignant transformation, and in somatic tumor cells from patients without the disease who develop colon cancer, while *MCC* mutations have thus far been detected only in somatic tumor tissue. Mutations in the gene cluster occur as initiating events in colon carcinogenesis, setting the stage for the complete malignant transformation that occurs with subsequent deletions or acquisitions of mutations in the tumor suppressor genes *p53* (located on chromosome 17p13) and "deleted in colon cancer" (*DCC* on chromosome 18q21) and in the *ras* oncogene, a paradigm of multistep carcinogenesis (102).

More recently, several groups of scientists have again tackled hereditary colon cancer, using genetic linkage analyses in families with histories of hereditary nonpolyposis colon cancer (HNPCC) to uncover a cancer susceptibility gene (1, 74, 95), and have thus also developed a revolutionary concept that links the inheritance of genetic instability and a predisposition to cancer, as distinct from the predisposition that is mediated by the inheritance of conventional tumor suppressor gene mutations. The susceptibility to cancer in HNPCC families is not relegated to the colon, as there appears to be a propensity to develop cancers at numerous organ sites (see below) (1), a situation perhaps reminiscent of the Li-Fraumeni syndrome (60). Genetic linkage and mapping studies of HNPCC kindreds demonstrate germ line and tumor cell alterations in the short repetitive DNA sequences (microsatellite DNA). HNPCC is a disease associated with widespread microsatellite instability, which is manifested by expansion or deletion of repeat elements (dinucleotide and trinucleotide) in neoplastic tissues. This genetic instability is thought to arise from inherited and somatic mutations in a set of DNA mismatch repair genes (*hMSH2, hMLH1, hPMS1,* and *hPMS2*) (71).

HNPCC accounts for up to 13% of colon cancers (or perhaps more, as HNPCC may be indistinguishable from sporadic cancers) and is thus far more common than familial adenomatous polyposis, which accounts for only about 1% of colon cancers (1). Furthermore, HNPCC patients are at elevated risk for developing cancers at other organ sites, namely, endometrium, stomach, pancreas, urinary tract, and perhaps breast and ovary (1, 95). Thus, the findings surrounding this cancer susceptibility gene have far-reaching implications that could touch large numbers of individuals at potential risk for development of several common epithelial malignancies, particularly in terms of genetic screening and early interventions for those at high risk. Further, these findings expand the boundaries of knowledge concerning the maintenance (or dysregulation) of genetic integrity and continue to serve as templates for investigations aimed at unraveling the complex sequence of events that preserve the balance between proliferation and differentiation or, when disrupted, culminate in the emergence and perpetuation of malignancy.

This line of research has taken on a fascinating twist. Microsatellite alterations are associated with small-cell lung cancer, head and neck tumors, bladder cancer, and most likely other tumors as well. Trinucleotide (and tetranucleotide) repeats are more prone to expansion or deletion in a variety of cancers. The mechanisms producing the altered alleles in such cancers are probably different from those described in HNPCC and most likely reflect a more subtle set of defects in DNA repair pathways, with some loci altered in a tumor-specific way. Indeed, Mao et al. (61) recently showed that microsatellite alterations can serve as markers to detect clonal expansions of tumor cells in pathologic specimens. This technique can provide an astonishing research tool for identifying surgical margins of resection and for identifying tumor cells in the corresponding fluids (e.g., urine in the case of bladder cancer and sputum in the case of lung cancer). It is likely that in the near future, these strategies will provide better diagnostic and staging information for established cancers and might even be adapted as a more generalized screening approach to common cancers.

AT, *p53*, AND GROWTH ARREST DNA DAMAGE-INDUCIBLE (*GADD*) GENES: A RARE CONDITION TEACHES US ABOUT COMMON CANCERS

It is worth focusing on the autosomal recessive disease ataxia-telangiectasia (AT) for several reasons. First, there is a growing body of knowledge regarding the mechanisms by which the DNA-binding protein product of the *p53* tumor suppressor gene, located on the short arm of chromosome 17 (17p13), exerts its growth-inhibitory activity (43, 58). Upregulation of p53 protein in human hemato-poietic cells in response to certain types of DNA damage (notably, UV light and other types of gamma irradiation) prolongs the G_1 phase of the cell cycle and gives the cell time to repair that damage before it enters S phase (50). Cells with aberrant net *p53* gene expression fail to undergo this G_1 arrest following irradiation-induced DNA damage and consequently enter S phase and replicate damaged DNA. The direct relationship between *p53* and DNA damage-induced inhibition of cell cycle progression has now been demonstrated by transfecting wild-type *p53* genes into cells lacking endogenous *p53* (57). Alternatively, the introduction of mutant *p53* genes into cells with the wild-type gene leads to the production of abnormal p53 proteins which act in a dominant negative fashion to inhibit normal p53 function and thus nullify the cell's ability to undergo DNA damage-induced G_1 arrest (42, 59). Second, a *p53*-knockout mouse model has been used to demonstrate that the isolated loss of functional p53 is sufficient to abrogate the G_1 arrest that would otherwise occur in response to gamma irradiation (51, 57).

The central mechanistic roles of the *p53* gene and its protein product in the pathway of repair of irradiation-induced DNA damage have been further eluci-cated. The p53 protein binds to and enhances the transcription of a specific gene in the family of five growth arrest DNA damage-inducible (*GADD*) genes (51). In fact, the p53 protein binds to a specific, highly conserved region of one such *GADD* gene (*GADD45*), which in turn produces a protein that effects G_1 arrest. *GADD* gene expression is particularly prominent in lymphoblasts and fibroblasts after exposure to irradiation. This is an important point, and I will return to it.

The p53-based response to DNA, including the induction of *GADD45* expres-sion, appears to be activated by the product(s) of genes that are defective in patients with AT, an inherited neurological disorder with substantial abnormalities of T-cell immunity that arise from mutations in a recessive gene on chromosome 11q22-23 (32). Several laboratories are trying to isolate the gene.

Patients with AT exhibit hypersensitivity to DNA-damaging agents (especially ionizing radiation), an impaired ability to repair that damage, marked DNA fragility with chromosomal instability, and a striking propensity to develop lymphoprolifer-ative malignancies (103). The theoretical clinical intersection of AT and *p53* is substantiated by molecular studies demonstrating that lymphoblasts and fibro-blasts from AT patients fail to show a radiation-induced increase in p53 protein and, further, do not demonstrate the consequent induction of *GADD45* gene expression.

The incidence of full-blown "homozygous" AT has been estimated from ret-rospective analyses to be between 1 in 20,000 and 1 in 100,000. However, the heterozygous (or carrier) state may be remarkably common, perhaps approxi-mately 1 to 2% of the population in the United States (32, 93). This is a potentially important statistic, since women who are heterozygous for AT, particularly those under the age of 60, appear to have an approximately fivefold-increased risk of developing breast cancer (93). From this it has been estimated that perhaps up to 9% of all breast cancer cases occur in AT heterozygotes, with a propensity for early onset (93). The known radiation hypersensitivity, documentable in skin fibroblasts and/or peripheral blood lymphocytes from AT heterozygotes, could contribute significantly to breast carcinogenesis in some of these women. The key point is that the genetic abnormality has a substantial impact on cancer in the general population (92).

It has recently been recognized that women who have received "mantle" radiation therapy (neck, axillae, mediastinum) for Hodgkin's disease are at increased risk for developing breast cancer at a median 15 years (with a range extending from 4 to 22 years in one series and 8 to 34 years in another) after radiation (38, 104). This risk is confined to women who have received radiation before the age of 30 and especially before the age of 15. In these younger age groups, the risk of developing breast cancer is strikingly high, with the relative risk being roughly 15 for women irradiated between ages 20 and 29, up to 40 for girls receiving radiation between ages 10 and 19, and over 100 if radiation was received before age 15 (38). Importantly, there is no detectable increased risk for women who receive radiation after age 30. The majority of these tumors appear to arise within or at the margin of the radiotherapy field, often in the medial portion of the breast and with some incidence of bilateral breast tumors at presentation. Histopathologically, these tumors are most commonly infiltrating ductal carcinomas. The propensity for ductal involvement could perhaps reflect a higher proliferative rate and/or a greater number of epithelial stem cells (relatively undifferentiated) in the breast tissue of adolescents and young women and thus a greater sensitivity to radiation-induced damage in the terminal ducts (104).

Given the clinical observations that link increased breast cancer risk with abnormalities in both *AT* and *p53* genes, it is logical to speculate that a lesion along the *AT-p53-GADD* gene pathway might have a formative role in the pathogenesis of radiation-induced breast cancers arising in the setting of previous Hodgkin's disease. Furthermore, lymphomas, like breast cancer, are also associated with defective expression of both *AT* (103) and *p53* (30). For instance, lymphomas are the most common tumors detected in the *p53*-knockout (or Li-Fraumeni) mouse (24), and breast cancers frequently arise in patients with Li-Fraumeni syndrome (60). All things considered, it is conceivable that the presence of such underlying germ line (inherited) lesions serves to predispose these individuals to both the primary and the secondary malignancies, as postulated for abnormalities in *AT* and/or *p53* expression. In this light, detection of absent or abnormal expression of the molecular components of this pathway could serve as markers for defining adolescent girls and young women with Hodgkin's disease who might be at particularly high risk for subsequent development of radiogenic (radiation-induced) breast cancer. From many vantage points, this clinical setting presents a unique group of high-risk women for whom the timing and magnitude of risk exposure can be quantitated, in whom longitudinal changes could be assessed for the purposes of identification and clinical application of new biomarkers, and who could participate in developmental prevention clinical trials.

Thus, the *AT* and *p53* genes appear to be functionally linked on both the molecular and the clinical level. These elegant studies define a signal transduction pathway that is induced in response to DNA damage caused by ionizing radiation, with *p53* as a central component: the initial activation of the *AT* gene(s) leads in some way to increased p53 protein levels, which in turn enhance *GADD45* gene transcription and expression (51).

p53 aberrations are among the most common genetic abnormalities found in a broad spectrum of malignancies. For example, roughly 50% of non-small-cell lung cancers have somatically acquired mutations in *p53,* and in small-cell lung cancer, the percentage is about 80% (21). Thus, disruption of this pathway is likely to be an important mechanism underlying the growth dysregulation and genetic instability that lead to commonly observed malignancies. In turn, this *AT-p53-GADD* signal transduction pathway could provide new molecular targets toward which to direct therapeutic and perhaps prevention strategies for some common cancers.

NEW INSIGHTS IN LEUKEMOGENESIS: FOCUS ON CHROMOSOME 11

Lymphohematopoietic malignancies are frequently accompanied by nonrandom chromosomal abnormalities that represent a disruption in the normal structure and function of genes that control the balance of proliferation and differentiation. As with many solid malignancies, duplications (e.g., trisomy 8) or losses of whole chromosomes (-5, -7) or chromosome segments (5q-, 7q-) and gene point mutations (e.g., in tumor suppressor genes *RB* and *p53*) are often associated with both lymphoid and myeloid malignancies. In addition, many leukemias and lymphomas of diverse lineages are typified by specific gene rearrangements and reciprocal chromosome translocations (70). Such translocations can juxtapose two growth-promoting oncogenes, e.g., the Philadelphia chromosome of chronic myelogenous leukemia (CML), t(9;22)(q34;q11), which fuses the *bcr* and *abl* oncogenes. Alternatively, an oncogene can recombine with a gene whose products encode factors influencing differentiation, as in acute progranulocytic leukemia, where t(15;17)(q11;q11.2-12 or q21.1) fuses a rearranged *PML* oncogene encoding a transcription factor composed of both a zinc finger and a leucine zipper region with a rearranged and truncated gene encoding retinoic acid receptor α (6, 12, 33). The resultant chimeric oncogenes in such translocations produce tumor-specific fusion mRNAs and proteins which in turn lead to dysregulated cellular proliferation and/or disorderly (or blocked) differentiation.

A segment of the long arm of chromosome 11 (11q23) is rearranged and reciprocally translocated to a variety of genes of diverse chromosomes in acute leukemias of both lymphoid and myeloid (particularly monocytic) lineages (49, 78). The 11q23 segment contains several genes involved in distinct aspects of lymphohematopoietic cell growth and differentiation. Genes in this segment encode proteins involved in intercellular communication, namely, the gamma and delta chains (subunits) of the CD3 T-cell antigen receptor complex and the cell adhesion molecules THY-1 and N-CAM (78). Also located on 11q23 are the *ets* oncogene, encoding a DNA-binding transcription factor (22), and the gene encoding the interleukin-1β (IL-1β) converting enzyme, a cysteine protease that cleaves the IL-1β precursor to produce active IL-1β (the form of IL-1 synthesized and released predominantly by monocytes) (13, 97). Provocatively, the gene(s) encoding several of the complementation groups for *AT* is localized to 11q23 and may overlap with some or all of the genes mentioned above (32).

Lessons from *Drosophila melanogaster*: a Gene Regulating Embryogenic Differentiation Is Involved in Stem Cell Leukemias

It has been recognized for some time that acute leukemias associated with 11q23 abnormalities often exhibit biphenotypic features that reflect both lymphoid and monocytic differentiation on the molecular and morphologic levels (14, 49, 78). The specific lineage expressed by an individual leukemia depends to a great extent on the particular recipient gene or chromosomal segment involved in the reciprocal translocation with 11q23. A high proportion of acute leukemias in infants demonstrate 11q23 translocations, most commonly to chromosome 4q21 [t(4; 11)] but also with other gene segments such as 19p13, 6q27, and 9p21 (14). Most of the 11q23 breakpoints in these various translocations appear to be clustered in a circumscribed region near but not necessarily within the CD3 gene cluster, although some heterogeneity has been detected in t(11;19)(q23;p13) (67). The malignant phenotype is usually that of B-cell acute lymphoblastic leukemia, especially in t(4;11) but often with concomitant monocytic markers (14).

The 11q23 breakpoint involved in infant acute lymphoblastic leukemia bearing t(4;11) or t(9;11) occurs within a gene (so-called *ALL-1*) (17, 36) that is highly homologous to the *Drosophila trithorax* (*trx*) gene (23, 98). *ALL-1* (also called human trithorax, or *HRX*) and *trx* encode strikingly similar zinc finger DNA-binding transcription factors (36). In *D. melanogaster,* the *trx* gene product regulates embryological differentiation (specifically, the process of segmentation) (23, 98). By extrapolation, it seems likely that the human counterpart could have a pivotal role in controlling differentiation in primitive pluripotent cells, for instance, hematopoietic stem cells. Disruption of the *ALL-1* (or *HRX*) gene in stem cells, such as occurs in t(4;11), t(9;11), and t(11;19), may lead to defective or altered differentiation along the lymphoid and myeloid pathways (45, 98). The resultant dysregulation of stem cell differentiation may account for the multilineage phenotypes and heterogeneous histopathologic and clinical features expressed among these 11q23-associated acute leukemias.

The rearrangement and fusion of *ALL-1* with another gene (in particular, one encoding a transcription factor or other growth-promoting protein) might confer a growth dysregulation that in turn would promote leukemic transformation, especially in the setting of disordered differentiation. This mechanism of leukemogenesis, linked to the aberrant activity of the unique fusion protein produced by the hybrid gene, appears to operate in CML and acute progranulocytic leukemia. A somewhat different mechanism, which leads to the constitutive expression of a growth-promoting protein that is normally temporally restricted in its expression (rather than the production of a novel, tumor-specific protein), probably operates in translocations involving c-*myc* and *bcl-2* oncogenes. In fact, recent molecular analyses of t(4;11) indicate that *ALL-1* fuses with a specific gene on chromosome 4 (called the *ALL-1*-fused gene from chromosome 4, or *AF-4*) to produce a specific chimeric protein (36). It is tempting to speculate that the AF-4 protein may itself be a transcription factor or may, by virtue of the *ALL-1–AF-4* gene fusion, alter the conformation and thus the DNA-binding and transcriptional activities of the ALL-1 protein. Along these lines, recent studies demonstrate that a gene on 4q21 at the translocation locus encodes a highly charged basic protein that contains large amounts of the amino acids serine and proline (68). This unusual protein, called FEL, is a nuclear protein that is normally expressed in T and B lymphocytes. Whether FEL and AF-4 are identical or disparate proteins has not yet been determined.

Similarly, t(11;19) culminates in the production of a unique, chimeric DNA-binding protein composed of a truncated *HRX* (or *ALL-1*) component fused to the protein product encoded by the juxtaposed gene on chromosome 19. The gene on 19p13 that is involved in t(11;19) also encodes a serine/proline-rich nuclear protein, so-called ENL (98). Intriguingly, transcriptional activator proteins often contain domains rich in serine and proline, suggesting that both FEL and ENL might function normally as transcriptional regulators (68, 98). When fused with HRX sequences, HRX-FEL or HRX-ENL protein might exhibit deregulation of DNA binding by the HRX portion and transcriptional activity by the FEL or ENL portion of the molecule.

Chromosome 11q23: a Locus for DNA Damage by Drugs That Target Topo II

While the t(4;11) acute lymphoblastic leukemia appears to arise de novo in all age groups, other 11q23-associated leukemias frequently exhibit a predominantly myeloid phenotype, especially with a monocytic component, and arise in the clinical setting of cytotoxic chemotherapy for a prior malignancy. The therapy-related leukemias involve deletions, breaks, rearrangements, and inversions or translocations of 11q23 to other chromosomal sites, most commonly 9p21 but also 19p13 (45, 76, 83). Of note is that 9p21 is the site of the alpha interferon (IFN-α) gene and has recently been found to be deleted in some kindreds with familial melanoma, suggesting that this locus may contain a tumor suppressor gene (perhaps IFN-α or another gene) that determines inherited genetic susceptibility to melanoma in some cases (10). Recent studies of children and adults with diverse malignancies demonstrate a relationship between exposure to the epipodophyllin VP-16 or VM-26 (and, to a lesser degree, certain anthracyclines in combination with other DNA-damaging agents, for instance, cisplatin and alkylating agents) and development of therapy-related acute myelogenous leukemias (AMLs) of the monocytic or myelomonocytic phenotype with 11q23 abnormalities (73, 76, 85). Both the epipodophyllins (especially VP-16) and the anthracyclines cause DNA damage by targeting the intranuclear enzyme topoisomerase II (topo II), stabilizing and "freezing" the DNA-topo II complex and inducing single- and double-stranded breaks (8). Perhaps the 11q23 lesion induced by VP-16 confers a defect in enzymes involved in the process of orderly DNA repair, including recombinase enzymes, which in turn leads to inappropriate translocations and malignant transformation (16). Along these lines, it is provocative that the *AT* gene is located on 11q23. While the AT gene is probably not directly damaged by the topo II-directed drugs, its structure and function could conceivably be altered by lesions in proximate genes. Such alterations could, at least in theory, confer defects similar to those detected in *AT,* i.e., pronounced DNA fragility, chromosomal instability, and hypersensitivity to certain DNA-damaging agents coupled with an impaired ability to repair that damage (especially DNA strand breaks).

The constellation of molecular findings surrounding 11q23-associated acute leukemias is germane to the pathogenesis and diagnosis of leukemias and to a diversity of treatment issues such as the ability to monitor response to antileukemic treatment by the molecular detection of minimal residual disease or the ability to design antisense molecules or other targeted inhibitors directed against the leukemia-specific fusion mRNAs and proteins. Perhaps of particular relevance is the ability to circumvent secondary (therapy-induced) AMLs. It may be feasible to monitor normal hematopoietic precursors for their inherent sensitivity to VP-16-induced DNA damage at critical 11q23 gene loci. The detection of such damage and/or the presence of defective repair of such induced lesions could define those individuals at particular risk for leukemogenesis by VP-16 (or perhaps certain anthracyclines plus cisplatin or alkylating agents) and determine a need to avoid such agents during the treatment of the primary malignancy.

PROGRAMMED CELL DEATH (APOPTOSIS), CANCER THERAPY, AND HUMAN IMMUNODEFICIENCY VIRUS INFECTION

Programmed cell death, or apoptosis, is a normal physiologic process by which the cellular life span is controlled and certain cells are programmed for elimination. Structurally, specific morphologic changes typify apoptosis: cell shrinkage, membrane blebbing, chromatin and cytoplasmic condensation, and, perhaps most characteristic, fragmentation of nuclear DNA in a nucleosomal stepladder pattern (11, 63, 84). Biochemically, the process represents the activation of a cascade of specific "death genes" that in turn leads to the induction of particular signal transduction pathways involving a calcium and magnesium (Ca^{2+}-Mg^{2+})-dependent endonuclease (which cleaves nuclear DNA into the characteristic oligonucleosomal ladder) and transglutaminases that cross-link cellular proteins and thereby cause cytoplasmic coagulation. This orderly process of cell death occurs in diverse cell types: hormonally sensitive normal and malignant prostate and breast cells undergo apoptosis in response to hormonal deprivation, and immune cells (especially activated T cells and natural killer cells) and neurons undergo apoptosis as a mechanism for negative selection (63, 77, 84). Indeed, programmed cell death is probably a critical mechanism for preserving homeostatic balance between cell birth and death in a given cell population. The net activity of the apoptosis signaling pathways can be modulated by several factors, including intracellular calcium level, the provision or deprivation of growth factors (including hormones), and the expression of various growth-related genes. The ability to selectively induce apoptosis in tumor cells could provide a uniquely potent mechanism for antitumor activity. On the other hand, there are settings in which overactivity of apoptosis pathways can lead to imbalanced losses of critical cell populations.

Regulation of Cellular Life Span: Roundworms, bcl-2, and c-myc

The dissection of the genetic bases for programmed cell death and its antidote has come from embryologic studies in *Caenorhabditis elegans*, a simple round-worm with fewer than 1,100 cells (29, 63, 101). In this experimental model, genes known as *ced-3* and *ced-4* encode proteins that prompt programmed cell death, a series of events that is countered by expression of the *ced-9* gene. Intriguingly, *ced-9* shares significant structural and functional homology with the mammalian apoptosis-inhibiting oncogene *bcl-2* (101). By extrapolation, this suggests that *ced-3* and *ced-4* could be at least partially homologous to mammalian death genes and raises the question of whether *bcl-2*, like *ced-9*, directly inhibits death genes in mammalian cells. The evolutionary conservation between nematode *ced-9* and mammalian *bcl-2* underscores the importance of programmed cell death as a basic cellular control mechanism, much like the highly conserved cyclins and other components of cell cycle regulators (for instance, the *p53* and c-*mos* genes, which exert their effects at critical cell cycle checkpoints).

The *bcl-2* oncogene, located on chromosome 18q21, encodes a protein that localizes to the inner mitochondrial membrane and has recently also been detected in the nuclear envelope and endoplasmic reticulum (56). Like the *C. elegans ced-9* gene product, the *bcl-2*-encoded protein blocks apoptosis and enhances cell longevity (56, 101). These findings, first described for immunoglobulin-secreting memory B cells and then T cells, have now been extended to other cell lineages, in particular, sympathetic neurons (31), in which *bcl-2* overexpression prevents the programmed cell death induced by deprivation of nerve growth factor. Provocatively, *bcl-2* is also expressed in neurons throughout the nervous system during embryogenesis and early postnatal development in mice (31, 77), a finding that could have enormous implications for a future therapeutic approach to various neurodegenerative diseases.

Abrogation of programmed cell death may be a pivotal lesion in the emergence and/or progression of certain malignancies. Toward this end, the *bcl-2*-encoded protein is overexpressed when *bcl-2* is juxtaposed to the immunoglobulin heavy-chain gene in t(14;18), a gene rearrangement often associated with follicular lymphomas (56) and more recently detected in patients with Hodgkin's disease (37, 91). The fusion protein produced by the translocation between the *bcl-2* oncogene and 14q32 exaggerates the *bcl-2*-encoded block of apoptosis and perpetuates tumor cell survival. Further, t(14;18)(q32;q21) permits clonal progression to a more aggressive variant to occur through additional genetic changes such as rearrangements or mutations in the growth-promoting c-*myc* oncogene (7, 28, 56). This multistep evolution from an indolent lymphoproliferative condition to a high-grade diffuse large-cell (or lymphoblastic) lymphoma has been related to acquisition of c-*myc* gene rearrangements both in a transgenic mouse model (in which mice have undergone germ line transfections with the human hybrid genes that typify follicular lymphoma in humans) and in human lymphomagenesis (56). The continuing molecular dissection of *bcl-2*-related pathways may allow design of specific therapies aimed at inactivating or circumventing this natural inhibitor of apoptosis.

Other genes under various conditions may regulate the net activity of the apoptosis pathway. For instance, wild-type *p53* induces apoptosis when it is transfected into murine AML cells, whereas mutant *p53* is capable of blocking certain types of apoptosis in growth-activated cells (84, 105). Paradoxically, expression of the growth-promoting c-*myc* gene induces apoptosis in serum-deprived fibroblasts (27) and in T cells activated through the T-cell receptor (89). The c-*myc*-encoded protein is a DNA-binding transcription factor that promotes cell proliferation, and the deregulation of c-*myc* expression is fundamental to the full development of several distinct lymphohematopoietic and epithelial malignancies. However, when deregulation occurs in the setting of a growth-arresting factor (e.g., wild-type *p53*, growth factor deprivation, cell overcrowding, or cytotoxic drugs), the result is a potent activation of programmed cell death (18, 27, 59). This dichotomy suggests that the proliferative and apoptotic functions of the Myc protein are tightly linked, a concept that is supported biochemically by the finding that the domains of the Myc molecule responsible for apoptosis overlap with the domains necessary for DNA binding and transcriptional activity, cotransformation, and inhibition of differentiation (27). Thus, for full c-*myc*-induced malignant transformation to ensue, there needs to be one or more additional survival mechanisms acting to prevent c-*myc*-induced apoptosis: cooperation with another mutant gene such as *ras* or *p53* (27, 84), a synergistic interaction with *bcl-2* (7, 28, 85), and/or the continuous presence of growth factors that are capable of abrogating apoptosis, as in the case of IL-6 overcoming wild-type *p53*-induced apoptosis in the murine AML cell line mentioned above (84, 105).

Mechanisms of Drug-Induced Apoptosis

Poly(ADP) ribosylation is a major mechanism that alters chromatin structure and appears to be central to the host cell's ability to respond to and repair DNA damage (9, 86). Like the formation of microtubules from tubulin, poly(ADP) ribosylation is, mechanically speaking, a polymerization process. The effector enzyme, poly(ADP) ribosylation polymerase, catalyzes the polymerization of NAD onto proteins (in particular, histone proteins that are bound to DNA) and, by so doing, reconfigures the histones in a way that transiently inhibits DNA metabolism. The polymerase is recruited into action by DNA strand breaks, especially 5′ breaks and particularly double-stranded breaks. The consequences of NAD polymerization at the strand break sites are twofold: (i) a shutdown in DNA metabolism that permits excisional repair of the damaged DNA locus and (ii) "chromatin opening," which facilitates the action of repair enzymes (e.g., DNA ligase) on the damaged template and may subsequently (and seemingly paradoxically) promote transcription (86). On the other hand, extensive DNA damage can translate into extensive NAD polymerization, which can lead to depletion of intracellular pools of NAD and eventual cell death (9). Further, poly(ADP) ribosylation polymerase activates the Ca^{2+}-Mg^{2+}-dependent endonuclease responsible for apoptosis. Taken together, these activities suggest that the process of poly(ADP) ribosylation may play a critical role in apoptosis.

The activation of poly(ADP) ribosylation and endonuclease pathways in response to DNA strand breakage may be a central mechanism by which several diverse chemotherapeutic agents induce apoptosis and, as a consequence, exert cytotoxicity. The induction of apoptosis via these pathways may explain at least in part the cytotoxic activities of cell cycle-active, antileukemic nucleoside drugs (cytosine arabinoside, fludarabine, 2-chlorodeoxyadenosine) against noncycling cells in addition to the cycle-dependent mechanisms that target the proliferative fraction of the cell population (5, 66, 81). The induction of DNA strand breaks and thus the activation of apoptosis pathways are quite possibly a common mechanistic motif for drugs that interact with the topoisomerases, for example, the camptothecins (targeting topo I) (90) and the anthracyclines and epipodophyllins (targeting topo II) (8), and perhaps for agents that complex with DNA (alkylators, cisplatin) and cause genotoxic damage (26). In fact, it is reasonable to speculate that activation of apoptosis pathways serves as a common final pathway for many cytotoxic agents irrespective of the primary mechanism of action.

As mentioned previously, the removal of growth factors from factor-dependent cells is another mechanism that triggers the biochemical pathways culminating in apoptosis. This mode of cytotoxicity is exemplified by the action of hormone antagonists (or hormone withdrawal) in hormone-sensitive breast and prostate cancers. However, it is well known that prostate cancer cells can become refractory to the tumor-suppressive effects of the antiandrogen flutamide (52, 88). Recent studies suggest that this refractoriness may relate to the emergence of a mutation in the androgen receptor that recognizes flutamide and other steroids as androgenic stimuli. A clinical trial in metastatic hormone-refractory prostate cancer demonstrates that flutamide withdrawal in conjunction with adrenal suppression is associated with salutary clinical responses and dramatic decreases in prostate-specific antigen levels (69a). It is intriguing to speculate that the withdrawal of flutamide in this setting (where flutamide is paradoxically acting as a growth stimulus) results in activation of programmed cell death (52). A similar mechanism might operate in hormone-refractory breast cancer, where withdrawal of the antiestrogen tamoxifen can result in clinical tumor regressions in some instances (44, 48).

Drugs that interrupt the normal process of signal transduction may exert cytotoxicity, at least in part, by inducing programmed cell death. Many of these drugs may directly or indirectly target the functional integrity of the *ras* family of oncogenes. *ras* oncogenes encode membrane-associated proteins that control cell signaling by modulating the phosphorylation state of so-called G proteins present in the cell cytoplasm. *ras* mutations and abnormal expression are detected in such diverse malignancies as colon, prostate, bladder, and brain cancers; myeloid leukemias; and lung cancer. (Indeed, in smoking-associated lung cancers, roughly one-third have somatic mutations in K-*ras*.) In this light, the mevalonate pathway is responsible for the synthesis of cholesterol and several other compounds critical to cell membrane structure and function. This pathway provides at least three enzymatic targets for cancer therapy and possibly for cancer prevention (53). Lovastatin, a drug currently used to treat hypercholesterolemia, inhibits the enzyme responsible for the synthesis of mevalonate, an early and rate-limiting precursor in sterol biosynthesis. Phenylacetate blocks the next step in this pathway, namely, the conversion of mevalonate pyrophosphate to isopentyl pyrophosphate. Finally, limonene (citrus oil), a plant monoterpene and a normal dietary constituent, specifically blocks the process known as isoprenylation, the most downstream reaction in the mevalonate pathway. The attachment of a farnesyl group (a type of isoprenyl group) to p21^{ras} proteins (the critical signal transduction proteins encoded by the *ras* oncogene family) is critical to Ras protein anchorage to the cell membrane and thus to Ras activity (53). Limonene and other terpenes such as farnesol (essence of lilac) inhibit farnesyl transferase, the enzyme that converts farnesyl pyrophosphate precursors (derived from isopentyl pyrophosphate) into active moieties. The inhibition of farnesyl transferase prevents the necessary post-translational modification of Ras proteins that would allow attachment to the cell membrane, which in turn inhibits *ras* family peptide activity (47, 55). Thus, each of these agents converges on the isoprenylation pathway, targeting different enzymatic sites to achieve antitumor activity by inhibiting *ras*-based signal transduction and *ras*-driven cellular proliferation. The cytotoxic effects of each agent are ultimately mediated by activation of programmed cell death. These agents are currently undergoing clinical development for prostate cancer and glioblastoma. On the basis of their *ras*-targeted activities, they could have a significant impact on several common malignancies that are frequently associated with *ras* mutation and overexpression, namely, bladder, pancreatic, and non-small-cell lung cancers.

PROGRESS TOWARD IMPLEMENTATION OF GENE-DIRECTED THERAPIES

The ability to directly modulate gene expression presents a surpassingly important challenge for rapid clinical development. The implementation of sophisticated gene-directed technologies in clinical practice stands as a paradigm for the effective transfer of information from the laboratory to the bedside. Indeed, gene transfection technology (gene therapy) in its present and future ramifications intersects all avenues of prevention and therapy alike: vaccine development predicated on gene transfection of tumor cells, immunomodulation via tumor-infiltrating lymphocyte modifications, and gene correction or replacement.

Antisense ODNs

Antisense oligodeoxynucleotides (ODNs) selectively target and inhibit the net expression of the targeted tumor cell gene, tumor cell proliferation, and survival without affecting normal cell counterparts. Examples of this exciting technology are now entering the clinical arena, with a particular focus on the tumor suppressor gene *p53*, the gene most commonly mutated in a widely diverse spectrum of lymphohematopoietic and epithelial malignancies. To this end, a clinical trial of phosphorothioate-modified antisense ODN targeting *p53* sequences in exon 10, the so-called OL-1-*p53*, is being conducted in patients with refractory and relapsing AML (3, 4). Phosphorothioates modify the phosphate backbone of the

ODN, thereby making the molecule resistant to destructive nucleases and permitting, at least in theory, the achievement and maintenance of effective intracellular levels by the process of diffusion (rather than transfection) (41, 79). In vitro data for primary cultures of fresh human AML cells demonstrate that antisense *p53* inhibits the proliferation of all 40 AML populations tested thus far, inducing cell fragmentation and disintegration compatible with apoptosis (3). Normal marrow cells, in contradistinction, do not exhibit such apoptotic changes. Further, the inhibitory responses are, at least to some degree, sequence specific and restricted to *p53* antisense constructs. Intriguingly, the incubation of long-term AML marrow cultures with antisense *p53* results in selective eradication of AML precursors (3), suggesting that this antisense molecule may have a special future role in the purging of autologous bone marrow or peripheral blood for stem cell reconstitution following marrow-ablative therapy with curative intent for AML in complete remission. A phase I study of 10-day continuous infusion of OL-1-*p53* in patients with refractory AML shows no evidence of antisense-induced rapid tumor lysis or other toxicity at the initial dose levels but does show interindividual variability in *p53* mRNA and protein production in the presence of antisense *p53* (4). The molecular pharmacology of antisense infusion is complex, and the measurements of various pharmacologic determinants in relation to clinical response are likely to be confounded by diverse cellular and molecular factors. Such factors might include the type of defect in net *p53* expression (genomic mutation versus posttranslational conformational change in the p53 protein) and/or the cellularity, vascularity, or presence of marrow necrosis in the leukemic marrow that in turn might influence the effective dose of antisense ODN at the cellular level. This trial provides an important model for attempting to give antisense ODNs in a systemic fashion to effect sequence-specific suppression of gene expression and will provide a basis for identifying and measuring important laboratory correlates of response.

A variety of approaches are being refined for potential use in CML for in vivo systemic treatment and/or for ex vivo selective purging of leukemic progenitors from bone marrow or peripheral blood stem cells of CML patients. One such approach involves the use of a diffusible, phosphorothioate-modified antisense ODN complementary to sequences in the c-*myb* oncogene (79), which produces a DNA-binding transcriptional factor. When given by multiple-day continuous infusion to severe combined immunodeficiency mice transplanted with the human CML blast crisis cell line K562, antisense *myb* prolongs survival and reduces tumor burden even at sites of sequestration where tumor cells are relatively protected from the effects of systemic therapies (i.e., central nervous system and ovaries) (79). The K562 cell line expresses both c-*myb* and the chimeric *bcr-abl* gene. In vivo treatment with antisense *myb* abrogates c-*myb* expression but has no impact on *bcr-abl* expression (79). Interestingly, in an in vitro model with fresh human CML cells, antisense *myb* was able to block the expression of *bcr-abl,* an activity perhaps related to the finding that MYB binds to the *bcr-abl* promoter region and suggesting that MYB may regulate *bcr-abl* transcription (80). Other gene-directed approaches under development for CML include antisense oligomers complementary to the aberrant mRNA encoded by the fusion *bcr-abl* gene (94), a retrovirally inserted antisense ODN directed against *bcr-abl* gene sequences

(62), and the potential use of antisense to the c-*myc* gene or a dominant negative MYC protein (87), the last based on the finding that the MYC protein is essential for full leukemic transformation in a transgenic mouse model for CML (20).

Gene-directed strategies using antisense ODNs are under preclinical development for eventual clinical application in the treatment of brain tumors. Recent studies suggest that such an approach is both feasible and effective. Insulin-like growth factor-1 (IGF-1) is a growth-promoting factor that is produced by and helps drive the proliferation of both human and rat glioblastoma cells (and many other malignant cell types). A rat glioblastoma model is being used to test an antisense construct directed against the gene encoding IGF-1 (100). Glioblastoma cells are transfected with antisense IGF-1 coupled to an episomal (circular) Epstein-Barr virus (EBV)-based vector that incorporates EBV replicative machinery. EBV is a neurotropic virus and as such may be especially well suited as a vector for transfecting brain tumor cells. When subsequently administered to tumor-bearing rats, the antisense-gene-manipulated cells elicit tumor-specific cytotoxic T lymphocytes (CTLs) that effectively eradicate both subcutaneous and intracerebral glioblastoma cells. In addition, these antisense-gene-transfected cells can prevent the development of glioblastoma by the same immune-based mechanism. This experimental design may serve as one basis for developing a vaccine for the purposes of treatment of active disease, secondary prevention (following surgical extirpation), or, possibly, primary prevention in individuals at high risk for development of brain tumors (100).

The modification of antisense constructs, for example, the development of triple-helix analogs that target double-stranded DNA or of high-affinity aptamers that target surface or extracellular molecules (41), may provide alternative approaches with high degrees of specificity for those genes or gene products whose aberrant expressions are central to the process of malignant transformation. An expansion of this theme is the development of DNA-targeting peptide nucleic acids, which invade duplex DNA at a specific site and displace a single strand of DNA (15, 39); these nucleic acids also bind avidly to RNA and in this way are capable of interrupting both transcriptional and translational processes. Still another technologic derivative is the construction of circular DNA ODNs that strongly bind single-stranded DNA and RNA by forming triple-helix complexes (41, 75). The development of clinically testable ODN constructs that focus their activities on tumor-specific genes or their transcribed products and spare the normal counterpart is an exceedingly important direction for continued development.

Update of Basic and Clinical Advances in Gene Therapy

This past year (1994) has seen a tremendous expansion in the design and implementation of gene transfer and gene therapy clinical trials for a broad spectrum of diseases. Gene transfer trials involve the transduction of the neomycin resistance marker gene for the purpose of determining gene-transduced cell migration, localization, and survival patterns, whereas gene therapy trials involve the transduction of genes (or antisense constructs, as discussed earlier) to confer a new function or to alter the expression of certain genes with therapeutic intent.

At present, more than 30 clinical protocols have been approved or are pending approval by the National Institutes of Health Recombinant DNA Advisory Committee, and over half of them have been developed within the last year. More than 75% of these trials target malignant diseases (melanoma; renal cell, small-cell and non-small-cell lung, ovarian, and breast cancers; neuroblastoma; glioblastoma; and hematopoietic malignancies including AML and CML, chronic lymphocytic leukemias, and multiple myeloma). The diverse approaches to these malignancies include the transfection of tumor-infiltrating lymphocytes, tumor cells, fibroblasts, and bone marrow cells with marker genes, antisense constructs, or genes for immunostimulatory cytokines, viral enzymes, or drug resistance (65, 96).

The concept of suicide vectors involves the selective transfer of genes into tumor cells to confer a new function that in turn can be exploited for antitumor therapy (19). The initial clinical application of this approach targets brain tumors. Exciting results have been obtained in an animal brain tumor model with gene therapy that transfers into the tumor cells a suicide vector, so-called because it allows subsequent selective killing of the gene-transfected cells by treatment with a cytotoxic agent that targets the transfected gene. Scientists at the National Cancer Institute and the National Institute of Neurological Disorders and Stroke have collaborated to design a clinical treatment protocol that combines stereotactic surgery with this form of gene therapy in patients with primary and metastatic brain tumors (19). This groundbreaking protocol, approved by the National Institutes of Health Recombinant DNA Advisory Committee and the Food and Drug Administration, began in December 1992. In humans (as in the mouse model), fibroblasts are transfected with the herpes simplex virus (HSV) thymidine kinase (TK) gene (*HSV-TK*) by use of a retroviral vector. These gene-altered fibroblasts are then injected directly into the brain tumors of patients by stereotactic implantation. The implanted fibroblasts continuously produce the disabled mouse viruses that transfer the *HSV-TK* gene (via a plasmid-like mechanism) into surrounding tumor cells, which consequently become sensitive to the anti-HSV drug ganciclovir. Ganciclovir is toxic specifically to cells containing *HSV-TK* and therefore should target only those cells containing and expressing the transfected *TK* gene. In addition, it seems that the incorporation of the *HSV-TK* gene generates CTLs, perhaps in response to the tumor cell producing a new immunogenic TK.

The suicide vector model serves as one template for further development in other cancers as well, especially those compartmentalized in relatively sequestered sites, e.g., intraperitoneal ovarian cancer or primary central nervous system lymphoma, an opportunistic cancer associated with AIDS. As an extension of this strategy, very recent animal studies demonstrate that the addition of the gene for type IV collagenase (an enzyme involved in tumor cell invasion) along with the *HSV-TK* gene facilitates the ability of the retrovirus to infect neighboring tumor cells by breaking down the basement membrane surrounding the tumor. This innovative approach is being expanded to other suicide vector genes for tumor cell transfections such as cytosine deaminase, which converts 5-flucytosine to 5-fluorouracil, a known cytotoxic agent (69).

Efforts to improve the effectiveness of immune responses to tumor antigens continue to focus on the use of gene transfer techniques to directly deliver a variety of naturally secreted cytokines to the target tumor cell and on strategies to overcome the tendency for immune cells to develop tolerance to malignant cells. Cytokines such as ILs, colony-stimulating factors (CSFs), IFNs, and tumor necrosis factors (TNFs) enhance tumor cell immunogenicity by upregulating the expression of major histocompatibility complex antigens on tumor cell surfaces and, in consequence, promoting the generation of tumor-targeted CTLs. These immunobiologic principles are the basis for serial clinical trials in patients with advanced melanoma that use autologous, TNF gene-modified tumor-infiltrating lymphocytes as immunotherapy or melanoma cells that have been transfected with either the TNF or the IL-2 gene to immunize patients against their tumors (2, 82). Diseases targeted for innovative trials of cytokine gene-transfected tumor cells include breast, ovarian, renal, and colon cancers, and development of their use against melanoma continues.

The principles established with an IL-4-transfected renal-cell mouse cancer vaccine, which demonstrated both immediate local tumor rejection and long-lasting systemic antitumor immunity (34), are now being translated into clinical testing of cytokine gene transfections in human renal cell carcinoma. Using a novel high-efficiency retroviral vector (MFG) for cytokine gene transfer, investigators have used diverse types of irradiated animal and primary explant human tumor cells to define the optimal cytokines for generating the responsible tumor-specific immune effector cells (25, 46). Of 10 immunostimulatory molecules tested, only granulocyte-macrophage CSF (GM-CSF) consistently showed superior antitumor activity in all cytokine gene-modified cancer cell lines (melanoma, renal, sarcoma, lung, and colon) (46). Intriguingly, GM-CSF is superior to IL-2, IL-4, IL-6, IFN-γ and TNF-α in terms of inducing long-lasting and specific antitumor immunity. The marked immunostimulatory capacity of GM-CSF may be related at least in part to its particular ability to quantitatively increase, attract, and activate non-lymphoid (as well as lymphoid) antigen-presenting cells, in particular monocytes/macrophages, eosinophils, and dendritic cells (25). The precise mechanisms by which GM-CSF exerts such immunostimulation will be pursued through the use of GM-CSF knockout mice created through homologous recombination technology. Gene transfer with MFG can be achieved in primary cell explants from various human cancers without the need for long-term culture or use of a selection marker, with impressive cellular transduction (roughly 60% efficiency) being detected for ovarian, colon, pancreatic, and prostate (as well as renal) tumors and with significant transduction efficiencies (20 to 40%) detected in small numbers of breast cancer cell populations (46). Importantly, tumor cell irradiation following GM-CSF gene transfer does not diminish GM-CSF secretion (25). Taken together, these provocative findings provide the foundation for an innovative phase I study of irradiated (nonreplicating) autologous tumor cell injections with cells prepared with or without GM-CSF gene transduction. The clinical trial of GM-CSF-transduced autologous tumor vaccine is currently being implemented in patients with metastatic renal cell carcinoma, with patient entry beginning in fall 1993 (89a); it will serve as a template for a similar approach to be tested in patients with other malignancies.

Dissection of the critical components of T-cell response to an antigenic stimulus has led to the finding that even with unimpeded antigen recognition by the T-cell receptor on a $CD8^+$ T cell, a costimulatory signal is required for full T-cell activation. Such costimulation results from the binding of a molecule known as B7 to the CD28 receptor on the T-cell surface (40, 99). Provocatively, while several types of antigen-presenting cells (e.g., dendritic cells, macrophages, activated B cells) express B7, epithelial cells do not (99). This lack of B7 has recently been offered as one explanation for the relative inability of epithelial cell cancers to generate effective CTL responses. This hypothesis has been tested and proven in recent experiments that demonstrate that transfection of the B7 gene into a murine melanoma cell line causes direct activation of $CD8^+$ T cells, which in turn inhibits the growth of established tumors, abrogates primary tumor development, and protects against subsequent tumor challenge by non-B7-transfected tumor cells (99). Importantly, $CD8^+$ activation by B7-transfected cells proceeds without invoking or recruiting $CD4^+$ T cells, thus apparently bypassing the need for $CD4^+$ cooperation in the generation of cellular immunity (40). The implications of these findings are significant for all epithelial tumors and may provide one mechanism with which to maximize the efficacy of cancer vaccine constructs. Further, the ability to bypass $CD4^+$ cells may suggest that B7 could have a special role in the therapy and prevention of human immunodeficiency syndrome-associated malignancies (where profound depletion of $CD4^+$ lymphocytes has a formative role in the emergence and progression of AIDS cancers, perhaps most especially AIDS-related lymphomas).

BUILDING ON NEW KNOWLEDGE

Molecular medicine continues to uncover the broad principles that govern the interplay between internal and external factors that modulate net gene expression and cellular function for many cell types. The principles provide a way of unifying our knowledge that cuts across multiple levels of biology to inform and advance the multiple disciplines of cancer investigation. The rapidly evolving understanding of molecular mechanisms brings with it unprecedented opportunities for advances in detection, therapy, and ultimately prevention of disease by detection of specific molecules or intervention at the molecular level. Indeed, as a result of revolutionary technological advances, molecular knowledge is now moving into the realm of clinical testing at an explosive pace. Expansion of investigations aimed at discerning the molecular foundations of cancer will continue to result in the development and application of new technologies to the broad spectrum of cancers that arise as a consequence of inherited or acquired molecular dysregulation and facilitate the transfer of those technologies from the laboratory to the clinic.

SUMMARY

An understanding of the molecular basis for cancer is gaining ground in many areas. If we maintain the tradition of excellence and the balance between creativity and scientific rigor exemplified by the work of Howard Temin, we will assuredly make progress against cancer.

REFERENCES

1. **Aaltonen, L. A., P. Peltomaki, F. S. Leach, P. Sistonen, L. Pylkkanen, J. P. Mecklin, H. Jarvinen, S. M. Powell, J. Jen, S. R. Hamilton, G. M. Petersen, K. W. Kinzler, B. Vogelstein, and A. D. L. Chapelle.** 1993. Clues to the pathogenesis of familial colorectal cancer. *Science* **260:**812–816.

2. **Asher, A. L., J. J. Mule, A. Kasid, N. P. Restifo, J. C. Salo, C. M. Reichert, G. Jaffe, B. Fendly, M. Kreigler, and S. A. Rosenberg.** 1991. Murine tumor cells transduced with the gene for tumor necrosis factor-alpha. *J. Immunol.* **146:**3227.

3. **Bayever, E., K. M. Haines, P. L. Iversen, R. W. Ruddon, S. J. Pirruccello, C. P. Mountjoy, M. A. Arneson, and L. J. Smith.** Selective cytotoxicity to human myeloblasts produced by oligodeoxyribonucleotides directed to p53 nucleotide sequences. *Leukemia Lymphoma,* in press.

4. **Bayever, E., P. L. Iversen, M. R. Bishop, J. G. Sharp, H. K. Tewary, M. A. Arneson, S. J. Pirruccello, R. R. Ruddon, A. Kessinger, G. Zon, and J. O. Armitage.** Systemic administration of a phosphorothioate oligonucleotide with a sequence complementary to p53 for acute myelogenous leukemia and myelodysplastic syndrome: initial results of a phase I trial. *Antisense Res. Dev.,* in press.

5. **Bhalla, K., C. Tang, A. M. Ibrado, S. Grant, E. Tourkina, C. Holladay, M. Hughes, M. E. Mahoney, and Y. Huang.** 1992. Granulocyte-macrophage colony-stimulating factor/interleukin-3 fusion protein (pIXY 321) enhances high-dose ara-C-induced programmed cell death or apoptosis in human myeloid leukemia cells. *Blood* **80:**2883–2890.

6. **Biondi, A., A. Rambaldi, M. Alcalay, P. P. Pandolfi, F. Lo Coco, D. Diverio, V. Rossi, A. Mencarelli, D. L. Longo, D. Zangrilli, G. Masera, T. Barbui, F. Mandelli, F. Grignani, and P. G. Pelicci.** 1991. RAR-α gene rearrangements as a genetic marker for diagnosis and monitoring in acute promyelocytic leukemia. *Blood* **77:**1418–1422.

7. **Bissonette, R. P., F. Echeverri, A. Mahboubi, and D. R. Green.** 1992. Apoptotic cell death induced by *c-myc* is inhibited by *bcl-2*. *Nature* (London) **359:**552–554.

8. **Bodley, A. L., and L. F. Liu.** 1988. Topoisomerases as novel targets for cancer chemotherapy. *Bio/Technology* **6:**1315–1319.

9. **Boulikas, T.** 1991. Relation between carcinogenesis, chromatin structure and poly(ADP-ribosylation). *Anticancer Res.* **11:**489–528.

10. **Cannon-Albright, L. A., D. E. Goldgar, L. J. Meyer, C. M. Lewis, D. E. Anderson, J. W. Fountain, M. E. Hegi, R. W. Wiseman, E. M. Petty, A. E. Bale, O. I. Olopade, M. O. Diaz, D. J. Kwiatkowski, M. W. Piepkorn, J. J. Zone, and M. H. Skolnick.** 1992. Assignment of a locus for familial melanoma, MLM, to chromosome 9p13-p22. *Science* **258:**1148–1152.

11. **Carson, D. A., and J. M. Ribeiro.** 1993. Apoptosis and disease. *Lancet* **341:**1251–1254.

12. **Castaigne, S., N. Bailitrand, H. de Thé, A. Dejean, L. Degos, and C. A. Chomienne.** 1992. PML/retinoic acid polymerase chain reaction in acute promyelocytic leukemia. *Blood* **79:**3110–3115.

13. **Cerretti, D. P., C. J. Kozlosky, B. Mosley, N. Nelson, K. Van Ness, T. A. Greenstreet, C. J. March, S. R. Kronheim, T. Druck, L. A. Cannizzaro, K. Huebner, and R. A. Black.** 1992. Molecular cloning of the interleukin-1β converting enzyme. *Science* **256:**97–100.

14. **Chen, C.-S., P. H. B. Sorensen, P. H. Domer, G. H. Reaman, S. J. Korsmeyer, N. A. Heerema, G. D. Hammond, and J. H. Kersey.** 1993. Molecular rearrangements on chromosome 11q23 predominate in infant acute lymphoblastic leukemia and are associated with specific biologic variables and poor outcome. *Blood* **81:**2386–2393.

15. **Cherny, D. Y., B. P. Belotserkovskii, M. D. Frank-Kamenetskii, M. Egholm, O. Buchardt, R. H.**

Berg, and P. E. Nielsen. 1993. DNA unwinding upon strand-displacement binding of a thymine-substituted polyamide to double-stranded DNA. *Proc. Natl. Acad. Sci. USA* **90:**1667–1670.

16. **Chiron, M., C. Demur, V. Pierson, J. P. Jaffrezou, C. Muller, S. Saivin, C. Bordier, C. Bousquet, N. Dastugue, and G. Laurent.** 1992. Sensitivity of fresh acute myeloid leukemia cells to etoposide: relationship with cell growth characteristics and DNA single-strand breaks. *Blood* **80:**1307–1315.

17. **Cimino, G., D. T. Moir, O. Canaani, K. Williams, W. M. Crist, L. Katzav, L. Cannizzaro, B. Lange, P. C. Nowell, C. M. Croce, and E. Canaani.** 1991. Cloning of ALL-1, the locus involved in leukemias with the t(4;11)(q21;q23),t(9;11)(p22;q23), and t(11;19)(q23;p13) chromosome translocations. *Cancer Res.* **51:**6712–6714.

18. **Clarke, A. R., C. A. Purdie, D. J. Harrison, R. G. Morris, C. C. Bird, M. L. Hooper, and A. H. Wyllie.** 1993. Thymocyte apoptosis induced by p53-dependent and independent pathways. *Nature* (London) **362:**849–852.

19. **Culver, K. W., Z. Ram, S. Wallbridge, H. Ishii, E. H. Oldfield, and R. M. Blaese.** 1992. In vivo gene transfer with retroviral vector-producer for treatment of experimental brain tumors. *Science* **256:**1550–1552.

20. **Daley, G. Q., R. A. Van Etten, and D. Baltimore.** 1990. Induction of chronic myelogenous leukemia in mice by the P210$^{bcr/abl}$ gene of the Philadelphia chromosome. *Science* **247:**824–830.

21. **D'Amico, D., D. Carbone, T. Mitsudomi, M. Nau, J. Fedorko, E. Russell, B. Johnson, D. Buchhagen, S. Bodner, R. Phelps, A. Gazdar, and J. D. Minna.** 1992. High frequency of somatically acquired p53 mutations in small-cell lung cancer cell lines and tumors. *Oncogene* **7:**339–346.

22. **Delattre, O., J. Zucman, B. Plougastel, C. Desmaze, T. Melot, M. Peter, H. Kovar, I. Joubert, P. de Jong, G. Rouleau, A. Aurias, and G. Thomas.** 1992. Gene fusion with an *ETS* DNA-binding domain caused by chromosome translocation in human tumours. *Nature* (London) **359:**162–165.

23. **Djabali, M., L. Selleri, P. Parry, M. Bower, B. D. Young, and G. A. Evans.** 1992. A trithorax-like gene is interrupted by chromosome 11q23 translocations in acute leukaemias. *Nature Genet.* **2:**113–118.

24. **Donehower, L. A., M. Harvey, B. L. Slagle, M. J. McArthur, C. A. Motgomery, Jr., J. S. Butel, and A. Bradley.** 1992. Mice deficient for p53 are developmentally normal but susceptible to spontaneous tumours. *Nature* (London) **356:**215–221.

25. **Dranoff, G., E. Jaffee, A. Lazenby, P. Golumbek, H. Levitsky, K. Brose, V. Jackson, H. Hamada, D. Pardoll, and R. C. Mulligan.** 1993. Vaccination with irradiated tumor cells engineered to secrete murine granulocyte-macrophage colony-stimulating factor stimulates potent-specific, and long-lasting anti-tumor immunity. *Proc. Natl. Acad. Sci. USA* **90:**3539–3543.

26. **Epstein, R. J.** 1990. Drug-induced DNA damage and tumor chemosensitivity. *J. Clin. Oncol.* **8:** 2062–2084.

27. **Evan, G. I., A. H. Wyllie, C. S. Gilbert, T. D. Littlewood, H. Land, M. Brooks, C. M. Waters, L. Z. Penn, and D. C. Hancock.** 1992. Induction of apoptosis in fibroblasts by c-Myc protein. *Cell* **69:**119–128.

28. **Fanidi, A., E. A. Harrington, and G. I. Evan.** 1992. Cooperative interaction between *c-myc* and *bcl-2* proto-oncogenes. *Nature* (London) **359:**554–556.

29. **Futreal, P. A., Q. Liu, D. Shattuck-Eidens, C. Cochran, K. Harshman, S. Tavtigian, L. M. Bennett, A. Haugen-Strano, J. Swensen, Y. Miki, K. Eddington, M. McClure, C. Frye, J. Weaver-Feldhaus, W. Ding, Z. Gholami, P. Soderkvist, L. Terry, S. Jhanwar, A. Berchuck, J. D. Iglehart, J. Marks, D. G. Ballinger, J. C. Barrett, M. H. Skolnick, A. Kamb, and R. Wiseman.** BRCA1 mutations in primary breast and ovarian carcinomas. *Science* **266:**120–122.

30. **Gaidano, G., P. Ballerini, J. Z. Gong, G. Inghirami, A. Neri, E. W. Newcomb, I. T. Magrath, D. M. Knowles, and R. Dalla-Favera.** 1991. p53 mutations in human lymphoid malignancies: association with Burkitt lymphoma and chronic lymphocytic leukemia. *Proc. Natl. Acad. Sci. USA* **88:**5413–5417.

31. **Garcia, I., I. Martinou, Y. Tsujimoto, and J.-C. Martinou.** 1992. Prevention of programmed cell death of sympathetic neurons by the *bcl-2* proto-oncogene. *Science* **258:**302–304.

32. **Gatti, R. A., E. Boder, H. V. Vinters, R. S. Sparkes, A. Norman, and K. Lange.** 1991. Ataxia-telangiectasia: an interdisciplinary approach to pathogenesis. *Medicine* **70:**99–117.

33. **Goddard, A. D., J. Borrow, P. S. Freemont, and E. Solomon.** 1991. Characterization of a zinc finger gene disrupted by t(15;17) in acute promyelocytic leukemia. *Science* **254:**1371–1374.

34. **Golumbek, P. T., A. J. Lazenby, H. I. Levitsky, L. M. Jaffee, H. Karasuyama, M. Baker, and D. M. Pardoll.** 1991. Treatment of established renal cell cancer by tumor cells engineered to secrete interleukin-4. *Science* **254:**713.

35. **Groden, J., A. Thliveris, W. Samowitz, M. Carlson, L. Gelbert, H. Albertson, G. Joslyn, J. Stevens, L. Spirio, M. Robertson, L. Sargeant, K. Krapcho, E. Wolff, R. Burt, J. P. Hughes, J. Warrington, J. McPherson, J. Wasmuth, D. Le Paslier, H. Abderrahim, D. Cohen, M. Leppert, and R. White.** 1991. Identification and characterization of the familial adenomatous polyposis coli gene. *Cell* **66:**589–600.

36. **Gu, Y., T. Nakamura, H. Alder, R. Prasad, O. Canaani, G. Cimino, C. M. Croce, and E. Canaani.** 1992. The t(4;11) chromosome translocation of human acute leukemias fuses the ALL-1 gene, related to drosophila trithorax, to the AF-4 gene. *Cell* **71:**701–708.

37. **Gupta, R. K., J. S. Whelan, T. A. Lister, B. D. Young, and J. G. Bodmer.** 1992. Direct sequence analysis of the t(14;18) chromosomal translocation in Hodgkin's disease. *Blood* **79:**2084–2088.

38. **Hancock, S. L., M. A. Tucker, and R. T. Hoppe.** 1993. Breast cancer after treatment of Hodgkin's disease. *J. Natl. Cancer Inst.* **85:**25–31.

39. **Hanvey, J. C., N. J. Peffer, J. E. Bisi, S. A. Thomson, R. Cadilla, J. A. Josey, D. J. Ricca, C. F. Hassman, M. A. Bonham, K. G. Au, S. G. Carter, D. A. Bruckenstein, A. L. Boyd, S. A. Noble, and L. E. Babiss.** 1992. Antisense and antigene properties of peptide nucleic acids. *Science* **258:**1481–1485.

40. **Harding, F. A., and J. P. Allison.** 1993. CD28-B7 interactions allow the induction of CD8 cytotoxic T lymphocytes in the absence of exogenous help. *J. Exp. Med.* **177:**1791–1796.

41. **Helene, C.** 1991. Rational design of sequence-specific oncogene inhibitors based on antisense and antigene oligonucleotides. *Eur. J. Cancer* **27:**1466–1471.

42. **Herskowitz, I.** 1987. Functional inactivation of genes by dominant negative mutations. *Nature* (London) **329:**219–222.

43. **Hollstein, M., D. Sidransky, B. Vogelstein, and C. C. Harris.** 1991. p53 mutations in human cancers. *Science* **253:**49–53.

44. **Howell, A., D. J. Dodwell, H. Anderson, and J. Redford.** 1992. Response after withdrawal of tamoxifen and progestogens in advanced breast cancer. *Ann. Oncol.* **3:**611–617.

45. **Hunger, S. P., D. C. Tkachuk, M. D. Amylon, M. P. Link, A. J. Carroll, J. L. Welborn, C. L. Willman, and M. L. Cleary.** 1993. HRX involvement in de novo and secondary leukemias with diverse chromosome 11q23 abnormalities. *Blood* **81:**3197–3203.

46. **Jaffee, E. M., G. Dranoff, L. K. Cohen, K. M. Hauda, S. Clift, F. F. Marshall, R. C. Mulligan, and D. M. Pardoll.** 1993. High efficiency gene transfer into primary human tumor explants without cell selection. *Cancer Res.* **53:**2221–2226.

47. **James, G. L., J. L. Goldstein, M. S. Brown, T. E. Rawson, T. C. Somers, R. S. McDowell, C. W. Crowley, B. K. Lucas, A. D. Levinson, and J. D. Marsters, Jr.** 1993. Benzodiazepine peptidomimetics: potent inhibitors of ras farnesylation in animal cells. *Science* **260:**1937–1942.

48. **Jiang, S.-Y., S. M. Langan-Fahey, A. L. Stella, R. McCague, and V. C. Jordan.** 1992. Point mutation of estrogen receptor (ER) in the ligand-binding domain of antiestrogens in ER-negative breast cancer cells stably expressing complementary DNAs for ER. *Mol. Endocrinol.* **6:**2167–2174.

49. **Kaneko, Y., N. Maseki, N. Takasaki, M. Sakuria, Y. Hayashi, S. Nakazawa, T. Mori, M. Sakurai, T. Takeda, T. Shikano, and Y. Hiyoshi.** 1986. Clinical and hematologic characteristics in acute leukemia with 11q23 translocations. *Blood* **67:**484–491.

50. **Kastan, M. B., O. Onyekwere, D. Sidransky, B. Vogelstein, and R. W. Craig.** 1991. Participation of p53 protein in the cellular response to DNA damage. *Cancer Res.* **51:**6304–6311.

51. **Kastan, M. B., Q. Zhan, W. S. El-Deiry, F. Carrier, T. Jacks, W. V. Walsh, B. S. Plunkett, B. Vogelstein, and A. J. Fornace, Jr.** 1992. A mammalian cell cycle checkpoint pathway utilizing p53 and GADD45 is defective in ataxia-telangiectasia. *Cell* **71:**587–597.

52. **Kelly, W. K., and H. I. Scher.** 1993. Prostate specific antigen decline after antiandrogen withdrawal. *J. Urol.* **149:**607–609.

53. **Khosvari-Far, R., A. D. Cox, K. Kato, and C. J. Der.** 1992. Protein prenylation: key to ras function and cancer intervention? *Cell Growth Diff.* **3:**461–469.

54. **Kinzler, K. W., M. C. Nilbert, L.-K. Su, B. Vogelstein, T. M. Bryan, D. B. Levy, K. J. Smith,**

A. C. Preisinger, P. Hedge, D. McKechnie, R. Finniear, A. Markham, J. Groffen, M. S. Boguski, S. F. Altschul, A. Horii, H. Ando, Y. Miyoshi, Y. Miki, I. Nishisho, and Y. Nakamura. 1991. Identification of FAP locus genes from chromosome 5q21. *Science* **253**:661–665.

55. **Kohl, N. E., S. D. Mosser, S. J. deSolms, E. A. Giuliani, D. L. Pompliano, S. L. Graham, R. L. Smith, E. M. Scolnick, A. Oliff, and J. B. Gibbs.** 1993. Selective inhibition of *ras*-dependent transformation by a farnesyltransferase inhibitor. *Science* **260**:1934–1937.

56. **Korsmeyer, S. J.** 1992. Bcl-2 initiates a new category of oncogenes: regulators of cell death. *Blood* **80**:879–886.

57. **Kuerbitz, S. J., B. S. Plunkett, W. V. Walsh, and M. B. Kastan.** 1992. Wild-type p53 is a cell cycle checkpoint determinant following irradiation. *Proc. Natl. Acad. Sci. USA* **89**:7491–7495.

58. **Levine, A. J., J. Momand, and C. A. Finlay.** 1991. The p53 tumour suppressor gene. *Nature* (London) **351**:453–455.

59. **Lowe, S. W., E. M. Schmitt, S. W. Smith, B. A. Osborne, and T. Jacks.** 1993. p53 is required for radiation-induced apoptosis in mouse thymocytes. *Nature* (London) **362**:847–849.

60. **Malkin, D., F. P. Li, L. C. Strong, J. F. Fraumeni, Jr., C. E. Nelson, D. H. Kim, J. Kassel, M. A. Gryka, F. Z. Bischoff, M. A. Tainsky, and S. H. Friend.** 1990. Germ line p53 mutations in a familial syndrome of breast cancer, sarcomas, and other neoplasms. *Science* **250**:1233–1238.

61. **Mao, L., D. J. Lee, M. S. Tockman, Y. S. Erozan, F. Askin, and D. Sidransky.** 1994. Microsatellite alterations as clonal markers for the detection of human cancer. *Proc. Natl. Acad. Sci. USA* **91**: 9871–9875.

62. **Martiat, P., P. Lewalle, A. S. Taj, M. Phillipe, Y. Larondelle, J. L. Vaerman, C. Wildmann, J. M. Goldman, and J. L. Michaux.** 1993. Retrovirally transduced antisense sequences stably suppress P210$^{BCR-ABL}$ expression and inhibit the proliferation of BCR/ABL-containing cell lines. *Blood* **81**:502–509.

63. **Marx, J.** 1993. Cell death studies yield cancer clues. *Science* **259**:760–761.

64. **Miki, Y., J. Swensen, D. Shattuck-Eidens, P. A. Futreal, K. Harshman, S. Tavtigian, Q. Liu, C. Cochran, L. M. Bennett, W. Ding, R. Bell, J. Rosenthal, C. Hussey, T. Tran, M. McClure, C. Frye, T. Hattier, R. Phelps, A. Haugen-Strano, H. Katcher, K. Yakumo, Z. Gholami, D. Shaffer, S. Stone, S. Bayer, C. Wray, R. Bogden, P. Dayananth, J. Ward, P. Tonin, S. Narod, P. K. Bristow, F. H. Norris, L. Helvering, P. Morrison, P. Rosteck, M. Lai, J. C. Barrett, C. Lewis, S. Neuhausen, L. Cannon-Albright, D. Goldgar, R. Wiseman, A. Kamb, and M. H. Skolnick.** 1994. A strong candidate for the breast and ovarian cancer susceptibility gene BRCA1. *Science* **266**: 66–71.

65. **Miller, A. D.** 1992. Human gene therapy comes of age. *Nature* (London) **357**:455–460.

66. **Miyashita, T., and J. C. Reed.** 1993. Bcl-2 oncoprotein blocks chemotherapy-induced apoptosis in a human leukemia cell line. *Blood* **81**:151–157.

67. **Morgan, G. J., F. Cotter, F. E. Katz, S. A. Ridge, P. Domer, S. Korsmeyer, and L. M. Wiedemann.** 1992. Breakpoints at 11q23 in infant leukemias with the t(11;19) (q23;p13) are clustered. *Blood* **80**:2172–2175.

68. **Morrissey, J., D. C. Tkachuk, A. Milatovich, U. Francke, M. Link, and M. L. Cleary.** 1993. A serine/proline-rich protein is fused to HRX in t(4;11) acute leukemias. *Blood* **81**:1124–1131.

69. **Mullen, C. A., M. Kilstrup, and R. M. Blaese.** 1992. Transfer of the bacterial gene for cytosine deaminase to mammalian cells confers lethal sensitivity to 5-fluorocytosine: a negative selection system. *Proc. Natl. Acad. Sci. USA* **89**:33–37.

69a. **Myers, C.** Personal communication.

70. **Nichols, J., and S. D. Nimer.** 1992. Transcription factors, translocations, and leukemia. *Blood* **80**:2953–2963.

71. **Nicolaides, N. C., N. Papadopoulos, B. Liu, Y. F. Wei, K. C. Carter, S. M. Ruben, C. A. Rosen, W. A. Haseltine, R. D. Fleischmann, C. M. Fraser, M. D. Adams, J. C. Venter, M. G. Dunlop, S. R. Hamilton, G. M. Petersen, A. de la Chapelle, B. Vogelstein, and K. W. Kinzler.** 1994. Mutations of two PMS homologues in hereditary nonpolyposis colon cancer. *Nature* (London) **371**:75–80.

72. **Nishisho, I., Y. Nakamura, Y. Miyoshi, Y. Miki, H. Ando, A. Horii, K. Koyama, J. Utsunomiya, S. Baba, P. Hedge, A. Markham, A. J. Krush, G. Petersen, S. R. Hamilton, M. C. Nilbert, S. B. Levy, T. M. Bryan, A. C. Preisinger, K. J. Smith, L.-K. Su, K. W. Kinzler, and B. Vogelstein.**

1991. Mutations of chromosome 5q21 genes in FAP and colorectal cancer patients. *Science* **253:** 665–669.

73. **Pedersen-Bjergaard, J., T. C. Sigsgaard, D. Nielson, S. B. Gjedde, P. Philip, M. Hansen, S. O. Larsen, M. Rorth, H. Mouridsen, and P. Dombernowsky.** 1992. Acute monocytic or myelomonocytic leukemia with balanced chromosome translocation to band 11q23 after therapy with 4-epi-doxorubicin and cisplatin or cyclophosphamide for breast cancer. *J. Clin. Oncol.* **10:** 1444–1451.

74. **Peltomaki, P., L. A. Aaltonen, P. Sistonen, L. Pylkkanenm, J. P. Mecklin, H. Jarvinen, J. S. Green, J. R. Jass, J. L. Weber, F. S. Leach, G. M. Petersen, S. R. Hamilton, A. D. L. Chapelle, and B. Vogelstein.** 1993. Genetic mapping of a locus predisposing to human colorectal cancer. *Science* **260:**810–812.

75. **Prakash, G., and E. T. Kool.** 1991. Molecular recognition by circular oligodeoxynucleotides. Strong binding of single-stranded DNA and RNA. *J. Chem. Soc. Chem. Commun.* **1991:** 1161–1163.

76. **Pui, C.-H., R. C. Ribeiro, M. L. Hancock, G. K. Riveria, W. E. Evans, S. C. Raimondi, D. R. Head, F. G. Behm, M. H. Mahmoud, J. T. Sandlund, and W. M. Crist.** 1991. Acute myeloid leukemia in children treated with epipodophyllotoxins for acute lymphoblastic leukemia. *N. Engl. J. Med.* **325:**1682–1687.

77. **Rabizadeh, S., J. Oh, L. Zhong, J. Yang, C. M. Bitler, L. L. Butcher, and D. E. Bredesen.** 1993. Induction of apoptosis by the low-affinity NGF receptor. *Science* **261:**345–348.

78. **Raimondi, S. C., S. C. Peiper, G. R. Kitchingman, F. G. Behm, D. L. Williams, M. L. Hancock, and J. Mirro, Jr.** 1989. Childhood acute lymphoblastic leukemia with chromosomal breakpoints at 11q23. *Blood* **73:**1627–1634.

79. **Ratajczak, M. Z., N. Hijiya, L. Catani, K. DeRiel, S. M. Luger, P. McGlave, and A. Gewirtz.** 1992. Acute- and chronic-phase chronic myelogenous leukemia colony-forming units are highly sensitive to the growth inhibitory effects of *c-myb* antisense oligodeoxynucleotides. *Blood* **79:** 1956–1961.

80. **Ratajczak, M. Z., J. A. Kant, S. M. Luger, N. Hijiya, J. Zhang, G. Zon, and A. M. Gewirtz.** 1992. *In vivo* treatment of human leukemia in a *scid* mouse model with c-myb antisense oligodeoxynucleotides. *Proc. Natl. Acad. Sci. USA* **89:**11823–11827.

81. **Roberston, L. E., S. Chubb, R. E. Meyn, M. Story, R. Ford, W. N. Hittleman, and W. Plunkett.** 1993. Induction of apoptotic cell death in chronic lymphocytic leukemia by 2-chloro-2'-deoxyadenosine and 9-b-D-arabinosyl-2-fluoroadenine. *Blood* **81:**143–150.

82. **Rosenberg, S. A., P. Aebersold, K. Cornetta, A. Kasid, R. A. Morgan, R. Moen, E. M. Karson, M. T. Lotze, J. C. Yang, S. L. Topalian, M. J. Merino, K. Culver, A. D. Miller, R. M. Blaese, and W. F. Anderson.** 1990. Gene transfer into humans—immunotherapy of patients with advanced melanoma, using tumor-infiltrating lymphocytes modified by retroviral gene transduction. *N. Engl. J. Med.* **323:**570–578.

83. **Rubin, C. M., D. C. Arthur, W. G. Woods, B. J. Lange, P. C. Nowell, J. D. Rowley, J. Nachman, B. Bostrom, E. S. Baum, C. R. Suarez, N. R. Shah, E. Morgan, H. S. Maurer, S. E. McKenzie, R. A. Larson, and M. M. Le Beau.** 1991. Therapy-related myelodysplastic syndrome and acute myeloid leukemia in children: correlation between chromosomal abnormalities and prior therapy. *Blood* **78:**2982–2988.

84. **Sachs, L., and J. Lotem.** 1993. Control of programmed cell death in normal and leukemic cells: new implications for therapy. *Blood* **82:**15–21.

85. **Sandoval, C., C. H. Pui, L. C. Bowman, D. Heaton, C. A. Hurwitz, S. C. Raimondi, F. G. Behm, and D. R. Head.** 1993. Secondary acute myeloid leukemia in children previously treated with alkylating agents, intercalating topoisomerase II inhibitors, and irradiation. *J. Clin. Oncol.* **11:** 1039–1045.

86. **Satoh, M. S., and T. Lindahl.** 1992. Role of poly(ADP-ribose) formation in DNA repair. *Nature* (London) **356:**356–358.

87. **Sawyers, C. L., W. Callahan, and O. N. Witte.** 1992. Dominant negative MYC blocks transformation by ABL oncogenes. *Cell* **70:**901–910.

88. **Scher, H., T. Curley, S. Yeh, J. M. Iversen, M. O'Dell, and S. M. Larson.** 1992. Therapeutic alternatives for hormone-refractory prostatic cancer. *Semin. Urol.* **10:**55–64.

89. **Shi, Y., J. M. Glynn, L. J. Guilbert, T. G. Cotter, R. P. Bissonette, and D. R. Green.** 1992. Role for c-myc in activation-induced apoptotic cell death in T cell hybridomas. *Science* **257**:212–214.

89a. **Simons, J. W., E. M. Jaffee, and D. M. Pardoll.** Personal communication.

90. **Slichenmeyer, W. J., E. K. Rowinsky, R. C. Donehower, and S. H. Kaufmann.** 1993. The current status of camptothecin analogues as antitumor agents. *J. Natl. Cancer Inst.* **85**:27.

91. **Stetler-Stevenson, M.** 1992. The t(14;18) translocation in Hodgkin's disease. *J. Natl. Cancer Inst.* **84**:1770–1771.

92. **Swift, M.** Ionizing radiation, breast cancer, and ataxia-telangiectasia. *J. Natl. Cancer Inst.*, in press.

93. **Swift, M., D. Morrell, R. B. Massey, and C. L. Chase.** 1991. Incidence of cancer in 161 families affected by ataxia-telangiectasia. *N. Engl. J. Med.* **325**:1831–1836.

94. **Szczylik, C., T. Skorski, N. C. Nicolaides, L. Manzella, L. Malaguarnera, D. Venturelli, A. M. Gewirtz, and B. Calabretta.** 1991. Selective inhibition of leukemia cell proliferation by BCR-ABL antisense oligodeoxynucleotides. *Science* **253**:562–565.

95. **Thibodeau, S. N., G. Bren, and D. Schaid.** 1993. Microsatellite instability in cancer of the proximal colon. *Science* **260**:816–819.

96. **Thompson, L.** 1992. At age 2, gene therapy enters a growth phase. *Science* **258**:744–746.

97. **Thornberry, N. A., H. G. Bull, J. R. Calaycay, K. T. Chapman, A. D. Howard, M. J. Kostura, D. K. Miller, S. M. Molineaux, J. R. Weidner, J. Aunins, K. O. Elliston, J. M. Ayala, F. J. Casano, J. Chin, G. J.-F. Ding, L. A. Egger, E. P. Gaffney, G. Limjuco, O. C. Palyha, S. M. Raju, A. M. Rolando, J. P. Salley, T.-T. Yamin, T. D. Lee, J. E. Shively, M. MacCross, R. A. Mumford, J. A. Schmidt, and M. J. Toccil.** 1992. A novel heterodimeric cysteine protease is required for interleukin-1β processing in monocytes. *Nature* (London) **356**:768–774.

98. **Tkachuk, D. C., S. Kohler, and M. L. Cleary.** 1992. Involvement of a homolog of Drosophila trithorax by 11q23 chromosomal translocations in acute leukemias. *Cell* **71**:691–700.

99. **Townsend, S. E., and J. P. Allison.** 1993. Tumor rejection after direct costimulation of CD8 T cells by B7-transfected melanoma cells. *Science* **259**:368–370.

100. **Trojan, J., T. R. Johnson, S. D. Rudin, J. Ilan, M. L. Tykocinski, and J. Ilan.** 1993. Treatment and prevention of rat glioblastoma by immunogenic C6 cells expressing antisense insulin-like growth factor I RNA. *Science* **259**:94–96.

101. **Vaux, D. L., I. L. Weissman, and S. K. Kim.** 1992. Prevention of programmed cell death in Caenorhabditis elegans by human bcl-2. *Science* **258**:1955–1957.

102. **Vogelstein, B., E. R. Fearon, S. R. Hamilton, S. E. Kern, A. C. Preisinger, M. Leppert, Y. Nakamura, R. White, A. M. M. Smits, and J. L. Bos.** 1988. Genetic alterations during colorectal-tumor development. *N. Engl. J. Med.* **319**:525–532.

103. **Waldmann, T. A., J. Misiti, D. L. Nelson, and K. H. Kraemer.** 1983. Ataxia-telangiectasia: a multisystem hereditary disease with immunodeficiency, impaired organ maturation, x-ray hypersensitivity, and a high incidence of neoplasia. *Ann. Intern. Med.* **99**:367–379.

104. **Yahalom, J., J. A. Petrek, P. W. Biddinger, S. Kessler, D. D. Dershaw, B. McCormick, M. P. Osborne, D. A. Kinne, and P. P. Rosen.** 1992. Breast cancer in patients irradiated for Hodgkin's disease: a clinical and pathologic analysis of 45 events in 37 patients. *J. Clin. Oncol.* **10**:1674–1681.

105. **Yonisch-Rouach, E., D. Resnitzky, J. Lotem, L. Sachs, A. Kimchi, and M. Oren.** 1991. Wild-type p53 induces apoptosis of myeloid leukaemic cells that is inhibited by interleukin-6. *Nature* (London) **352**:345–347.

PART III

VIROLOGY

The DNA Provirus: Howard Temin's Scientific Legacy
Edited by G. M. Cooper, R. Greenberg Temin, and B. Sugden
© 1995 American Society for Microbiology, Washington, DC 20005

Chapter 11

Reprint of Shimotohno, Mizutani, and Temin's 1980 Paper on the Similarity of the Structures of Retroviral Proviruses to That of Transposable Elements

The latter half of the 1970s witnessed a blossoming of the development and application of techniques to analyze nucleic acids by introducing, identifying, and amplifying them in prokaryotic hosts. Howard Temin and his colleagues capitalized on these developments to address two questions central to the study of retroviruses and to his provirus hypothesis: what are the structures of the products of retroviral replication, and what is the mechanism by which retroviruses replicate? Their answers to the first question allowed them to begin to infer an answer to the second question.

In the following paper by Shimotohno, Mizutani, and Temin, the authors describe the identification, cloning, and sequencing of the long terminal repeats (LTRs) and the abutting cellular DNA of infectious proviruses of spleen necrosis virus. They had previously isolated clones of these proviruses in lambda vectors developed in 1977 and 1978 by Fred Blattner, working in the neighboring Genetics Department in Madison. The LTRs of retroviruses are direct repeats of sequences of DNA at the ends of the provirus. The 3′ terminus of the 5′ LTR should therefore be identical to the 3′ terminus of the intact provirus; similarly, the 5′ terminus of the 3′ LTR should be identical to the 5′ terminus of the intact provirus. A knowledge of this structure along with the determined DNA sequences allowed Howard and his colleagues to identify the junctions of proviral DNA and the cellular DNA into which the proviral DNA had integrated. They made two striking observations: the proviral DNA lost two nucleotides from each end prior to its integration, and five nucleotides of the cellular DNA were duplicated at the site of integration.

These two observations contributed substantively to the basis for their hypothesis that retroviruses have evolved from cellular movable genetic elements that often share these features. It is now known that retroviruses share facets of the mechanism of their replication with a wide variety of cellular movable genetic elements. Howard Temin continued to study and elucidate the mechanism of retroviral replication through genetic and biochemical dissections of the products of replication during a single cycle of infection. These studies allowed him to validate his provirus hypothesis and to reconstruct pathways by which retroviruses assimilate proto-oncogenes.

Bill Sugden

Reprinted from Nature, Vol. 285, No. 5766, pp. 550-554, June 19 1980
© *Macmillan Journals Ltd., 1980*

Sequence of retrovirus provirus resembles that of bacterial transposable elements

Kunitada Shimotohno, Satoshi Mizutani & Howard M. Temin

McArdle Laboratory for Cancer Research, University of Wisconsin-Madison, Madison, Wisconsin 53706

The nucleotide sequences of the terminal regions of an infectious integrated retrovirus cloned in the modified λ phage cloning vector Charon 4A have been elucidated. There is a 569-base pair direct repeat at both ends of the viral DNA. The cell–virus junctions at each end consist of a 5-base pair direct repeat of cell DNA next to a 3-base pair inverted repeat of viral DNA. This structure resembles that of a transposable element and is consistent with the protovirus hypothesis that retroviruses evolved from the cell genome.

RETROVIRUSES are a family of RNA viruses that infect animals and replicate through a DNA intermediate, the provirus[1]. The protovirus hypothesis suggests that retroviruses evolved from the cell genome, in particular from a portion of the cell genome involved in normal transfer of genetic information in and between cells[2].

The recent availability of recombinant DNA techniques for cloning eukaryotic genes and of rapid DNA sequencing techniques has enabled us to isolate several infectious proviruses of spleen necrosis virus and their surrounding cellular sequences[3] and to sequence several hundred bases including the cell–virus junctions of one of these proviruses. (Spleen necrosis virus is an avian retrovirus that has genus-specific relationships to mammalian type C retroviruses[4–6].) The regions sequenced include the two cell–virus junctions, the complete terminal repeats, the positions of origin and terminus of viral RNA, and a binding site for a putative primer tRNA. In particular, the cell–virus junction sequences seem to have a

5-base pair direct repeat of cellular DNA next to an inverted repeat of viral DNA. This structure resembles that of bacterial transposable elements[7–12] and is consistent with the protovirus hypothesis that retroviruses have evolved from part of the normal cell genome.

Sequencing of terminal regions of an integrated spleen necrosis virus

14-44 is a recombinant clone of the modified phage λ cloning vector, Charon 4A, containing chicken DNA with an infectious provirus of spleen necrosis virus (SNV)[3]. 14-44 was selected from phage packaged with Charon 4A arms ligated to EcoRI-digested DNA isolated from chicken cells 4 days after infection (acute infection). Restriction endonuclease cleavage sites in 14-44 have been mapped previously, as indicated in Figs 1 (top) and 2a (see ref. 3). In particular, the virus–cell junctions were located just left of the 5′ SacI site, and approximately 600 base pairs right of the 3′ SacI site and 200 base pairs left of the 3′ EcoRI site by comparison with restriction enzyme digests of 10 cloned proviruses and of unintegrated viral DNA, and also by electron microscopic heteroduplexing[3].

Subclones containing the junction regions were isolated as described in Fig. 1 legend. Fine structure restriction enzyme maps (Fig. 2b) were prepared for the subclones containing each end of the provirus using the method of Smith and Birnstiel[13] with labelling of the SalI end or the EcoRI end, respectively. Starting from the left end of both subclones, after

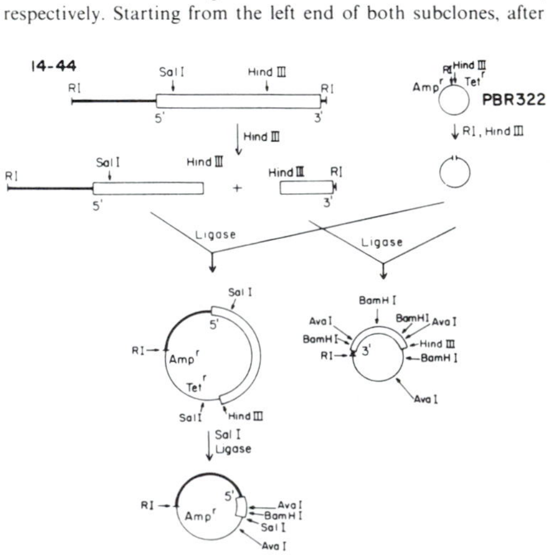

Fig. 1 Construction of subclones of 14-44 DNA in pBR322. DNA (100 µg) of clone 14-44 was digested with HindIII and EcoRI, and the fragments separated by gel electrophoresis in 1% agarose. The DNA bands of 14-44 were visualized with the aid of ethidium bromide, were cut out, and the agarose removed by soaking the crushed gel in 1 M NaCl, 1 mM EDTA, 0.1% SDS, 20 mM Tris-HCl (pH 8.0) at 37 C for 20 h followed by ethanol precipitation. Each DNA fragment was then separately ligated with T4 DNA ligase at 4 C for 20 h to the large fragment of EcoRI-HindIII-digested pBR322 DNA. Calcium chloride-treated competent Escherichia coli cells were transformed with the ligated DNA and the cells plated on medium containing 70 µg each of tetracycline and ampicillin. The colonies which appeared were transferred to nitrocellulose filters for colony hybridization[49]. The filter was treated with 0.1 M NaOH for 1 min, neutralized with 1 M Tris-HCl (pH 7.2), 0.6 M NaCl, and dried at 80 C under vacuum to immobilize the DNA on it. The filter was hybridized with ³²P-labelled SNV cDNA (specific activity about 10⁸ c.p.m. per µg) at 63 C for 20 h and washed to remove unhybridized cDNA. Autoradiographs were made after drying. The subcloned DNA which carried the 5′ half EcoRI-HindIII fragment of 14-44 DNA was further digested with SalI and self-ligated with T4 DNA ligase as described above. Calcium-treated competent E. coli cells were transformed, and cells were plated on Petri dishes containing 50 µg ampicillin. An aliquot of each colony which appeared was transferred to agar-medium containing 50 µg tetracycline, and plates were kept at 37 C for 24 h. Colonies which could not grow on the medium containing tetracycline contained the 5′ EcoRI-SalI DNA fragment of 14-44. Plasmid DNA was purified from chloramphenicol-treated bacteria which contained the subcloned plasmids as follows: The bacteria which contained the plasmids were grown to late log phase. Chloramphenicol was added to a final concentration of 20 µg per ml of culture medium. Incubation was continued another 5 h. Cells were collected by centrifugation and lysed with EDTA, lysozyme and Brij 58 (ref. 50). Cellular DNA was pelleted by centrifugation at 27,000 r.p.m. for 60 min. The supernatant was taken, CsCl added, and the refractive index adjusted to 1.395. Ethidium bromide was also added to a final concentration of 50 µg per ml of solution. The lysates were centrifuged at 35,000 r.p.m. for 72 h using a Beckman 50 Ti rotor. The lower DNA band was collected and centrifuged at the same speed for 48 h. The DNA band was taken out, the ethidium bromide was removed by extraction with isobutyl alcohol, and the DNA was dialysed and ethanol precipitated.

Fig. 2 Structure of recombinant clone 14-44 and subclones containing 5' and 3' ends of SNV. *a*, Restriction map of the inserted DNA of clone 14-44, containing an infectious provirus of SNV[3]. 5' And 3' are defined in terms of the ends of viral RNA[14]. The formation of subclones containing the 5' and 3' junction sequences is described in Fig. 1 legend. Mapping of the cleavage sites of restriction endonucleases in the subclones (*b*) was done as follows: The purified subcloned DNAs in pBR322 were digested with *Sal*I or *Eco*RI. After removal of phosphate residues from the 5' end of the fragments with alkaline phosphatase, the 5' ends were labelled using polynucleotide kinase in the presence of [γ-³²P]ATP. Labelled DNAs were further digested with *Eco*RI or *Hind*III, respectively, and were separated by agarose gel electrophoresis. The DNAs were extracted from the agarose as described in Fig. 1 legend and purified by ethanol precipitation. Labelled DNAs were partially digested with restriction endonuclease (2 min, 4 min, 8 min and 15 min incubation at 37 °C in 50 µl of reaction mixture containing 1 µg DNA and 1 unit restriction enzyme) and fragments were separated by gel electrophoresis in 1.5% agarose at 1.5 V cm⁻¹ for 8 h. Gels were dried and autoradiographs made. The molecular weight of each fragment was calculated from the extent of migration of labelled restriction fragments. End-labelled *Hind*III and *Taq*I fragments of pBR322 DNA were used as markers gel electrophoresis. *c*, *d* The extent of sequence determined from restriction enzyme fragments labelled at their 5' ends with ³²P as described by

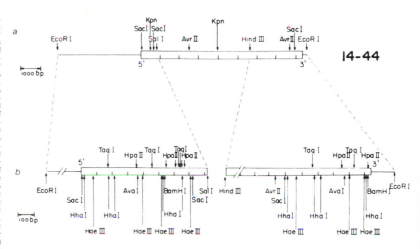

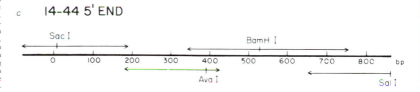

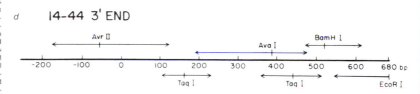

Maxam and Gilbert[51]. Labelled termini are indicated by (⊢) and the direction and extent of sequencing with arrows. Fragments for sequencing were isolated using agarose or polyacrylamide gel electrophoresis. In each case, after 5' labelling, molecules were digested with another restriction enzyme as indicated below. (Numbering is from the beginning of the terminal repeat (see Fig. 4).) *3' end. Eco*RI: digested with *Hind*III and the 2.6-kilobase-pair fragment was isolated. *Bam*HI: The 4-kilobase pair fragment was isolated and digested with *Eco*RI and the 160-pair fragment was isolated. The 1.7-kilobase-pair fragment was isolated and digested with *Ava*II and the 570-base pair fragment was isolated. *Ava*I: The 3.6-kilobase-pair fragment was isolated and digested with *Eco*RI and the 300-base pair fragment isolated. The 2.5-kilobase pair fragment was isolated and digested with *Ava*II and the 420-base pair fragment was isolated. *Avr*II: digested with *Eco*RI and the 700-base pair and 6.7-kilobase pair fragments were isolated. *Avr*II to *Eco*RI 750-base pair fragment was digested with *Taq*I. The 260-base pair fragment was digested with *Bam*HI and the 170-base pair fragment was isolated. Half the 250-base pair fragment was digested with *Ava*I and the 200-base pair fragment was isolated. The other half of the 250-base pair fragment was digested with *Hha*I and the 210-base-pair fragment was isolated. *5' end. Sal*I: digested with *Eco*RI and the 6-kilobase pair fragment was isolated. *Bam*HI: digested with *Eco*RI and both fragments were isolated. *Ava*I: digested with *Eco*RI and both fragments were isolated. *Sac*I: digested with *Eco*RI and *Ava*I and the 4.2-kilobase-pair and 400-base-pair fragments were isolated.

the *Avr*II site present only in the subclone with the 3' end of SNV, the same cleavage sites are seen: *Sac*I, *Hha*I, *Hae*III, . . . up to the *Hae*III, *Hha*I, *Bam*HI and *Hpa*II sites. Then the maps diverge. These data indicate that the terminal regions in the SNV provirus are repeated, as they are in the unintegrated SNV DNA[14], and that the terminal repeats are near to or at the ends of the viral DNA, as defined by comparison with the restriction cleavage sites in 10 cloned proviruses and in unintegrated DNA and by electron microscopic heteroduplexing of several of the cloned proviruses[3].

DNAs of the pBR322 subclones containing the 5' and the 3' cell virus junctions were digested with different restriction endonucleases, and the fragments were isolated, labelled, and further digested by restriction enzymes as described in Fig 2*c*, *d*. The fragments were then sequenced as illustrated in Fig. 3.

Nucleotide sequences of terminal regions of provirus in 14-44

The sequences of the 890 bases at the 5' end of SNV in 14-44 and the 850 bases at the 3' end of 14-44 are given in Fig. 4. These sequences contain all of the restriction endonuclease sites mapped at the ends of the virus DNA (Fig. 2*b*), as well as the *Alu*I and *Hinf*I restriction endonuclease sites which were also mapped in the plasmid DNA (data not shown).

Comparing the sequences of the 5' and 3' ends directly, one sees that a 569-base pair sequence from nucleotides 1 to 569 at the 5' end is the same as the nucleotide sequence from nucleotides 1 to 569 at the 3' end. The sequences to either side diverge. Therefore, the terminal repeat in viral DNA is 569 base pairs. This is similar in size to that of the terminal repeat

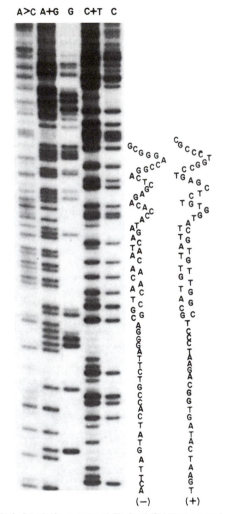

A>C A+G G C+T C

(−)   (+)

Fig. 3 Example of a sequencing gel. The 5' end of 14-44 was sequenced from the *Sal*I site (Fig. 2*d*) using the Maxam–Gilbert method[51]. Five base-specific cleavages were used (A>C, A+G, G, C+T, and C). Electrophoresis was carried out in thin 20″, 10″, 8″, polyacrylamide-7 M urea gels[52]. The bases marked are from nucleotides 815 to 750 of 14-44 5' end (see Fig. 4). (−) Represents strand which is complementary to sense of viral RNA (+).

14-44 5' end, approximately 25 nucleotides after the sequence TATAAG and after the sequence TTGCT (boxed in Fig. 4). These sequences have been found in other animal DNAs 5' to the beginning of RNAs[17–20]. Viral RNA ends at approximately nucelotide 473 of the 14-44 3' end, 20 nucleotides after the sequence AATAAA (AAUAAA in viral RNA) (boxed in Fig. 4). This sequence has been found near the 3' end of mRNAs 5' to the poly(A)[17,21–24].

The terminal repetition in the RNA, called R (ref. 25), is about 80 nucleotides long. As was the case with the DNA repeat, R is larger than the 21 base-pair repeat in avian leukosis/sarcoma virus RNA, and the approximately 60-base pair repeat in murine leukaemia virus RNA[26–31].

One consequence of the proviral nucleotide sequence is that viral RNA synthesis starting at the 5' end of viral DNA progresses through a stop sequence at the 5' end of viral DNA. In addition, there is a second start signal at the 3' end of viral DNA. Thus, RNA synthesis might begin also at the 3' end of viral DNA.

tRNA^{Pro} has been reported to be the primer for REV-A, a member of the same retrovirus species as SNV[54]. A sequence complementary to the 18 3' nucleotides of tRNA^{Pro} (ref. 32) is found from nucleotides 572 to 589 of the 14-44 5' end (Fig. 4). This sequence includes the complement to the tRNA terminal CCA. Presumably this sequence represents the primer binding site (PBS) in viral RNA. The size of 'strong stop' DNA (from PBS to the 5' end of viral RNA) and of U5 (from PBS to R)[25] can then be calculated as 180 and 100 base pairs, respectively. These are also somewhat larger than for avian sarcoma and murine leukaemia viruses[26,28,29,33].

We have previously shown that some sequences from the 3' end of SNV RNA are present at the 5' end of SNV DNA[14], called U3 (ref. 25). Therefore, we compared the sequence at the 5' end of viral DNA, as defined above, with the homologous sequence at the 3' end of viral RNA (or DNA) (found in the 14-44 3' end, Fig. 4) (Fig. 5). From 5' (reading backwards) to TGT (nucleotides 1 to 3 in the 14-44 5' and 3' ends) the sequences diverge. We propose that this divergence marks the 5' junction of cellular and viral sequences.

Similarly, we compared the sequence at the 3' end of viral DNA, as defined above, to the homologous sequence at the 5' end of viral RNA (or DNA) (Fig. 5). From 3' (reading forwards) to ACA (nucleotides 567–569 in the 14-44 5' and 3' ends) the sequences diverge. We propose that this divergence marks the 3' junction of viral and cellular sequences. It is 2 base pairs 5' to the primer binding site and the start of viral DNA synthesis[54].

Resemblances to transposable elements

The sequence of AAAAT in cellular DNA is repeated at both proposed junctions of the viral sequences (Fig. 5). A 5-base pair repeat of cellular DNA has been found at the end of several transposable elements, IS2, Tn3, phage Mu, a 200-base pair sequence in pSC101, and δγ (refs 34–39). Although our data do not permit the unambiguous conclusion that the cell DNA is repeated as a result of the integration process, we suggest that it is. (Further clones containing proviruses are being sequenced to check this hypothesis.) Furthermore, the 3-base pair sequence of viral DNA next to the 5-base pair repeat of cellular DNA is an inverted repeat, TGT and ACA. However, the sequence of the provirus, especially the location of the putative primer binding site, leads us to propose that the structure of the precursor to integration is that shown in Fig. 6, that is a 5-base pair inverted repeat at the ends of linear unintegrated viral DNA with integration occurring 2 bases from the terminus. (This hypothesis is being checked by sequencing cloned unintegrated linear viral DNA.) The putative 'insertion sequence' ACATT probably has to be terminal, as the sequence ACATT is also found in the terminal repeat starting at nucleotide 409. Note that the analogous sequence, next to the primer binding site, in avian sarcoma and murine leukaemia virus RNAs is TCATT[26,29,33].

of unintegrated SNV DNA (J. J. O'Rear, personal communication), and considerably larger than the 300-base pair repeat reported for avian sarcoma virus DNA[15,16]. There are no ATG codons in the only reading frame open for a large distance, nucleotides 344 to 664 of the 14-44 5' end. Therefore, apparently no polypeptide is coded by the terminal repeat.

We have also sequenced portions of the ends of SNV RNA (manuscript in preparation). The RNA sequence starting approximately 10 bases from the 5' cap of viral RNA is the same as the proviral DNA sequence from nucleotides 405 to 425 of the 14-44 5' end (Fig. 4). The RNA sequence for 150 bases from the 3'-polyadenylated end is the same as the proviral DNA sequence from nucleotides 470 to 320 of the 14-44 3' end (Fig. 4).

Therefore, viral RNA begins at about nucleotide 394 of the

14-44 5' End

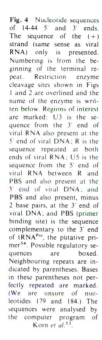

Fig. 4 Nucleotide sequences of 14-44 5' and 3' ends. The sequence of the (+) strand (same sense as viral RNA) only is presented. Numbering is from the beginning of the terminal repeat. Restriction enzyme cleavage sites shown in Figs 1 and 2 are overlined and the name of the enzyme is written below. Regions of interest are marked: U3 is the sequence from the 3' end of viral RNA also present at the 5' end of viral DNA; R is the sequence repeated at both ends of viral RNA; U5 is the sequence from the 5' end of viral RNA between R and PBS and also present at the 3' end of viral DNA; and PBS and also present, minus 2 base pairs, at the 3' end of viral DNA; and PBS (primer binding site) is the sequence complementary to the 3' end of tRNA^Pro, the putative primer[54]. Possible regulatory sequences are boxed. Neighbouring repeats are indicated by parentheses. Bases in these parentheses not perfectly repeated are marked. (We are unsure of nucleotides 179 and 184.) The sequences were analysed by the computer program of Korn et al.[53].

Chicken cell
AAGCA AAATAAAAGA TACAAAAAAT

U3-SNV
TGTGGGAGGG AGCTCTGGGG GAAATAGCGC TGGCTCGCAA CTGCTATATT AGCTTCTGTA CTCATGTCTG CTTGCCTGGC CACTAACCGC
Sac I   Hha I   Hae III

CATATTAGCT TCTGTACACA TGCTTGCTTG CCGTAGCCGC CATTGTACTT GATATGCCAT TTCTCGGAAT CGGCATCAAG TTTCGCTTCT

CGAGAGCAAG CCCACAAACC ACAAAAGGAA ACGCGCACCG AAGGCAAGCA TCAGACCACT TGCGCCATCC AATCATGAAC GGACACGAGA
TaqI   HhaI   HhaI

TCGGACTATC ATACTGGAGC CAATGGTTGT AAAGGGCAGA TGCTACTCTC CAATGAGGGA AAATGTCATG TAACACCTG TAAGCTGTAA

GCGGCTATAT AAGCCGGGTA CATCTCTTGC TCGGGGTCGC CGTCCTGCAC ATTGTTGTTG TGACGTGCGG CCCAGATTCG AATCTGTAAT
         HpaII   AvaI   HaeIII   TaqI

AAAACTTTTT CTTCTGAATC CTCAGATTGG CAGTGAGAGG AGATTTTGTT CGTGGTGTTG GCTGGCCTAC TGGGTGGGCG CAGGGATCCG
R-U5   HaeIII   HhaI   BamHI Hpa

GACTGAATCC GTAGTACTTC GGTACAACAT TTGGGGGCTC GTCCGGGGTA CCCTCCCCAT CGGCAGAGGT GCCAACTGCT TCTTCGAACT)
II   U5|PBS   PBS   HpaII   TaqI

(TTCTTCGAAC TCCCGGCGCG GTGAGTTAAG TACTTGATTT TGGTACCTCG CGAGGGTTTG GGAGGATCGG AGTGGTGGCG GGACGCTGCC
    TaqI   HpaII HhaI HpaII   KpnI   Hpa

GGGAAGCTCC ACCTCCGCTC AGCAGGGGAC GCCCTGGCCT AGCTCTGTG GTATCTGATT GTTGTTGAGC CGTCCCTAAG ACGGTGATAC
II   HaeIII SacI

TAAGTCGTGG (CTTGTGTGTT TGTTTGT)TGC (CTTGTGTTTG TTCGT)CGTTT GTCGAC
                                                        SalI

14-44 3' End

G AAGGCCCTTG ACTCTGCATT ATCCATTGAC AAAATGCAGG CAGTAAAAAT CCTAGCACTA GTCCCACAAT

ACAAGCCACT CCCACAAATAC AAGCCACTCC CAACAGAGAT GGATACCTTA GGTCAATGAT TTGACCAGAA TGTACAAGAG CAGTGGGGAA
U3   AvrII

TGTGGGAGGG AGCTCTGGGG GAAATAGCGC TGGCTCGCAA CTGCTATATT AGCTTCTGTA CTCATGTCTTG CTTGCCTGGC CACTAACCGC
         SacI   HhaI   HaeIII

CATATTAGCT TCTGTACACA TGCTTGCTTG CCGTAGCCGC CATTGTACTT GATATGCCAT TTCTCGGAAT CGGCATCAAG TTTCGCTTCT

CGAGAGCAAG CCCACAAACC ACAAAAGGAA ACGCGCACCG AAGGCAAGCA TCAGACCACT TGCGCCATCC AATCATGAAC GGACACGAGA
TaqI   HhaI   HhaI

TCGGACTATC ATACTGGAGC CAATGGTTGT AAAGGGCAGA TGCTACTCTC CAATGAGGGA AAATGTCATG TAACACCTG TAAGCTGTAA

GCGGCTATAT AAGCCGGGTA CATCTCTTGC TCGGGGTCGC CGTCCTGCAC ATTGTTGTTG TGACGTGCGG CCCAGATTCG AATCTGTAAT
         HpaII   AvaI   HaeIII   TaqI
R-U5
AAAACTTTTT CTTCTGAATC CTCAGATTGG CAGTGAGAGG AGATTTTGTT CGTGGTGTTG GCTGGCCTAC TGGGTGGGCG CAGGGATCCG
U5-SNV|CHICKEN CELL   HaeIII   HhaI   BamHI Hpa

GACTGAATCC GTAGTACTTC GGTACAACAA AAATCCGATT TACCCAGGCA ATTCTCCAAA AAAGTTTAAA AAAGAGGGGG TGGGAAGTGG
II

ATTTTTATGAA GGTACTATGG AGGCTGGCGA AAGCAAGGAA TGAATGCCTA GAAGAAAAGA AAGAATTC
                                                                    EcoRI

Fig. 5 Portions of sequencing gels with the ends of the terminal repeats of viral DNA. Portions of the sequencing gels used to secure the sequences shown in Figs 4 and 6 are presented. NT, nucleotides. 14-44 3' end NT −11 to 25 is from a gel labelled at the AvrII cleavage site; 14-44 5' end NT 4 to −17 is from a gel labelled at the SacI cleavage site; and 14-44 5' end NT 564 to 584 and 14-44 3' end 548 to 574 are from gels labelled at the BamHI cleavage site. Arrows mark the last nucleotide in common in the analogous sequences (gels placed next to each other). (14-44 5' end NT 4 to −17 is read in the opposite direction (3' to 5') to the other gels (5' to 3').) (−) Represents strand which is complementary to sense of viral RNA (+).

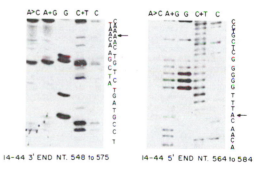

14-44 5' END NT. 4 to −17   (−) (+)

14-44 3' END NT. −11 to 25

14-44 3' END NT. 548 to 575

14-44 5' END NT. 564 to 584

Fig. 6 Proposed sequences of viral and cell precursors to provirus. The viral genome is that of a linear un-integrated DNA with terminal repeats at both ends starting at the end of the primer binding site and the analogous nucleotide on the 5′ (U3) end. Integration occurs two bases from each end in the centre of a 5-base pair inverted terminal repeat. The cell DNA is nucleotides −24 to 0 of 14-44 5′ end (Fig. 4) and nucleotides 570 to 698 of 14-44 3′ end (Fig. 4) with an overlap of five bases. Nucleotide 1 corresponds to nucleotide −24 of 14-44 5′ end; and nucleotide 30 corresponds to nucleotide 579 of 14-44 3′ end. Arrows in the cell DNA represent possible cleavage sites of a putative nuclease. Alternatively, a 5′ overhang could be formed.

```
                              8.3 kbp
        3' T T A C A ········· T G T A A 5'
Virus
        5' A A T G T ········· A C A T T 3'

                              ↓
     3' ···T T T C T A T G T T T T T T A G G C T A A A T G G·····5'
Cell ···A A A G A T A C A A A A A A A T C C G A T T T A C C····3'
     5'                                                      3'
                              ↑

        10        20         30        40         50        60
   AAGCAAAATA AAAGATACAA AAAATCCGAT TTACCCAGGC AATTCTCCAA AAAAGTTTAA

        70        80         90       100        110       120
   AAAAGAGGGG GTGGGAAGTG GATTTTATGA AGGTACTATG GAGGCTGGCG AAAGCAAGGA

       130       140
   ATGAATGCCT AGAAGAAAAG AAAGAATTC
```

Transposable elements also end with inverted repeats, although usually they are over 20 base pairs long[7-12, 35-42].

The integration of viral DNA may require only one or two specific nucleases and a ligase. The presence of DNase and DNA ligase activities in purified virions of avian retroviruses has been reported[43-45]. These findings support the hypothesis that retrovirus virions contain all the proteins necessary for provirus synthesis[44].

The sequences in the provirus next to the 5′ end and near the 3′ end are G+C rich. This base composition may be significant.

The sequence of cell DNA around the site of insertion may be reconstructed (Fig. 6). It is very A+T rich (73% A+T; chicken DNA is 58% A+T) and has no stop codons in one reading frame. Further sequencing of proviral clones will be required to determine if these characteristics or the exact sequence at the integration site have any specificity or if they are correlated with proviral infectivity[14,46]. The sites of insertion of Tn3 seem to be A+T rich[47].

The resemblance of the provirus to transposable elements indicates that retroviruses may have evolved from such elements. Evolution might possibly have involved the appearance of start and stop sequences for RNA synthesis in an insertion element[48]; appearance of a sequence that could bind a tRNA in the insertion element proximal to the site of the start sequence for RNA; the presence of a DNA polymerase in cells that could use this tRNA as a primer and the RNA as a template; formation of a transposon by integration of two of these insertion sequences around the coding sequences for this polymerase; and addition of nucleotide sequences for the rest of the present virion proteins. As the transposon transposed through an RNA intermediate, a long inverted repeat would not be needed to mark its end, as the end would also be the end of a molecule. Therefore, the size of the inverted terminal repeat would have decreased.

Since submission of this article, the same pattern of a 5-base pair repeat of cellular DNA, different in every case, next to the 3-base pair inverted repeat of viral DNA, the same in every case, has been found in five other clones of SNV proviruses.

This work was supported by grants (CA-07173 and CA-22443) from the NCI. H.M.T. is an American Cancer Society Research Professor. We thank I. Chen, C. Gross, J. Mertz, J. O'Rear and B. Sugden for comments.

Received 19 February; accepted 16 April 1980.

1. Temin, H. M. Science 192, 1075–1080 (1976).
2. Temin, H. M. J. natn. Cancer Inst. 46, III–VII (1971).
3. O'Rear, J. J., Mizutani, S., Hoffman, G., Fiandt, M. & Temin, H. M. Cell 20, 423–430 (1980).
4. Hunter, E., Bhown, A. S. & Bennet, J. C. Proc. natn. Acad. Sci. U.S.A. 75, 2708–2712 (1978).
5. Barbacid, M., Hunter, E. & Aaronson, S. A. J. Virol. 30, 508–514 (1979).
6. Bauer, G. & Temin, H. M. J. Virol. 34, 168–177 (1980).
7. Bukhari, A-I. A. Rev. Genet. 10, 389–412 (1976).
8. Cohen, S. Nature 263, 731–738 (1976).
9. Starlinger, P. & Saedler, H. Curr. Topics Microbiol. Immun. 75, 111–152 (1976).
10. Nevers, P. & Saedler, H. Nature 206, 109–115 (1977).
11. Kleckner, N. Cell 11, 11–23 (1977).
12. Bukhari, A. I., Shapiro, J. A. & Adhya, S. (eds) DNA Insertion Elements, Plasmids and Episomes (Cold Spring Harbor Laboratory, New York, 1977).
13. Smith, H. O. & Birnstiel, M. L. Nucleic Acids Res. 3, 2387–2398 (1976).
14. Keshet, E., O'Rear, J. J. & Temin, H. M. Cell 16, 51–61 (1979).
15. Shank, P. R. et al. J. Virol. 15, 1383–1395 (1978).
16. Sabran, J. L. et al. J. Virol. 29, 170–178 (1979).
17. Konkel, D. A., Tilghman, S. M. & Leder, P. Cell 15, 1124–1132 (1978).
18. Goldberg, M. thesis, Stanford Univ. (1979).
19. Ziff, E. B. & Evans, R. M. Cell 15, 1463–1475 (1978).
20. Nishioka, Y. & Leder, P. Cell 18, 875–882 (1979).
21. Proudfoot, N. J. J. molec. Biol. 197, 491–525 (1976).
22. Hamlyn, P. H., Brownlee, G. G., Cheng, C. C., Gait, M. J. & Milstein, C. Cell 15, 1067–1075 (1978).
23. Seeburg, P. H., Shine, J., Martial, J. A., Baxter, J. D. & Goodman, H. Nature 270, 486–494 (1977).
24. Tucker, P. W., Marcu, K. B., Slightom, J. L. & Blattner, F. R. Science 206, 1299–1303 (1979).
25. Coffin, J. M. J. gen. Virol. 42, 1–26 (1979).
26. Haseltine, W. A., Maxam, A. M. & Gilbert, W. Proc. natn. Acad. Sci. U.S.A. 74, 989–993 (1977).
27. Schwartz, D. W., Zamecnik, P. C. & Weith, H. L. Proc. natn. Acad. Sci. U.S.A. 74, 994–998 (1977).
28. Stoll, E., Billeter, M. A., Palmenberg, A. & Weissmann, C. Cell 12, 57–72 (1977).
29. Shine, J., Czeinilofsky, A. P., Friederick, R., Bishop, J. M. & Goodman, H. M. Proc. natn. Acad. Sci. U.S.A. 74, 1473–1477 (1977).
30. Coffin, J. M. & Haseltine, W. A. Proc. natn. Acad. Sci. U.S.A. 74, 1908–1912 (1977).
31. Coffin, J. M., Hageman, T. C., Maxam, A. M. & Haseltine, W. A. Cell 13, 761–773 (1978).
32. Harada, F., Peters, G. & Dahlberg, J. G. J. biol. Chem. 254, 10979–10985 (1979).
33. Haseltine, W. A., Kleid, D. G., Panet, A., Rothenberg, E. & Baltimore, D. J. molec. Biol. 106, 109–131 (1976).
34. Rosenberg, M., Court, D., Shimatake, H., Brady, C. & Wulff, D. L. Nature 272, 414–423 (1978).
35. Ghosal, D., Sommer, H. & Saedler, H. Nucleic Acids. Res. 6, 1111–1122 (1979).
36. Ohtsubo, H., Ohmori, H. & Ohtsubo, E. Cold Spring Harb. Symp. quant. Biol. 43, 1269–1277 (1978).
37. Allet, B. Cell 16, 123–129 (1979).
38. Ravetich, J. V. et al. Proc. natn. Acad. Sci. U.S.A. 76, 2195–2198 (1979).
39. Reed, R. R., Young, R. A., Steitz, J. A., Grindley, N. D. F. & Guyer, M. S. Proc. natn. Acad. Sci. U.S.A. 76, 4882–4885 (1979).
40. Grindley, N. D. F. Cell 13, 419–426 (1978).
41. Calos, M., Johnsnud, L. & Miller, J. H. Cell 13, 411–418 (1978).
42. Ohtsubo, A. & Ohtsubo, E. in DNA Insertion Elements, Plasmids and Episomes (eds Bukhari, A. I., Shapiro, J. A. & Adhya, S.) 591–593 (Cold Spring Harbor Laboratory, New York, 1977).
43. Mizutani, S., Boettiger, D. & Temin, H. M. Nature 228, 424–427 (1970).
44. Mizutani, S., Temin, H. M., Kodama, M. & Wells, R. D. Nature new Biol. 230, 232–235 (1971).
45. Golomb, M. & Grandgenett, D. P. J. biol. Chem. 254, 1606–1613 (1979).
46. Keshet, E. & Temin, H. M. Proc. natn. Acad. Sci. U.S.A. 75, 3372–3376 (1978).
47. Tu, C.-P. D. & Cohen, S. N. Cell 19, 151–160 (1980).
48. Benton, W. D. & Davis, R. W. Science 196, 180–182 (1977).
50. Clewell, D. B. & Helinski, D. R. Proc. natn. Acad. Sci. U.S.A. 62, 1159–1166 (1969).
51. Maxam, A. M. & Gilbert, W. Proc. natn. Acad. Sci. U.S.A. 74, 560–565 (1977).
52. Sanger, F. & Coulson, A. R. FEBS Lett. 87, 197–110 (1978).
53. Korn, L. J., Queen, C. L. & Wegman, M. N. Proc. natn. Acad. Sci. U.S.A. 74, 4401–4405 (1977).
54. Peters, G. G. & Glover, C. J. Virol. 33, 708–716 (1980).

The DNA Provirus: Howard Temin's Scientific Legacy
Edited by G. M. Cooper, R. Greenberg Temin, and B. Sugden
© 1995 American Society for Microbiology, Washington, DC 20005

Chapter 12

Beyond the Provirus: from Howard Temin's Insights on Rous Sarcoma Virus to the Study of Epstein-Barr Virus, the Prototypic Human Tumor Virus

Bill Sugden

Howard Temin contributed to the science I shall describe in two ways: he was a master builder of much of the foundation of retrovirology in particular and tumor virology in general, and he was a close colleague for 19 years. In the summer of 1970, Howard Temin and Tom Benjamin taught the animal virology course at Cold Spring Harbor, where I was doing my graduate work. At the end of the 6-week course, Howard eyed the 15 or so students in it and said, "If you have understood what we taught you about RNA tumor viruses, then we have doubled the number of people in the world who do." This estimate may have been a bit pessimistic. However, it occurred at an instant in time before which retrovirology was a somewhat arcane and dubious pursuit and after which it grew exponentially and, with the discovery of the human immunodeficiency virus, became the central focus of virology. Howard Temin contributed fundamentally to the beginning, the takeoff, and the exponential expansion of retrovirology, as is evidenced by his own work and the tributes to him in this volume. He contributed to my work with Epstein-Barr virus (EBV), a human tumor virus, by providing the sustained interest and critical appreciation of a demanding scientific colleague and the nurturing of a friend.

Here I shall relate briefly the development of our understanding of carcinogenesis mediated by avian retroviruses, to which Howard Temin contributed so much. This understanding, coupled with the associated study of retroviruses in general, has been essential to dealing effectively with human disease. For example, it prepared virologists to isolate, identify, and study the human immunodeficiency virus. It has also been essential for the identification of proto-oncogenes as mutational targets of chemical carcinogens active in people.

I shall also outline our current appreciation of human tumor viruses, using EBV as a model. Human tumor viruses are less efficient pathogens than most of

Bill Sugden • McArdle Laboratory for Cancer Research, University of Wisconsin, 1400 University Avenue, Madison, Wisconsin 53706.

the well-studied avian oncogenic retroviruses. The roles of the human host and the host's environment therefore dramatically influence the outcome of infection with human tumor viruses. These influences have made it difficult to molecularly define the contributions of viruses to human cancer.

RSV

Howard Temin refined a transformation assay for Rous sarcoma virus (RSV) (117) such that it became a standard method for detecting and measuring the transforming abilities of different viruses in cell culture. A dose-response curve for transformation of chick embryo fibroblasts that was developed by using this assay indicated that a single particle of virus is sufficient to infect a cell and to change its morphology (117). This finding reflects the fact that RSV is both replication competent and transforming. The early characterization of RSV facilitated the subsequent recognition that most rapidly transforming retroviruses are replication incompetent and require helper viruses to mediate their propagation.

RSV infects susceptible strains of chickens efficiently and can induce sarcomas in infected animals rapidly (88). The infected animals can become viremic (88), and this viremia is presumably facilitated by the replication competence of the virus. Tumors develop in 3 to 5 days after a sufficient inoculum is injected into the wing webs of newborn chicks (88). This rapid onset of disease limits the development of an effective humoral or cellular immune response to the virus or to virally infected cells.

Experimental conditions can be manipulated, however, to detect an immune response to RSV. In natural infections, infection with a related avian leukosis virus (ALV) can provide immunity to structural proteins of RSV (88). Laboratory inoculations of chickens with derivatives of a sarcoma virus that express protein fusions of short stretches of the amino terminus of *env* with $p60^{src}$ yield sarcomas that regress in a few weeks (34). Inoculation of chickens with these viruses establishes immunity such that a subsequent challenge with RSV leads to no tumors or to tumors that regress (34). This protection is dependent on active recognition of the immunizing *src* protein (35). Despite the ability of chickens to mount a protective response to the *src* protein, inoculation of susceptible chicks with high titers of RSV can yield sarcomas and rapid death of the host (88). These observations underscore the notion that the pathogenicity of RSV is in part derived from the rapidity with which it transforms cells to proliferate in vivo. Many well-studied animal tumor viruses share with RSV the ability to induce tumors rapidly. This common feature of rapid tumor induction has inhibited some investigators from considering human viruses as potentially oncogenic, because human tumors do not arise soon after viral infections.

RSV induces tumors efficiently not only because it is replication competent and able to outstrip the immune response but also because it introduces into cells the dominantly acting viral oncogene v-*src*. v-*src* has apparently evolved from the cellular proto-oncogene c-*src* via a mechanism outlined by Howard Temin in 1974 (116). The structural differences between the proteins encoded by the v-*src* and c-*src* genes have profound effects on their enzymatic activities and on the regulation of those activities. The *src* proteins constitute a family of nonreceptor tyrosine kinases that localize to the plasma membrane (95) and are regulated by their own state of phosphorylation at specific tyrosine residues (48). The *csk* tyrosine kinase mediates phosphorylation of the tyrosine residue at position 527 of c-*src*, which is instrumental in negatively regulating kinase activity (46, 82, 86). v-*src* is truncated at its carboxy terminus and lacks this regulatory residue (17). It is this lack that contributes to the increased tyrosine kinase activity of v-*src* and its transforming ability relative to that of c-*src* (15, 58, 90). The function of c-*src* in normal cells is yet to be elucidated; however, it is clear that efficient expression of its hyperactive cousin v-*src* by RSV in certain cell types is rate limiting in the rapid evolution of those infected cells into tumors.

The intellectual and experimental foundations for the formation of retroviral oncogenes that Howard Temin established culminated in the identification of v-*src* by Michael Bishop, Harold Varmus, and their colleagues as a relative of normal genes in the uninfected cell (110). Their findings led to great excitement and a widespread search for viral oncogenes transduced by rapidly transforming retroviruses and for the cellular proto-oncogenes that gave rise to these viral relatives. Approximately 100 such genes have been identified, and this understanding in the aggregate has led to the expectation that human tumor viruses might transduce dominantly acting oncogenes similar to those carried by rapidly transforming retroviruses.

ALVs

ALVs provide one example of weakly transforming viruses. These viruses formerly often infected commercial flocks of chickens, in which they were propagated either vertically or horizontally (88). Infection of susceptible chickens with ALV can lead to bursal lymphomas that develop and are fatal by 24 to 40 weeks after infection (88). ALVs do not contain a dominantly acting transforming gene and do not morphologically affect cells they infect in cell culture. The mechanism by which they contribute to lymphomagenesis remained enigmatic until Hayward et al. (43) demonstrated that the ALV provirus is inserted in the vicinity of the c-*myc* proto-oncogene and affects the expression of this gene in bursal lymphomas. This retroviral "promoter insertion" can be either 3' or 5' to the c-*myc* locus but consistently alters expression of the proto-oncogene (87).

The events that lead to this form of ALV-associated lymphoma are not thoroughly understood. However, they must be multiple, because transformed follicles arise in the bursa months before the lymphoma appears (6). It is also clear that efficient viral replication and spread are essential to the development of the disease, because they must provide sufficient independent infections so that the provirus is inserted near the c-*myc* locus in one cell destined to evolve into a tumor.

Tumors associated with weakly transforming retroviruses arise in other species (see chapter 13 in this volume). The study of weakly transforming retroviruses has allowed the identification of additional proto-oncogenes not associated with known viral oncogenes. Their study has also provided another model by which human viruses might be expected to contribute to tumorigenesis: viremia leading to promoter insertion, which affects expression of a proto-oncogene.

HUMAN TUMOR VIRUSES

Four human tumor viruses—EBV (a herpesvirus), hepatitis B virus (HBV), human papilloma virus types 16 (HPV-16), -18, -31, and -33, and human T-cell leukemia virus type 1 (HTLV-1; a retrovirus)—have been studied sufficiently to be considered here. A fifth, hepatitis C virus, has been identified, but virologic studies of it are only now beginning. These tumor viruses are associated with perhaps 10 to 20% of all human cancers. The vast majority of the remaining human cancers arise from chemical carcinogenesis, in which proto-oncogenes are often targets for carcinogen-induced somatic mutation. These four human tumor viruses do not contain dominantly acting transforming genes related to proto-oncogenes. They are not known in general to affect expression of proto-oncogenes by nearby insertion of their viral genomes. They do, however, affect the proliferation of the infected cell, as do many rapidly transforming retroviruses. Finally, the outcome of infection with some human tumor viruses can be profoundly affected by the immune response of the host. The contrasts between tumor induction by RSV, ALVs, and EBV are outlined in Fig. 1.

EBV, INFECTIOUS MONONUCLEOSIS, AND BURKITT'S LYMPHOMA

One way of briefly delineating EBV's contributions to different human diseases is to compare and contrast our reconstructions of the etiologies of those diseases. These reconstructions can then serve as the basis for developing a more

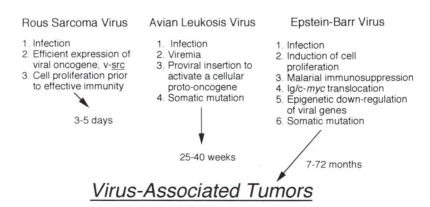

FIGURE 1. Synopsis of steps proposed to occur between infection with these different model tumor viruses and development of their associated tumors. The order in which the steps occur is consistent with current understanding of the etiologies of these virus-associated tumors but is hypothetical. Ig, immunoglobulin.

detailed understanding of individual contributions made by the virus. We now know little about EBV-associated nasopharyngeal carcinoma, but we know much about the origins of benign and malignant diseases associated with EBV infections of B lymphocytes. A description of these origins can highlight some of the conditions under which EBV is thought to be tumorigenic in its human host.

The heterophile-positive form of infectious mononucleosis is a self-limiting lymphoproliferative disease caused by EBV (24). This disease becomes demonstrable 3 to 4 weeks after primary infection and consists of EBV-infected B lymphoblasts that proliferate after infection with EBV (37, 96) and activated T cells that can specifically kill syngeneic EBV-positive B-cell blasts (115). Over time, the number of detectable EBV-positive B lymphocytes in the peripheral blood drops from a high of 1 to 10% of total immunoglobulin-positive cells to 0.01% or less (97, 131). These few EBV-positive cells are maintained throughout life and presumably provide the immunizing antigens that elicit immunity to subsequent EBV-associated disease in immunologically competent hosts.

From these and additional observations, we can reconstruct a brief likely scenario for EBV's contributions to infectious mononucleosis. EBV is transmitted orally; it wends its way to some source of B lymphocytes (perhaps the tonsils), infects B-lymphocytes, and induces them to proliferate. Studies in cell culture have demonstrated that EBV both infects primary B lymphocytes and induces them to proliferate extraordinarily efficiently (44, 113). The proliferating cells are latently infected: they do not support productive infection by the virus but instead maintain its DNA as a plasmid replicon and support expression of only a small subset of viral genes, termed "latent genes" (112). Many of the products of the latent genes are recognized by the humoral and cellular arms of the host immune response and elicit a robust T-cell-mediated cytoxic response (14, 53). This immunologic response quells the disease such that only a few EBV-positive cells remain in the peripheral blood, and the patient is healthy and immune to further EBV-associated disease. The state of viral gene expression within the surviving EBV-positive cells and their proliferative capacity in vivo are uncertain (65, 91, 118), but this small reservoir of virally infected cells persists for life and if transplanted to an immunologically compromised host can proliferate fatally (80, 105). The kernel, then, of EBV's contribution to infectious mononucleosis is its induction and maintenance of infected-B-lymphocyte proliferation. The disease-quelling response of the host is to specifically kill most virally infected cells.

This description of EBV and infectious mononucleosis is simple relative to that for EBV's relationship to Burkitt's lymphoma, but the contribution of the virus to these disparate diseases appears to be similar. Burkitt's lymphoma arises in young children 7 to 72 months after primary infection with EBV (21, 36). Children that have a particularly vigorous humoral response to this virus's capsid antigens have an increased risk of developing this lymphoma (21, 36). This risk may reflect a larger-than-normal viral load and a correspondingly increased number of EBV-infected proliferating B lymphoblasts. The tumor is endemic only in those regions of the world in which malaria is also endemic, and malaria is probably a cofactor in the risk for developing the tumor (13, 79). Malaria is known to suppress a host T-cell-mediated response to syngeneic EBV-positive B lymphoblasts (25,

128). Direct measurements in children from The Gambia, where Burkitt's lymphoma is endemic, indicate that the number of B cells infected with or releasing EBV is fivefold higher in the peripheral blood of youngsters acutely infected with malaria than in the same children during convalescence (60). These observations are consistent with a reconstruction in which a virulent infection of children with EBV and an immune suppression mediated via malaria combine to yield a large pool of EBV-infected proliferating B lymphoblasts. The vast majority of children with such an extreme infection do not, however, develop Burkitt's lymphoma. Several rare genetic events must occur for lymphomagenesis. It seems likely that the accumulation of these mutations in one cell and its progeny is favored in the replicating pool of EBV-infected B lymphoblasts.

These genetic events include chromosomal translocations involving the immunoglobulin and c-*myc* loci that lead to preferential expression of the translocated allele of the c-*myc* proto-oncogene (108). The translocation is found in more than 90% of EBV-positive Burkitt's lymphomas. Most Burkitt's lymphoma biopsy samples also include mutations in the tumor suppressor gene, *p53* (28, 33). These mutations have been found in lymphoma biopsy samples and in cell lines derived from the biopsy samples; in the former cases, the mutations are heterozygous (23). The contributions of these mutations to lymphomagenesis have not been elucidated, but they may permit an epigenetic change that allows the evolving tumor to escape any residual immune surveillance of the host.

EBV-positive B lymphoblasts, established either by infection in vitro or by explanting of infected cells from normal human hosts, express nine latent viral proteins, most of which are efficiently recognized by cytotoxic T cells. One of these latent proteins, EBNA-1, is not recognized by T cells and is required to maintain EBV DNA as a plasmid in proliferating cells (134). EBNA-1 is the only viral gene product detected in freshly explanted Burkitt's lymphoma tumor cells (100, 101). It seems likely that the epigenetic event that yields the shutdown of most of EBV's expression also provides the infected cell with a selective advantage in the presence of any residual immune response of the patient. Some of the shut-down viral genes formerly expressed products essential for proliferation of the EBV-positive cell (39, 49, 119). It is reasonable to hypothesize that the immunoglobulin–c-*myc* translocation, *p53* mutations, and perhaps other yet to be identified genetic changes in the EBV-infected cells support proliferation of those cells in the absence of formerly critical viral gene expression.

The brief reconstruction of the relationship between EBV and infectious mononucleosis indicates that the host immune response limits this disease. Similarly, the brief reconstruction of the relationship between EBV and Burkitt's lymphoma highlights the host immune response as the crucial determinant for whether or not an EBV-infected cell evolves into a lymphoma. The similarities and differences between these models for the development of these EBV-associated diseases are outlined in Fig. 2. For a full understanding of the etiology of EBV-associated Burkitt's lymphoma and how it differs from that of infectious mononucleosis, at least three uncertainties need to be resolved. (i) Why is Burkitt's lymphoma a disease of children? (ii) What are the contributions of the immune response to the development of Burkitt's lymphoma? (iii) What are

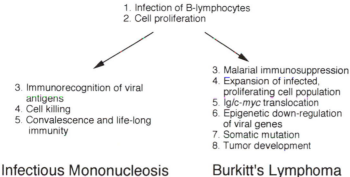

1. Infection of B-lymphocytes
2. Cell proliferation

3. Immunorecognition of viral antigens
4. Cell killing
5. Convalescence and life-long immunity

3. Malarial immunosuppression
4. Expansion of infected, proliferating cell population
5. Ig/c-*myc* translocation
6. Epigenetic down-regulation of viral genes
7. Somatic mutation
8. Tumor development

Infectious Mononucleosis   **Burkitt's Lymphoma**

FIGURE 2. Synopsis of the shared and different steps proposed to occur between infection with EBV and the development of a benign lymphoproliferative disease (infectious mononucleosis) or a malignant one (Burkitt's lymphoma). Ig, immunoglobulin.

the contributions of somatic mutations to the development of Burkitt's lymphoma? None of these questions has been answered fully, but we do have information that bears on them.

WHY IS BURKITT'S LYMPHOMA A DISEASE OF CHILDREN?

Two general observations provide background for the question of why Burkitt's lymphoma affects only children. First, most human cancers occur in adults, and the incidence of human cancers considered in the aggregate rises exponentially with the age of the population considered. This age dependence for the development of cancers is explained by a requirement for the accumulation of multiple genetic lesions in one cell and its progeny in order for the cell to evolve into a monoclonal tumor. Burkitt's lymphomas are monoclonal (29) and do contain multiple genetic lesions, but they arise rapidly in children (21, 36).

Second, EBV can induce lymphoproliferative disease in 3 to 4 weeks after primary infection in hosts of all ages. Infectious mononucleosis, a disease characterized by proliferation of infected lymphoid cells, usually occurs in adolescents upon primary infection with EBV (24); however, primary infection of adults also can yield infectious mononucleosis (11). Burkitt's lymphoma, on the other hand, does not occur upon primary infection with EBV but requires 7 to 72 months after infection to develop. This latent period, although dramatically brief for most human cancers, is long relative to that for the development of infectious mononucleosis. The length of the latent period for Burkitt's lymphoma indicates that the host immune response to primary infection by EBV must limit that infection. If EBV-associated lymphoproliferation characteristic of infectious mononucleosis were to expand unrestrictedly in children destined to develop Burkitt's lymphoma, they would die of polyclonal B-cell proliferations within a few months after primary infection (41). That they do not develop a lymphoma upon primary infection indicates that an infected cell takes many generations to evolve into a Burkitt's lymphoma; this evolution, however, can occur only in children. That Burkitt's lymphoma is limited to children is all the more striking when considered against the background that EBV does cause polyclonal B-cell lymphomas in immunosuppressed patients of all ages (19, 41).

Two simple explanations for Burkitt's lymphoma being a disease of children have probably been eliminated. This tumor has not been noted to be familial, as is bilateral retinoblastoma or Wilms' tumor; that is, there is no evidence for a highly penetrant germ line mutation in children that predisposes them to develop the tumor. Nor is there a strain dependence of EBV associated with Burkitt's lymphoma. Isolates of EBV from Burkitt's lymphoma biopsy samples are structurally and biologically equivalent to those isolated from patients with infectious mononucleosis (75, 111).

There are at least two global explanations that could explain the age dependence of Burkitt's lymphoma. Elaborating these explanations will require more information, and, if still considered plausible, they will need to be tested. One of these explanations is intrinsic to the host. It is possible that lymphoid development in children destined to develop Burkitt's lymphoma differs from that in adults. These children may harbor large numbers of target cells for Burkitt's lymphoma that are rare in adults, or their immune responses may differ critically from those of adults. There are no compelling data for either of these notions about contributory factors intrinsic to the host, but too little is known to exclude them.

The second explanation invokes factors extrinsic to the host. It is already known that both infection with EBV and infection with malaria are risk factors for the development of Burkitt's lymphoma. It is possible that a third environmental insult, be it a common viral or parasitic infection or a common dietary contaminant, affects the evolution of an EBV-infected cell into a Burkitt's lymphoma by supporting its survival and proliferation and/or by inhibiting the immune response to that infected cell. Such environmental insults have so far avoided detection. As long as the environmental insults are common, as, for example, aflatoxin B1 is, and act at the confluence of extreme infections with EBV and/or malaria, they will be difficult to identify. It is clear that the explanations for both intrinsic and extrinsic factors that favor Burkitt's lymphoma in children are ad hoc and insubstantial. We need a much deeper understanding of infections in children in Central Africa and New Guinea, where EBV-associated Burkitt's lymphoma is endemic, to appreciate why this disease develops in them.

WHAT ARE THE CONTRIBUTIONS OF THE IMMUNE RESPONSE TO THE DEVELOPMENT OF BURKITT'S LYMPHOMA?

A reflexive answer to the question concerning immune responses to Burkitt's lymphoma might be, "The immune response prevents the disease." However, a battery of findings indicates that Burkitt's lymphomas develop in children who retain at least portions of their immune response and that their immune responses may in fact shape their disease. First, infection of immunodeficient children and adults with EBV can yield virally induced B-cell lymphomas distinct from Burkitt's lymphomas (19). These tumors can be polyclonal, can lack the immunoglobulin–c-*myc* translocations characteristic of Burkitt's lymphoma, and do express a variety of viral latent genes (19, 41). Some immunodeficient hosts therefore support development of polyclonal EBV-positive lymphomas.

Second, at least three observations indicate that children who are destined to develop or who have developed Burkitt's lymphoma have vigorous humoral immune responses and may have effective cellular immune responses, too. The prospective seroepidemiologic survey that identified EBV as a risk factor for developing Burkitt's lymphoma found that children who would later develop the disease had significantly higher titers of antibodies for viral capsid antigens than did those who would not (21, 36). Burkitt's lymphomas are tumors strikingly susceptible to chemotherapeutic treatment. This sensitivity reflects the rapid proliferation of the tumor and may also indicate that the treated patient retains an immune response capable of eliminating residual, surviving tumor cells. Finally, if tumors do recur in treated patients, their recurrence is signaled by elevated titers of antibodies to viral early antigens (84). This signaling demonstrates that the patient does have both an active humoral response to viral antigens and the T-cell help required for that response.

Characteristics of the EBV-positive Burkitt's lymphoma tumor cell are consistent with its having evolved to escape an immune response directed against its viral antigens. These cells express only low levels of cell adhesion molecules required for efficient recognition and killing by cytotoxic T cells (38). They also detectably express only one virally encoded protein (100, 101) and fail to express six viral proteins that are expressed in EBV-positive cells isolated from patients with infectious mononucleosis and that are known to be targets for cytotoxic T cells (14, 81). It is possible that the Burkitt's lymphoma tumor cell has evolved from an uninfected B lymphocyte that is prevalent in young children, expresses only low levels of cell adhesion molecules, and can proliferate in the absence of all but one viral protein. Such cells have yet to be identified. It is also possible that Burkitt's lymphoma tumor cells have evolved from EBV-infected B lymphocytes of the type characterized from infectious mononucleosis patients. These latter cells express normal levels of cell adhesion molecules and 10 viral proteins. If Burkitt's lymphoma tumor cells represent selected descendants of infectious mononucleosis-like cells, then it is likely that the host immune response contributes to the selection of these tumor cells.

The hypothesis that a child's immune response may contribute to the development of Burkitt's lymphoma is consistent with several observations but is still too skeletal an idea to be satisfying. In particular, we need to know how the immune response selects for surviving cells in these children when it eliminates them in other hosts.

WHAT ARE THE CONTRIBUTIONS OF SOMATIC MUTATIONS TO THE DEVELOPMENT OF BURKITT'S LYMPHOMA?

Experimental and clinical data from chickens, mice, and people indicate that one mutation common to Burkitt's lymphomas is a tissue-specific transforming event. ALV-induced bursal lymphomas in chickens are associated with a proviral insertion near the c-*myc* proto-oncogene that increases the expression of this gene. This promoter insertion leads to B-cell proliferation but is insufficient for lymphomagenesis (6). BALB/c mice inoculated peritoneally with pristane develop plasmacytomas, which are tumors of immunoglobulin-secreting B cells (5). These tumors contain immunoglobulin–c-*myc* translocations that lead to preferential expression of the translocated c-*myc* allele (50, 85, 109). These tumors proliferate initially in the peritoneal cavity, which as a result of the pristane treatment provides a microenvironment rich in macrophages and their associated growth factors. The primary BALB/c plasmacytomas require these growth factors for proliferation and usually will not proliferate when explanted into cell culture. EBV-negative Burkitt's lymphomas arise throughout the world at about 1% of the frequency with which EBV-positive Burkitt's lymphomas arise in the regions to which they are endemic. EBV-negative Burkitt's lymphomas also have immunoglobulin–c-*myc* translocations in which the translocated allele is expressed, and they proliferate in vivo and in vitro upon explantation into cell culture. In all of these examples, mutations that affect expression of c-*myc* are associated with proliferation of a B-lymphoid cell and may predispose it to develop into a lymphoma.

EBV-positive Burkitt's lymphomas also have immunoglobulin–c-*myc* translocations, and on the basis of the available examples, this mutation would be expected to support proliferation of the affected cell in certain microenvironments. It is equally clear that infection with EBV supports proliferation of the infected cell in vivo and in vitro. Why, then, should an evolving tumor acquire two genetic alterations, each of which supports its proliferation? One answer to this query has already been outlined. Either an immunoglobulin–c-*myc* translocation eventually occurs in an EBV-infected B-lymphoid cell, or EBV infects a B-lymphoid cell that has already acquired the translocation. In either case, the virus provides the cell with a selective advantage, because the virus is retained as a plasmid throughout the cell's development into a lymphoma. In the presence of an effective immune response, however, at least six viral proteins are recognized by cytotoxic T-cells as targets. Only those progeny cells that can continue to proliferate and evade killing by turning off most viral gene expression and downregulating cell adhesion molecules survive to develop into a Burkitt's lymphoma. This survival is a rare event even in patients with Burkitt's lymphoma, because the tumor is clonal in origin. The mechanism for the downregulation of viral and cellular genes is not known. It may be the result of a normal but rare developmental switch such as, for example, the development of an activated B cell into a memory cell (57). It may also be an epigenetic event peculiar to tumor cells.

Mutations in addition to the immunoglobulin–c-*myc* translocations are common to EBV-associated Burkitt's lymphomas, but their potential contributions to the development of this tumor are obscure. One class of mutations is clustered in the second exon of the translocated c-*myc* allele (130). These mutations may affect transcriptional activation by this proto-oncogene. Mutations in the *p53* tumor suppressor gene have been detected in cell lines derived from Burkitt's lymphomas and directly from tumor biopsy samples. In the latter case, these mutations may be heterozygous, with the tumor cell also retaining the wild-type allele. Homozygous mutations in *p53* are associated with progression of a variety of human tumors and resistance to chemotherapy and radiotherapy (67), while heterozygous mutations may reflect a gain of function for the mutated allele. What this gain of function might be or how it might affect the EBV-positive Burkitt's lymphoma is unknown.

The age dependence of Burkitt's lymphomas and the roles of the immune response and somatic mutations in the development of this tumor are unresolved problems in our understanding of its etiology. The contribution of EBV to it and other associated lymphoproliferations is now well established, and the molecular details of that contribution are gradually being defined.

EBV INDUCES AND MAINTAINS PROLIFERATION OF INFECTED B-LYMPHOID CELLS

Much of our understanding of infection of cells by EBV has been gained from studies with human B lymphocytes in vitro. Explanted neonatal or adult B lymphocytes do not proliferate in cell culture and do not enlarge to form blasts or incorporate [^{3}H]thymidine under standard conditions. They can, however, be induced to form blasts and to divide by treatment with mitogens such as formaldehyde-fixed *Staphylococcus aureus,* but this induction is limited to a few divisions. Some subset of these cells (e.g., cells found in spleens or tonsils) can be induced to proliferate for a multiple but still limited number of generations in a complex mitogenic environment (9). These observations indicate that human B lymphocytes cannot readily be induced with known growth factors to proliferate in vitro. Either we lack the appropriate growth factors or the explanted B lymphocytes normally would not be destined to proliferate extensively in vivo either.

Infection of explanted human B lymphocytes with EBV is dramatic, because it efficiently yields proliferating cells. More than 50% of adult peripheral B lymphocytes can be infected with EBV in vitro, and at least 3% of the infected cells yield colonies when plated into semisolid medium. The plating efficiency of similar established cells in semisolid medium is approximately 3 to 10%, so that detection of 3% of the infected cells as proliferating colonies indicates that many or most of the infected cells have the capacity to proliferate but that the assay to measure their initial proliferation is inefficient.

Induction of proliferation by infection with EBV is also efficient with respect to the virus. A constant number of isolated human B lymphocytes have been exposed to decreasing amounts of EBV, and the number of resulting, proliferating cell colonies has been measured in two ways (44, 113). These dose-response studies indicate that one infectious particle of EBV is sufficient to yield a proliferating cell. The frequency of infectious particles in stocks of EBV has also been measured in related experiments and found to be approximately 1 in 30, which is similar to the frequency for other herpesviruses. All of these findings indicate that induction and maintenance of proliferation of infected cells by EBV is efficient for the target cell as well as for the infecting virus. It is, therefore, a wild-type B lymphocyte infected by a wild-type virus that yields proliferating infected B lymphoblasts.

There is no evidence that either mutant target cells or mutant viruses contribute in vitro to proliferation early after infection.

Proliferating B lymphoblasts infected with EBV in vitro efficiently yield immortalized cells, that is, cells that proliferate in perpetuity as long as they are adequately maintained in cell culture. This process may require somatic mutations, but if it does, the required mutations occur sufficiently frequently that between one in two and one in three of the infected cells that proliferate for 20 to 25 generations eventually yields some immortalized progeny. (The frequency with which Burkitt's lymphoma biopsy samples yield immortalized cells has not been documented. It is possible that not all of these clonal tumors survive in culture to yield cell lines or that the mutations they have accumulated in vivo ensure that they will proliferate in perpetuity in vitro.)

EBV contributes at least five genes absolutely required for the induction and/or maintenance of proliferation of infected cells in vitro; it also contributes genes that increase the efficiency of this process. These viral transforming genes and their functions can be described practically by recounting first their effects on the viral genome and subsequently their effects on the cellular genome.

The first viral genes to be expressed from an infecting EBV genome in a resting B lymphocyte are EBV nuclear antigens, termed EBNA-LP and EBNA-2 (3, 99), which appear to be translated from mRNAs initiated from the Wp promoter (107, 129). (These viral genes, the transcripts that encode their gene products, and the promoters from which the transcripts are initiated are depicted in Fig. 3.) They are detectably expressed 12 to 24 h after infection, i.e., after the linear viral DNA has entered the nucleus and been circularized. EBNA-LP is not absolutely required for induction and maintenance of proliferation, and what it might contribute to the regulation of viral gene expression is unknown (39). EBNA-2, on the other hand, is the initial determinant of viral gene expression. It positively

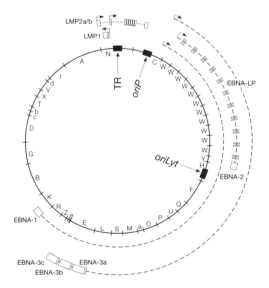

FIGURE 3. Map of the EBV genome, depicting transcripts expressed during the latent phase of the viral life cycle (adapted from reference 51). The circle represents the circular DNA of EBV joined at the terminal repeats (TR), as occurs in latently infected cells. The letters and slash marks in the circle denote fragments of EBV DNA generated by digestion with *Bam*HI endonuclease. *oriP* and *oriLyt* are the two origins of DNA replication of EBV; *oriP* is used during the latent phase of the viral life cycle, and *oriLyt* is used during the lytic phase. Arrows represent start sites for transcription used during the EBV latent phase and considered in this chapter. Open boxes following the arrows are coding sequences for translation of EBV latent genes (EBNA-LP is made up of the repeats within BamW, while EBNA-2 is made up of the single exon within BamY and BamH), and the dashed lines between them represent introns and untranslated exons.

regulates at least three viral promoters by associating with a cellular protein termed variously RBPJ$_K$ or CBF1, which binds site specifically 200 to 300 bp upstream of the start sites of transcription for those promoters (66, 137). These three EBNA-2-responsive promoters are LMP 2a/b, LMP 1, and Cp. They drive transcripts that are spliced (LMP 1) or differentially spliced (LMP 2a/b and Cp) to yield mRNAs that encode all the proteins expressed in latently infected cells, including EBNA-2 (107). The longest transcript initiated from the Cp promoter is close to 100,000 bases long. It is differentially spliced to yield several mature mRNAs, one of which is 3,700 bases long and encodes the EBNA-1 protein (12). The initial expression of EBNA-1 is clearly dependent on EBNA-2 function and is detectable 24 to 36 h after infection. However, once EBNA-1 is expressed, it assumes a central role in regulating EBV at the level of both transcription and DNA replication.

EBNA-1 binds to 20 adjacent sites, which make up the family of repeats that is part of the EBV origin of plasmid replication (*oriP*), which lies within the EBV *cis*-acting regulatory domain (92, 94, 132). The family of repeats bound by EBNA-1 is a potent transcriptional enhancer (93) for the C promoter, which lies about 3,000 bp from it, and for the LMP 1 promoter, which lies 10,000 bp from it (32, 114). The mechanism by which EBNA-1 bound to the family of repeats stimulates transcription is not known, but we do know that it does not absolutely require the site upstream of the LMP 1 promoter at which EBNA-2 acts to affect this promoter (32). The C promoter can yield mRNAs through differential splicing and varied use of polyadenylation sites that encode all six EBNAs, that is, EBNA-LP, EBNA-1, EBNA-2, EBNA-3a, EBNA-3b, and EBNA-3c (107). It is not known what EBNA-3a and EBNA-3b contribute to the regulation of viral gene expression, but some observations indicate that EBNA-3c is a transcriptional activator that can affect expression from the LMP 1 promoter (4, 124). EBNA-1, EBNA-2, and EBNA-3c contribute to the regulation of the LMP 1 promoter, which encodes a viral oncoprotein, the latent membrane protein, not known to affect viral gene expression. LMP 1 can first be detected 48 to 72 h after infection of resting B lymphocytes.

EBNA-1 also binds to four adjacent sites, known as the dyad symmetry element (92), which constitute the second element of *oriP* and are the site at which viral plasmid DNA synthesis initiates (31, 94). EBNA-1 is the only viral gene product required to mediate replication at *oriP* (134). This replication occurs once per cell cycle during S phase and must use cellular enzymatic machinery (2, 133). EBNA-1 is the origin-binding protein required for replication of EBV plasmid DNA during latent infection and is therefore central to the regulation of viral DNA synthesis and maintenance of viral information in proliferating cells. Viral DNA synthesis is assumed to first take place when cellular DNA synthesis occurs 72 to 96 h after infection of a resting B lymphocyte. The viral plasmid replicon is subsequently maintained in the proliferating cells. It is lost from cells on which it does not confer a selective advantage at rates of 2 to 3% per cell per generation (56, 132).

The third promoter regulated by EBNA-2, the LMP 2a/b promoter, has not been analyzed for its possible regulation by EBNA-1. It yields two mRNAs via alternative splicing that encode proteins found in the plasma membrane (62, 103). These membrane proteins can affect the frequency with which the latent phase of the EBV life cycle can convert to the lytic phase (74). These mRNAs appear to be the predominant transcripts expressed in latently infected cells found in the peripheral blood of healthy adults (91).

The combined actions of EBNA-2, EBNA-1, EBNA-3c, and possibly EBNA-LP regulate viral gene expression in the initially infected resting B lymphocyte and the proliferating lymphoblast. EBNA-1 mediates viral DNA replication as an extrachromosomal element. These genes of EBV, through their own activities and those of additional viral genes that they regulate, affect the infected host cell dramatically.

It is not known what the earliest influence of EBV on resting B lymphocyte is. It is possible that the virus particle affects the target cell before viral genes are transcribed either by binding to the cellular receptor CD21 (30, 83) or by introducing transcription factors from within the virion that act on cellular genes. However, such possible effects have not yet been documented. If they do occur, they are insufficient to induce the infected cell to incorporate [^{3}H]thymidine, because a break in virion duplex DNA inhibits induction of cellular DNA synthesis (70). A break in virion duplex DNA would not be expected to affect either the consequences of virus particle binding of its cellular receptor or the functioning of a virion-encapsidated transcription factor. It is clear that EBNA-2, one of the first two viral genes to be detectably expressed, positively regulates some cellular genes. CD23 is a cellular protein first mistakenly thought to be uniquely expressed on EBV-infected B cells because its expression is so efficiently induced by EBV in general and EBNA-2 in particular (18, 55, 123, 125). It is also positively regulated by LMP 1, which is in turn itself positively regulated by EBNA-2 (1, 26, 120, 124, 126). The expression of EBNA-2 in EBV-negative B lymphoblasts also correlates with that of additional cellular genes whose expression may therefore be positively affected by EBNA-2. Perhaps the most tantalizing insights into the contributions of EBNA-2 to the infected host cell have been glimpsed recently from studies with a derivative of EBV that contains a conditionally functional EBNA-2. Some clones of B lymphoblasts infected with this mutant are dependent on EBNA-2 for their own continued proliferation (52). When such clones are withdrawn from the cell cycle and then stimulated to reenter it by a resupply of functional EBNA-2, a variety of regulatory cellular genes, including *myc* and cyclin D2, are induced (52). It is not known what links EBNA-2 function to the regulation of these cellular genes. It is clear, however, that because these genes are cellular determinants of proliferation, the effects of EBNA-2 on them are central to EBV induction and maintenance of proliferation of the infected cell.

EBNA-LP is synthesized early in infection, along with EBNA-2, from an alternatively spliced transcript (98). Although the intact protein is not essential to inducing and maintaining proliferation of the infected cell, it contributes to the proliferative capacity of the infected cell (39, 69). Cells infected with a derivative of EBV that lacks the last two exons of EBNA-LP initially survive inefficiently in semisolid medium (39). It is not known what EBNA-LP contributes to its host cell that mediates the survival or proliferative capacity of the cell when it is plated at clonal densities.

EBNA-1 is a central regulator of viral genes but has not been shown to directly influence expression of cellular genes. One observation indicates, however, that it may do so. Biopsy samples of Burkitt's lymphomas apparently express only EBNA-1 (101). These tumor cells proliferate in vivo and in vitro and maintain the viral DNA as a plasmid. Experimental reconstructions have shown that viral plasmid replicons are lost from proliferating cells in the absence of selection (56, 94). If it is correct that EBNA-1 is the only viral gene product expressed in biopsy cells, then it is likely that it provides the cells the selective advantage that ensures maintenance of EBV; this advantage may be an effect on cellular gene expression. No sites to which EBNA-1 binds specifically in the human genome have been reported; however, two cellular proteins that bind to EBNA-1 binding sites in *oriP* have been identified and cloned (136). Presumably, these cellular, site-specific, DNA-binding proteins recognize similar sequences in the human host. These sequences may be sites to which EBNA-1 can bind to influence the host cell.

Like EBNA-2, the LMP 1 oncoprotein appears to be a major viral determinant of cellular phenotypes. LMP 1 is required to induce and/or maintain proliferation in the infected cell (49). It does not alone render EBV-infected B lymphoid cells oncogenic, but it does transform certain established rodent cell lines and is therefore characterized as an oncoprotein (8, 121). This viral protein rapidly homes to the plasma membrane, where it patches, is attached to the cytoskeleton, and turns over rapidly (7, 63, 64, 68, 72). It affects a variety of phenotypes in cells, although a biochemical activity intrinsic to it has not been identified. Several properties of LMP 1 are consistent with its acting at the plasma membrane to effect signaling across it, as a receptor does, or acting as a regulator of a cellular receptor. Of the LMP 1 synthesized during a 20-min labeling period, 90% is found at the plasma membrane at the end of the labeling period (71). It is degraded either at the membrane or immediately upon being internalized. The expression of LMP 1 in a variety of EBV-negative B-lymphoid cells correlates with an increased expression of the cellular genes *CD21*, *CD23*, *ICAM-1*, and *Bcl-2* (45, 73, 122, 124). It can also, when introduced into primary B lymphocytes under selected conditions, induce cellular DNA synthesis (89). These and other phenotypes (20, 27) all indicate that LMP 1 affects a host cell dramatically and in some cells can contribute to their proliferative capacities. In some cell lines, the introduction of LMP 1 rapidly leads to an induction of activity of the NF-κB family of transcription factors (40, 59, 76). Again, the mechanism by which LMP 1 induces this activity is not known, but the mechanism may underlie one or more of the cellular phenotypes associated with the expression of this viral oncoprotein.

The last viral gene known to affect the host cell is EBNA-3a. It is required for initiating proliferation of infected cells but apparently not for maintaining their proliferation (51, 119). A derivative of EBV that has a frameshift mutation in codon 304 of the EBNA-3a gene fails to initiate proliferation of cells but does so when complemented by an EBNA-2⁻ EBNA-3a⁺ variant of EBV (51). Two of 47 cell clones that arose from a mixed infection with these two variants of EBV eventually contained only the EBNA-3a⁻ derivative. This finding indicates that once cells are proliferating, EBNA-3a is not required for maintaining their proliferation (51). The mechanism of action EBNA-3a and what it contributes to the resting B lymphocyte to initiate cellular proliferation are not known.

At least five genes of EBV are required for and several more contribute to inducing and maintaining proliferation of the infected B lymphocyte. These genes act to regulate the expression of the B lymphocytes, to maintain the viral genome as an extrachromosomal replicon in the proliferating host cell, and to affect the expression of cellular genes. The net result of the action of these viral genes for EBV is twofold: first, the infected B lymphocyte is induced to proliferate, and second, infected progeny cells accumulate sufficiently to ensure that they are maintained lifelong in the human host and that they release sufficient infectious virus to spread infection to naive human beings. EBV is amazingly successful and infects more than 90% of all people by the time they are adults. Although EBV can contribute to certain human cancers, it is clear that it does so only under extraordinary conditions.

A CHARACTERISTIC COMMON TO HUMAN TUMOR VIRUSES

EBV, HBV, HPV, and HTLV-1 do not encode dominantly acting oncogenes akin to v-*src,* nor are they known to integrate near cellular proto-oncogenes to affect the expression of those genes, as ALV can do. Each, however, can affect proliferation of the cell types that make up the cancers with which the viruses are associated. EBV is a prototype for these other viruses because it was the first-identified human tumor virus and the first human virus shown to induce and maintain proliferation of infected cells. It is causally associated with polyclonal B-cell tumors in immunocompromised hosts and with Burkitt's lymphomas in inhabitants of certain areas of the world. It is also causally associated with nasopharyngeal carcinoma and perhaps with Hodgkin's lymphoma.

HBV is causally associated with primary hepatocellular carcinoma: it infects hepatocytes and apparently leads quite indirectly to their proliferation. The infected cells are efficiently recognized by the host immune response and are killed by cytotoxic T cells (78). Hepatocytes retain the capacity to proliferate, particularly in younger people, to replace the killed, infected cells. These repeated cycles of infection, cell killing, and proliferation to repopulate the liver are thought to promote primary hepatocellular carcinogenesis. A prospective epidemiological survey has demonstrated that chronic infection with HBV increases the risk of developing liver cancer by a factor of 100 relative to that in uninfected, age-matched controls (10). The HBV-associated cancers, which take 40 to 60 years to develop, are preceded by chronic cirrhosis, which is also a product of repeated cycles of infection and cell killing (10).

HPVs are associated with genital carcinomas and affect proliferation of squamous epithelia in vitro. This family of viruses, particularly its tumorigenic members, encode genes, termed *E6* and *E7,* whose products bind the cellular tumor suppressors p53 and Rb (22, 127). In the former case, the E6 viral protein enhances ubiquitin-mediated degradation of p53 (104). In the latter case, the E7 viral protein displaces cellular proteins normally sequestered by Rb (16). Each of these viral activities contributes to proliferation of infected cells. Both together can immortalize human epithelial cells in culture (42), and both together are tumorigenic when expressed as transgenes in epithelial cells in transgenic mice (61). Although prospective surveys have not been completed, retrospective surveys have found an association of HPV-16, -18, -31, and -33 with human cervical carcinomas, which probably take 15 to 30 years to develop (106). In biopsy samples from those tumors, the viral DNA is integrated apparently randomly with respect to the host cell's chromosomes but nonrandomly with respect to the viral genome. The HPV DNA, which is circular in the virion, is linearized upon integration such that the *E6* and *E7* genes are intact and expressed, and their products accumulate to high levels (102). It thus appears that HPV-16, -18, and -33 affect proliferation of infected epithelial cells and that in those rare instances in which the viral genome integrates such that E6 and E7 can accumulate to high levels, the host cell is predisposed to evolve into a carcinoma.

HTLV-1 is associated with adult T-cell lymphomas. It infects T lymphocytes and can both induce and maintain their proliferation in vitro (77). The virus affects expression of cellular genes, including those encoding interleukin 2 (IL-2) and the α chain of the IL-2 receptor (47). IL-2 is a growth factor for T cells, and the expression of this ligand and its receptor in one cell may be sufficient to mediate proliferation of that cell. It is clear that infection of T lymphocytes with HTLV-1 in vitro can yield immortalized infected cells, although the efficiency of this process has not been measured, in part because HTLV-1 cannot be readily isolated in a cell-free infectious form. Infection in vivo with HTLV-1 presumably also yields proliferating infected cells. As with EBV and Burkitt's lymphoma, HTLV-1-associated adult T-cell lymphomas fail to express much viral information (54). The events that must occur between primary infection and the development of lymphomas have not been identified. However, it has been established that lifelong infection with HTLV-1 is associated with development of adult T-cell lymphomas. Two to 5% of such infected people in Japan are thought to develop the tumor eventually (135).

These four human tumor viruses vary in the cell types they infect, in the times between infection and development of cancers, and in the mechanisms by which they induce and/or maintain proliferation of infected cells. However, they all share the capacity to affect proliferation of the cells they infect. In no case do these viral proliferative effects appear to be sufficient to cause a cancer, but it is likely that they are the contributions made by these viruses to their associated tumors.

LOOKING BACKWARD

No human tumor virus had been identified when Howard Temin began studying retroviruses in the 1950s. None was generally thought to exist. His work with RSV stimulated a wide effort in the 1970s to search for human retroviruses that are tumor viruses. This wide-ranging research, conducted by many virologists, laid the foundation for the eventual isolation of HTLV-1 in the early 1980s. Howard Temin's work in particular contributed profoundly to our recognition that retroviruses can transduce oncogenes and to an appreciation of the mechanisms by which they do so. These insights led to the identification in the 1970s and 1980s of many of the proto-oncogenes that are the targets for initiation of human cancers by chemical carcinogens. Howard Temin's work also endowed the study of tumor virology with an intellectual respectability that encouraged the search for and characterization of those other viruses now accepted as being risk factors in the development of at least seven different human cancers.

ACKNOWLEDGMENTS. I thank Toni Gahn, Tom Mitchell, Donata Oertel, and Ilse Riegel for helpful criticisms of this review.

I was supported by Public Health Service grants CA-22443 and CA-07175 from the National Cancer Institute.

REFERENCES

1. **Abbot, S. D., M. Rowe, K. Cadwallader, A. Ricksten, J. Gurdon, F. Wang, L. Rymo, and A. B. Rickinson.** 1990. Epstein-Barr virus nuclear antigen 2 induces expression of the virus-encoded latent membrane protein. *J. Virol.* **64:**2126–2134.

2. **Adams, A.** 1987. Replication of latent Epstein-Barr virus genomes in Raji cells. *J. Virol.* **61:** 1743–1746.

3. **Alfieri, C., M. Birkenbach, and E. Kieff.** 1991. Early events in Epstein-Barr virus infection of human B lymphocytes. *Virology* **181:**595–608.

4. **Allday, M. J., and P. J. Farrell.** 1994. Epstein-Barr virus nuclear antigen EBNA 3C/6 expression maintains the level of latent membrane protein 1 in G_1-arrested cells. *J. Virol.* **68:**3491–3498.

5. **Anderson, P. N., and M. Potter.** 1969. Induction of plasma cell tumors in BALB/c mice with 2,6,10,14-tetramethylpentadecone (pristane). *Nature* (London) **222:**994–995.

6. **Baba, T. W., and E. H. Humphries.** 1985. Formation of a transformed follicle is necessary but not sufficient for development of an avian leukosis virus-induced lymphoma. *Proc. Natl. Acad. Sci. USA* **82:**213–216.

7. **Baichwal, V. R., and B. Sugden.** 1987. Posttranslational processing of an Epstein-Barr virus-encoded membrane protein expressed in cells transformed by Epstein-Barr virus. *J. Virol.* **61:** 866–875.

8. **Baichwal, V. R., and B. Sugden.** 1989. Transformation of BALB 3T3 cells by the BNLF-1 gene of Epstein-Barr virus. *Oncogene* **2:**461–467.

9. **Banchereau, J., P. de Paoli, A. Vallé, E. Garcia, and F. Rousset.** 1991. Long-term human B cell lines dependent on interleukin-4 and antibody to CD40. *Science* **251:**70–72.

10. **Beasley, R. P.** Hepatitis B virus. *Cancer* **61:**1942–1956.

11. **Blacklow, N. R., B. K. Watson, G. Miller, and B. M. Jacobson.** 1971. Mononucleosis with heterophile antibodies and EB virus infection. Acquisition by an elderly patient in hospital. *Am. J. Med.* **51:**549–552.

12. **Bodescot, M., and M. Perricaudet.** 1986. Epstein-Barr virus mRNAs produced by alternative splicing. *Nucleic Acids Res.* **17:**7130–7134.

13. **Burkitt, D.** 1962. A children's cancer dependent upon climatic factors. *Nature* (London) **194:** 232–234.

14. **Burrows, S. R., J. Gardner, R. Khanna, T. Steward, D. J. Moss, S. Rodda, and A. Suhrbier.** 1994. Five new cytotoxic T cell epitopes identified within Epstein-Barr virus nuclear antigen 3. *J. Gen. Virol.* **75:**2489–2493.

15. **Cartwright, C. A., W. Eckhart, S. Simon, and P. L. Kaplan.** 1987. Cell transformation by $pp60^{c-src}$ mutated in the carboxy-terminal regulatory domain. *Cell* **49:**83–91.

16. **Chellappan, S., V. B. Kraus, B. Kroger, K. Munger, P. M. Howley, W. C. Phelps, and J. R. Nevins.** 1992. Adenovirus E1A, simian virus 40 tumor antigen, and human papillomavirus E7 protein share the capacity to disrupt the interaction between transcription factor E2F and the retinoblastoma gene product. *Proc. Natl. Acad. Sci. USA* **89:**4549–4553.

17. **Cooper, J. A., K. L. Gould, C. A. Cartwright, and J. Hunter.** 1986. Tyr 527 is phosphorylated in pp60$^{c\text{-}src}$: implications for regulation. *Science* **231:**1431–1434.

18. **Cordier, M., A. Calender, M. Billand, V. Zimber, G. Rousselet, O. Pavish, J. Banchereau, T. Tursz, G. Bornkamm, and G. M. Lenoir.** 1990. Stable transfection of Epstein-Barr virus (EBV) nuclear antigen 2 in lymphoma cells containing the EBV P3HRI genome induces expression of B-cell activation molecules CD21 and CD23. *J. Virol.* **64:**1002–1013.

19. **Craig, F. E., M. L. Gully, and P. M. Banks.** 1993. Post-transplantation lymphoproliferative disorders. *Am. J. Clin. Pathol.* **99:**265–276.

20. **Dawson, C. W., A. B. Rickinson, and L. S. Young.** 1990. Epstein-Barr virus latent membrane protein inhibits human epithelial cell differentiation. *Nature* (London) **344:**777–780.

21. **de-Thé, G., A. Geser, N. E. Day, P. M. Tukei, E. H. Williams, D. P. Beri, P. G. Smith, A. G. Dean, G. W. Bornkamm, P. Feorino, and W. Henle.** 1978. Epidemiological evidence for causal relationship between Epstein-Barr virus and Burkitt's lymphoma from Ugandan prospective study. *Nature* (London) **274:**756–761.

22. **Dyson, N., P. M. Howley, K. Munger, and E. Harlow.** 1989. The human papilloma virus-16 E7 oncoprotein is able to bind to the retinoblastoma gene product. *Science* **243:**934–937.

23. **Edwards, R. H., and N. Raab-Traub.** 1994. Alterations of the p53 gene in Epstein-Barr virus-associated immunodeficiency-related lymphomas. *J. Virol.* **68:**1309–1315.

24. **Evans, A. S.** 1982. The transmission of EB viral infections, p. 211–225. *In* J. J. Hooks and G. W. Jordan (ed.), *Viral Infections in Oral Medicine.* Elsevier/North Holland, New York.

25. **Facer, C. A., and J. H. L. Playfair.** 1989. Malaria, Epstein-Barr virus, and the genesis of lymphomas. *Adv. Cancer Res.* **55:**33–72.

26. **Fåhraeus, R., A. Jansson, A. Ricksten, A. Sjöblom, and L. Rymo.** 1990. Epstein-Barr virus-encoded nuclear antigen 2 activates the viral latent membrane protein promoter by modulating the activity of a negative regulatory element. *Proc. Natl. Acad. Sci. USA* **87:**7390–7394.

27. **Fåhraeus, R., L. Rymo, J. S. Rhim, and G. Klein.** 1990. Morphological transformation of human keratinocytes expressing the LMP gene of Epstein-Barr virus. *Nature* (London) **345:**447–449.

28. **Farrell, P. J., G. J. Allan, F. Shanahan, K. H. Vousden, and T. Crook.** 1991. p53 is frequently mutated in Burkitt's lymphoma cell lines. *EMBO J.* **10:**2879–2887.

29. **Fialkow, P. J., G. Klein, S. M. Gartler, and P. Clifford.** 1970. Clonal origin for individual Burkitt tumors. *Lancet* **i:**384–386.

30. **Fingeroth, J. D., J. J. Weis, T. F. Tedder, J. L. Strominger, P. A. Biro, and D. T. Fearon.** 1984. Epstein-Barr virus receptor of human B lymphocytes is the C3d receptor CR2. *Proc. Natl. Acad. Sci. USA* **81:**4510–4514.

31. **Gahn, T. A., and C. L. Schildkraut.** 1989. The Epstein-Barr virus origin of plasmid replication, *oriP,* contains both the initiation and termination sites of DNA replication. *Cell* **58:**527–535.

32. **Gahn, T. A., and B. Sugden.** 1995. An EBNA-1 dependent enhancer acts from a distance of 10 kilobase pairs to increase expression of the Epstein-Barr virus LMP gene. *J. Virol.* **69:**2633–2636.

33. **Gaidano, G., P. Ballerini, J. Z. Gong, G. Inghirami, A. Neri, E. W. Newcomb, I. T. Magrath, D. M. Knowles, and R. Dalla-Favera.** 1991. p53 Mutations in human lymphoid malignancies: association with Burkitt lymphoma and chronic lymphocytic leukemia. *Proc. Natl. Acad. Sci. USA* **88:**5413–5417.

34. **Gelman, I. H., and H. Hanafusa.** 1989. Suppression of Rous sarcoma virus-induced tumor formation by preinfection with viruses encoding *src* protein with novel N termini. *J. Virol.* **63:**2461–2468.

35. **Gelman, I. H., and H. Hanafusa.** 1993. *src*-specific immune regression of Rous sarcoma virus-induced tumors. *Cancer Res.* **53:**915–920.

36. **Geser, A. D., G. deThé, G. Lenoir, N. E. Day, and E. H. Williams.** 1982. Final case reporting from the Uganda prospective study of the relationship between EBV and Burkitt's lymphoma. *Int. J. Cancer* **29:**397–400.

37. **Giuliano, V. J., H. E. Jasin, and M. Ziff.** 1974. The nature of the atypical lymphocyte in infectious mononucleosis. *Clin. Immunol. Immunopathol.* **3**:90–98.

38. **Gregory, C. D., R. J. Murray, C. F. Edwards, and A. B. Rickinson.** 1988. Down-regulation of cell adhesion molecules LFA-3 and ICAM-1 in Epstein-Barr virus-positive Burkitt's lymphoma underlies tumor cell escape from virus-specific T cell surveillance. *J. Exp. Med.* **167**:1811–1824.

39. **Hammerschmidt, W., and B. Sugden.** 1989. Genetic analysis of immortalizing functions of Epstein-Barr virus in human B-lymphocytes. *Nature* (London) **340**:393–397.

40. **Hammerskjöld, M., and M. Simurda.** 1992. Epstein-Barr latent membrane protein transactivates the human immunodeficiency virus type 1 long terminal repeat through induction of NF-κB activity. *J. Virol.* **66**:6496–6501.

41. **Hanto, D. W., G. Frizzera, D. T. Purtilo, K. Sakamoto, J. L. Sullivan, A. K. Saemundsen, G. Klein, R. L. Simmons, and J. S. Najarian.** 1981. Clinical spectrum of lymphoproliferative disorders in renal transplant recipients and evidence for the role of the Epstein-Barr virus. *Cancer Res.* **41**:4253–4261.

42. **Hawley-Nelson, P., K. H. Vousden, N. L. Hubbert, D. R. Lowy, and J. T. Schiller.** 1989. HPV16 E6 and E7 proteins cooperate to immortalize human foreskin keratinocytes. *EMBO J.* **8**:3905–3910.

43. **Hayward, W. S., B. G. Neel, and S. M. Astrin.** 1981. Activation of a cellular *onc* gene by promoter insertion in ALV-induced lymphoid leukosis. *Nature* (London) **290**:475–480.

44. **Henderson, E., G. Miller, J. Robinson, and L. Heston.** 1977. Efficiency of transformation of lymphocytes by Epstein-Barr virus. *Virology* **76**:152–163.

45. **Henderson, S., M. Rowe, C. Gregory, D. Croom-Carter, F. Wang, R. Longnecker, E. Kieff, and A. Rickinson.** 1991. Induction of Bcl-2 expression by Epstein-Barr virus latent membrane protein 1 protects infected B cells from programmed cell death. *Cell* **65**:1107–1115.

46. **Imamoto, A., and P. Soriano.** 1993. Disruption of the *csk* gene, encoding a negative regulator of *src* family tyrosine kinases, leads to neural tube defects and embryonic lethality in mice. *Cell* **73**:1117–1124.

47. **Inoue, J., M. Seiki, T. Taniguchi, S. Tsuru, and M. Yoshida.** 1986. Induction of interleukin 2 receptor gene expression by p40 encoded by human T-cell leukemia virus type I. *EMBO J.* **5**:2883–2888.

48. **Jove, R., and H. Hanafusa.** 1987. Cell transformation by the viral *src* oncogene. *Annu. Rev. Cell Biol.* **3**:31–56.

49. **Kaye, K. M., K. M. Izumi, and E. Kieff.** 1993. Epstein-Barr virus latent membrane protein 1 is essential for B-lymphocyte growth transformation. *Proc. Natl. Acad. Sci. USA* **90**:9150–9154.

50. **Keath, E. J., A. Keleker, and M. Cole.** 1984. Transcriptional activation of the translocated c-*myc* oncogene in mouse plasmacytomas: similar RNA levels in tumor and proliferating normal cells. *Cell* **37**:521–528.

51. **Kempkes, B., D. Pich, R. Zeidler, B. Sugden, and W. Hammerschmidt.** 1995. Immortalization of human B lymphocytes by a plasmid containing 71 kilobase pairs of Epstein-Barr virus DNA. *J. Virol.* **69**:231–238.

52. **Kempkes, B., D. Spitkovsky, P. Jansen-Dunn, J. W. Ellwart, E. Kremmer, H.-J. Delecluse, C. Rottenberger, G. W. Bornkamm, and W. Hammerschmidt.** 1995. B-cell proliferation and induction of early G_1-regulating proteins by Epstein-Barr virus mutants conditional for EBNA2. *EMBO J.* **14**:88–96.

53. **Khanna, R., S. R. Burrows, M. G. Kurilla, C. A. Jacob, I. S. Misko, T. B. Sculley, E. Kieff, and D. J. Moss.** 1992. Localization of Epstein-Barr virus cytotoxic T cell epitopes using recombinant vaccinia: implications for vaccine development. *J. Exp. Med.* **176**:169–176.

54. **Kinoshita, T., M. Shimoyama, K. Tobinai, M. Ito, S. Ito, S. Ikeda, K. Tajima, K. Shimotohno, and T. Sugimura.** 1989. Detection of mRNA for the tax_1/rex_1 gene of human T cell leukemia virus type 1 in fresh peripheral blood mononuclear cells of adult T-cell leukemia patients and viral carriers by using the polymerase chain reaction. *Proc. Natl. Acad. Sci. USA* **86**:5620–5624.

55. **Kintner, C., and B. Sugden.** 1981. Identification of antigenic determinants unique to the surfaces of cells transformed by Epstein-Barr virus. *Nature* (London) **294**:458–460.

56. **Kirchmaier, A., and B. Sugden.** 1995. Plasmid maintenance of derivatives of *oriP* of Epstein-Barr virus. *J. Virol.* **69**:1280–1283.

57. **Klein, G.** 1994. Epstein-Barr virus strategy in normal and neoplastic B cells. *Cell* **77**:791–793.

58. **Kmiecik, T. E., and D. Shalloway.** 1987. Activation and suppression of pp60$^{c\text{-}src}$ transforming ability by mutation of its primary sites of tyrosine phosphorylation. *Cell* **49:**65–73.

59. **Laherty, C. D., H. M. Hu, A. W. Opipari, F. Wang, and V. M. Dixit.** 1992. The Epstein-Barr virus LMP1 gene product induces A20 zinc finger protein expression by activating nuclear factor κB. *J. Biol. Chem.* **34:**24157–24160.

60. **Lam, K. M. C., N. Syed, H. Whittle, and D. H. Crawford.** 1991. Circulating Epstein-Barr virus-carrying B cells in acute malaria. *Lancet* **337:**876–878.

61. **Lambert, P. F., H. Pan, H. C. Pitot, A. Liem, M. Jackson, and A. E. Griep.** 1993. Epidermal cancer associated with expression of human papillomavirus type 16 E6 and E7 oncogenes in the skin of transgenic mice. *Proc. Natl. Acad. Sci. USA* **90:**5583–5587.

62. **Laux, G., M. Perricaudet, and P. J. Farrell.** 1988. A spliced Epstein-Barr virus gene expressed in latently transformed lymphocytes is created by circularization of the linear viral genome. *EMBO J.* **7:**769–774.

63. **Leibowitz, D., R. Kopan, E. Fuchs, J. Sample, and E. Kieff.** 1987. An Epstein-Barr virus transforming protein associates with vimentin in lymphocytes. *Mol. Cell. Biol.* **7:**2299–2308.

64. **Leibowitz, D., D. Wang, and E. Kieff.** 1986. Orientation and patching of the latent infection membrane protein encoded by Epstein-Barr virus. *J. Virol.* **58:**233–237.

65. **Lewin, N., P. Åman, M. G. Masucci, E. Klein, G. Klein, B. Öberg, H. Strander, W. Henle, and G. Henle.** 1987. Characterization of EBV-carrying B-cell populations in healthy seropositive individuals with regard to density, release of transforming virus and spontaneous outgrowth. *Int. J. Cancer* **39:**472–476.

66. **Ling, P. D., D. R. Rawlins, and S. D. Hayward.** 1993. The Epstein-Barr virus immortalizing protein EBNA-2 is targeted to DNA by a cellular enhancer-binding protein. *Proc. Natl. Acad. Sci. USA* **90:**9237–9241.

67. **Lowe, S. W., S. Bodis, A. McClatchey, L. Remington, H. E. Ruley, D. E. Fisher, D. E. Housman, and T. Jacks.** 1994. p53 status and the efficacy of cancer therapy in vivo. *Science* **266:**807–810.

68. **Mann, K. P., D. Staunton, and D. A. Thorley-Lawson.** 1985. Epstein-Barr virus-encoded protein in plasma membranes of transformed cells. *J. Virol.* **55:**710–720.

69. **Mannick, J. B., J. Cohen, M. Birkenbach, A. Marchini, and E. Kieff.** 1991. The Epstein-Barr virus nuclear protein encoded by the leader of the EBNA RNAs (EBNA-LP) is important in B-lymphocyte transformation. *J. Virol.* **65:**6829–6837.

70. **Mark, W., and B. Sugden.** 1982. Transformation of lymphocytes by Epstein-Barr virus requires only one-fourth of the viral genome. *Virology* **122:**431–443.

71. **Martin, J. M., and B. Sugden.** 1991. The LMP oncoprotein resembles activated receptors in its properties of turnover. *Cell Growth Differ.* **2:**653–660.

72. **Martin, J. M., and B. Sugden.** 1991. Transformation by the oncogenic latent membrane protein correlates with its rapid turnover, membrane localization, and cytoskeletal association. *J. Virol.* **65:**3246–3258.

73. **Martin, J. M., D. Veis, S. J. Korsmeyer, and B. Sugden.** 1993. Latent membrane protein of Epstein-Barr virus induces cellular phenotypes independently of expression of Bcl-2. *J. Virol.* **67:**5269–5278.

74. **Miller, C. L., J. H. Lee, E. Kieff, and R. Longnecker.** 1994. An integral membrane protein (LMP2) blocks reactivation of Epstein-Barr virus from latency following surface immunoglobulin cross-linking. *Proc. Natl. Acad. Sci. USA* **91:**772–776.

75. **Miller, G., D. Coope, J. Niederman, and J. Pagano.** 1976. Biological properties and viral surface antigens of Burkitt lymphoma- and mononucleosis-derived strains of Epstein-Barr virus released from transformed marmoset cells. *J. Virol.* **18:**1071–1080.

76. **Mitchell, T., and B. Sugden.** 1995. Stimulation of NF-κB-mediated transcription by mutant derivatives of the latent membrane protein of Epstein-Barr virus. *J. Virol.* **69:**2968–2976.

77. **Miyoshi, I., I. Kubonishi, S. Yoshimoto, T. Akagi, Y. Ohtsuki, Y. Shiroishi, K. Nagata, and Y. Hinuma.** 1981. Type C virus particles in a cord T cell line derived by co-cultivating normal human cord leukocytes with human leukemic T cells. *Nature* (London) **294:**770–772.

78. **Moriyama, T., S. Guilhot, K. Klopchin, B. Moss, C. A. Pinkert, R. D. Palmiter, R. L. Brinster, O. Kanagawa, and F. V. Chisar.** 1990. Immunobiology and pathogenesis of hepatocellular injury in hepatitis B virus transgenic mice. *Science* **248:**361–364.

79. **Morrow, R. H., A. Kisuule, M. C. Pike, and P. G. Smith.** 1976. Burkitt's lymphoma in the Mengo districts of Uganda: epidemiologic features and their relationship to malaria. *J. Natl. Cancer Inst.* **56:**479–486.

80. **Mosier, D. E., S. M. Baird, M. B. Kirven, R. J. Gulizia, D. B. Wilson, R. Kubayashi, G. Picchio, J. L. Garnier, J. L. Sullivan, and T. J. Kipps.** 1990. EBV-associated B-cell lymphomas following transfer of human peripheral blood lymphocytes to mice with severe combined immunodeficiency. *Curr. Top. Microbiol. Immunol.* **166:**317–323.

81. **Murray, R. J., M. G. Kurilla, H. M. Griffin, J. M. Brooks, M. Mackett, J. R. Arrand, M. Rowe, S. R. Burrows, D. J. Moss, E. Kieff, and A. B. Rickinson.** 1990. Human cytotoxic T-cell responses against Epstein-Barr virus nuclear antigens demonstrated by using recombinant vaccinia viruses. *Proc. Natl. Acad. Sci. USA* **87:**2906–2910.

82. **Nada, S., T. Yagi, H. Takeda, T. Tokunaga, H. Nakagawa, Y. Ikawa, M. Okada, and S. Aizawa.** 1993. Constitutive activation of src family kinases in mouse embryos that lack csk. *Cell* **73:** 1125–1135.

83. **Nemerow, G. R., C. Mold, V. K. Schwend, V. Tollefson, and N. R. Cooper.** 1987. Identification of gp350 as the viral glycoprotein mediating attachment of Epstein-Barr virus (EBV) to the EBV/ C3d receptor of B cells: sequence homology of gp350 and C3 complement fragment C3d. *J. Virol.* **61:**1416–1420.

84. **Nkrumah, F., W. Henle, G. Henle, R. Herberman, V. Perkins, and R. Depue.** 1976. Burkitt's lymphoma: its clinical course in relation to immunologic reactivities to Epstein-Barr virus and tumor related antigens. *J. Natl. Cancer Inst.* **57:**1051–1056.

85. **Ohno, S., M. Babonits, F. Wiener, J. Spira, G. Klein, and M. Potter.** 1979. Non-random chromosome changes involving the 1g-gene carrying chromosomes 12 and 6 in pristane-induced mouse plasmacytomas. *Cell* **18:**1001–1007.

86. **Okada, M., and H. Nakagawa.** 1989. A protein tyrosine kinase involved in regulation of pp60$^{c\text{-}src}$ function. *J. Biol. Chem.* **264:**20886–20893.

87. **Payne, G. S., J. M. Bishop, and H. E. Varmus.** 1982. Multiple arrangements of viral DNA and an activated host oncogene in bursal lymphomas. *Nature* (London) **295:**209–214.

88. **Payne, L. N., and H. G. Purchase.** 1991. Leukosis/sarcoma group, p. 386–439. *In* B. W. Calnek (ed.), *Diseases of Poultry*, 9th ed. Iowa State University Press, Ames.

89. **Peng, M., and E. Lundgren.** 1992. Transient expression of the Epstein-Barr virus LMP1 gene in human primary B cells induces cellular activation and DNA synthesis. *Oncogene* **7:**1775–1782.

90. **Piwnica-Worms, H., K. B. Saunders, T. M. Roberts, A. E. Smith, and S. H. Cheng.** 1987. Tyrosine phosphorylation regulates the biochemical and biological properties of pp60$^{c\text{-}src}$. *Cell* **49:**75–82.

91. **Qu, L., and D. T. Rowe.** 1992. Epstein-Barr virus latent gene expression in uncultured peripheral blood lymphocytes. *J. Virol.* **66:**3715–3724.

92. **Rawlins, D. R., G. Milman, S. D. Hayward, and G. S. Hayward.** 1985. Sequence-specific DNA binding of the Epstein-Barr virus nuclear antigen (EBNA-1) to clustered sites in the plasmid maintenance region. *Cell* **42:**859–868.

93. **Reisman, D., and B. Sugden.** 1986. *trans* activation of an Epstein-Barr viral transcriptional enhancer by the Epstein-Barr viral nuclear antigen 1. *Mol. Cell. Biol.* **6:**3838–3846.

94. **Reisman, D., J. Yates, and B. Sugden.** 1985. A putative origin of replication of plasmids derived from Epstein-Barr virus is composed of two *cis*-acting components. *Mol. Cell. Biol.* **5:**1822–1832.

95. **Resh, M. D.** 1994. Myristylation and palmitylation of src family members: the fats of the matter. *Cell* **76:**411–413.

96. **Robinson, J., D. Smith, and J. Niederman.** 1980. Mitotic EBNA-positive lymphocytes in peripheral blood during infectious mononucleosis. *Nature* (London) **287:**334–336.

97. **Robinson, J. E., D. Smith, and J. Niederman.** 1981. Plasmacytic differentiation of circulating Epstein-Barr virus-infected B lymphocytes during acute infectious mononucleosis. *J. Exp. Med.* **153:**235–244.

98. **Rogers, R. P., M. Woisetschlaeger, and S. H. Speck.** 1990. Alternative splicing dictates translational start in Epstein-Barr virus transcripts. *EMBO J.* **9:**2273–2277.

99. **Rooney, C., J. G. Howe, S. H. Speck, and G. Miller.** 1989. Influences of Burkitt's lymphoma and primary B-cells on latent gene expression by the nonimmortalizing P3J-HR-1 strain of Epstein-Barr virus. *J. Virol.* **63:**1531–1539.

100. **Rowe, D. T., M. Rowe, G. I. Evan, L. E. Wallace, P. J. Farrell, and A. B. Rickinson.** 1986. Restricted expression of EBV latent genes and T-lymphocyte-detected membrane antigen in Burkitt's lymphoma cells. *EMBO J.* **5:**2599–2607.

101. **Rowe, M., D. T. Rowe, C. D. Gregory, L. S. Young, P. J. Farrell, H. Rupani, and A. B. Rickinson.** 1987. Differences in B cell growth phenotype reflect novel patterns of Epstein-Barr virus latent gene expression in Burkitt's lymphoma cells. *EMBO J.* **6:**2743–2751.

102. **Saewha, J., and P. Lambert.** 1995. Integration of HPV-16 DNA into the human genome leads to increased stability of E6/E7 mRNAs: implications for cervical carcinogenesis. *Proc. Natl. Acad. Sci. USA* **92:**1654–1658.

103. **Sample, J., D. Liebowitz, and E. Kieff.** 1989. Two related Epstein-Barr virus membrane proteins are encoded by separate genes. *J. Virol.* **63:**933–937.

104. **Scheffner, M., J. M. Huibregtse, R. D. Vierstra, and P. M. Howley.** 1993. The HPV-16 E6 and E6-AP complex functions as a ubiquitin-protein ligase in the ubiquitination of p53. *Cell* **75:**495–505.

105. **Schubach, W. H., R. Hackman, P. E. Neiman, G. Miller, and E. D. Thomas.** 1982. A monoclonal immunoblastic sarcoma in donor cells bearing Epstein-Barr virus genomes following allogenic marrow grafting for acute lymphoblastic leukemia. *Blood* **60:**180–187.

106. **Schwartz, E., U. K. Freese, L. Gissman, W. Mayer, B. Roggenbuck, A. Stremlau, and H. zur Hausen.** 1985. Structure and transcription of human papilloma virus sequences in cervical carcinoma cells. *Nature* (London) **314:**111–114.

107. **Speck, S. H., and J. L. Strominger.** 1989. Transcription of Epstein-Barr virus in latently infected, growth-transformed lymphocytes. *Adv. Viral Oncol.* **8:**133–150.

108. **Spencer, C. A., and M. Groudine.** 1991. Control of c-*myc* regulation in normal and neoplastic cells. *Adv. Cancer Res.* **56:**1–48.

109. **Stanton, L. W., R. Watt, and K. B. Marcu.** 1983. Translocation, breakage, and truncated transcripts of c-*myc* oncogene in murine plasmacytomas. *Nature* (London) **303:**401–406.

110. **Stehelin, D., H. E. Varmus, J. M. Bishop, and P. K. Vogt.** 1976. DNA related to the transforming gene(s) of avian sarcoma viruses is present in normal avian DNA. *Nature* (London) **260:**170–173.

111. **Sugden B.** 1977. Comparison of Epstein-Barr viral DNAs in Burkitt lymphoma biopsy cells and in cells clonally transformed *in vitro. Proc. Natl. Acad. Sci. USA* **74:**4651–4655.

112. **Sugden, B.** 1994. Latent infection of B-lymphocytes by Epstein-Barr virus. *Semin. Virol.* **5:**197–205.

113. **Sugden, B., and W. Mark.** 1977. Clonal transformation of adult human leukocytes by Epstein-Barr virus. *J. Virol.* **23:**503–508.

114. **Sugden, B., and N. Warren.** 1989. A promoter of Epstein-Barr virus that can function during latent infection can be transactivated by EBNA-1, a viral protein required for viral DNA replication during latent infection. *J. Virol.* **63:**2644–2649.

115. **Svedmyr, E., and M. Jondal.** 1975. Cytotoxic effector cells specific for B cell lines transformed by Epstein-Barr virus are present in patients with infectious mononucleosis. *Proc. Natl. Acad. Sci. USA* **72:**1622–1626.

116. **Temin, H. M.** 1974. On the origin of the genes for neoplasia: G. H. A. Clowes Memorial Lecture. *Cancer Res.* **34:**2835–2841.

117. **Temin, H. M., and H. Rubin.** 1958. Characteristics of an assay for Rous sarcoma virus and Rous sarcoma cells in tissue culture. *Virology* **6:**669–688.

118. **Tierney, R. J., N. Steven, L. S. Young, and A. B. Rickinson.** 1994. Analysis of viral gene transcription during primary infection and in the carrier state. *J. Virol.* **68:**7374–7385.

119. **Tomkinson B., E. Robertson, and E. Kieff.** 1993. Epstein-Barr virus nuclear proteins (EBNA) 3A and 3C are essential for B-lymphocyte growth transformation. *J. Virol.* **67:**2014–2025.

120. **Tsang, S.-F., F. Wang, K. M. Izumi, and E. Kieff.** 1991. Delineation of the *cis*-acting element mediating EBNA-2 transactivation of latent infection membrane protein expression. *J. Virol.* **65:**6765–6771.

121. **Wang, D., D. Liebowitz, and E. Kieff.** 1985. An EBV membrane protein expressed in immortalized lymphocytes transforms established rodent cells. *Cell* **43:**831–840.

122. **Wang, D., D. Liebowitz, F. Wang, C. Gregory, A. Rickinson, R. Larson, T. Springer, and E.**

Kieff. 1988. Epstein-Barr virus latent infection membrane protein alters the human B-lymphocyte phenotype: deletion of the amino terminus abolishes activity. *J. Virol.* **62**:4173–4184.

123. **Wang, F., C. D. Gregory, M. Rowe, A. B. Rickinson, D. Wang, M. Birkenbach, H. Kikutani, T. Kishimoto, and E. Kieff.** 1987. Epstein-Barr virus nuclear antigen 2 specifically induces expression of the B-cell activation antigen CD23. *Proc. Natl. Acad. Sci. USA* **84**:3452–3456.

124. **Wang, F., C. Gregory, C. Sample, M. Rowe, D. Liebowitz, R. Murray, A. Rickinson, and E. Kieff.** 1990. Epstein-Barr virus latent membrane protein (LMP1) and nuclear proteins 2 and 3C are effectors of phenotypic changes in B lymphocytes: EBNA-2 and LMP1 cooperatively induce CD23. *J. Virol.* **64**:2309–2318.

125. **Wang, F., H. Kikutani, S.-F. Tsang, T. Kishimoto, and E. Kieff.** 1991. Epstein-Barr virus nuclear protein 2 transactivates a *cis*-acting CD23 DNA element. *J. Virol.* **65**:4101–4106.

126. **Wang, F., S. Tang, M. G. Kurilla, J. Cohen, and E. Kieff.** 1990. Epstein-Barr virus nuclear antigen 2 transactivates latent membrane protein LMP1. *J. Virol.* **64**:3407–3416.

127. **Werness, B. A., A. J. Levine, and P. M. Howley.** 1990. Association of human papillomavirus types 16 and 18 E6 proteins with p53. *Science* **248**:76–79.

128. **Whittle, H. C., J. Brown, K. Marsh, B. M. Greenwood, P. Seidelin, H. Tighe, and L. Wedderburn.** 1984. T-cell control of Epstein-Barr virus-infected B cells is lost during *P. falciparum* malaria. *Nature* (London) **312**:449–450.

129. **Woisetschlaeger, M., C. N. Yandava, L. A. Furmanski, J. L. Strominger, and S. H. Speck.** 1990. Promoter switching in Epstein-Barr virus during the initial stages of infection of B lymphocytes. *Proc. Natl. Acad. Sci. USA* **87**:1725–1729.

130. **Yano, T., C. A. Sander, H. M. Clark, M. V. Dolezal, E. S. Jaffe, and M. Raffeld.** 1993. Clustered mutation in the second exon of the MYC gene in sporadic Burkitt's lymphoma. *Oncogene* **8**: 2741–2749.

131. **Yao, Q. Y., A. B. Rickinson, and M. A. Epstein.** 1985. A re-examination of the Epstein-Barr virus carrier state in healthy sero-positive individuals. *Int. J. Cancer* **35**:35–42.

132. **Yates, J., N. Warren, D. Reisman, and B. Sugden.** 1984. A *cis*-acting element from the Epstein-Barr viral genome that permits stable replication of recombinant plasmids in latently infected cells. *Proc. Natl. Acad. Sci. USA* **81**:3806–3810.

133. **Yates, J. L., and N. Guan.** 1991. Epstein-Barr virus-derived plasmids replicate only once per cell cycle and are not amplified after entry into cells. *J. Virol.* **65**:483–488.

134. **Yates J. L., N. Warren, and B. Sugden.** 1985. Stable replication of plasmids derived from Epstein-Barr virus in various mammalian cells. *Nature* (London) **313**:812–815.

135. **Yoshida, M.** 1994. Retroviruses (HTLVs), p. 929–943. *In* G. Stamatoyannopoulos, A. W. Nienhaus, P. W. Majerus, and H. Varmus (ed.), *The Molecular Basis of Blood Diseases*. The W. B. Saunders Co., Philadelphia.

136. **Zhang, S., and M. Nonoyama.** 1994. The cellular proteins that bind specifically to the Epstein-Barr virus origin of plasmid DNA replication belong to a gene family. *Proc. Natl. Acad. Sci. USA* **91**:2843–2847.

137. **Zimber-Strobl, U., E. Kremmer, F. Grässer, G. Marschall, G. Laux, and G. W. Bornkamm.** 1993. The Epstein-Barr virus nuclear antigen 2 interacts with an EBNA2 responsive *cis*-element of the terminal protein 1 gene promoter. *EMBO J.* **12**:167–175.

The DNA Provirus: Howard Temin's Scientific Legacy
Edited by G. M. Cooper, R. Greenberg Temin, and B. Sugden
© 1995 American Society for Microbiology, Washington, DC 20005

Chapter 13

Under the Influence: from the Provirus Hypothesis to Multistep Carcinogenesis

Harold Varmus

Unlike many other authors in this volume, I was neither one of Howard's students nor one of his fellows. Because I was rarely in Madison and he was rarely out of it, we met relatively infrequently over the 20-odd years of our friendship. Yet I think of him as one of the truly important people in my life.

There are a few simple reasons for this. My scientific life—and the scientific lives of most of the people who have worked with me—have been profoundly influenced by the provirus hypothesis. Reading Howard's papers in the late 1960s, while I was in training at the National Institutes of Health, inspired me to look for proviral DNA and, later, to try to understand its synthesis, structure, and function. Our group's interest in cancer genes, especially the cellular origins of retroviral oncogenes, had many sources, but important among them was one of my first long conversations with Howard on a bus trip between Mexico City and Cuernavaca in the fall of 1972, when his enthusiasm for our plan to seek a cellular version of the transforming gene of Rous sarcoma virus proved persuasive.

Over the ensuing years, our common interests in proviral integration, endogenous viruses, oncogenes, and reverse transcription gave us a language that permitted conversation to thrive even when the intervals between conversations were several months in length. As we came to know each other better, our talks spread beyond science to politics, families, history, and values. Because Howard was never shy about advancing his point of view, I hold him at least partly responsible for my current job through his advocacy with his friend (now my boss) Secretary of Health and Human Services Donna Shalala.

Howard's name will always be consubstantial with the provirus. Because most of my scientific work has been, directly or indirectly, in the service of the provirus, I have been indentured to him for nearly a quarter of a century. The diagram that I used to introduce new students to the major themes in my laboratory at the University of California at San Francisco (UCSF) illustrated vividly the extent to which our studies oriented themselves around the provirus. Part of our group at UCSF worked on the early events in retrovirus replication (virus entry through receptors, reverse transcription, and integration) that result in the estab-

Harold Varmus • Office of the Director, National Institutes of Health, Bethesda, Maryland 20892.

lishment of a provirus. Another part worked on aspects of proviral gene expression, especially mechanisms, such as ribosomal frameshifting, that permit synthesis of the viral enzymes (reverse transcriptase and integrase) required for forming a provirus. Yet another portion of the group studied the oncogenic consequences of proviral integration, including insertional activation and the capture of host proto-oncogenes. All of these components, arrayed around a schematic drawing of a provirus, revealed the depth of our commitment to Teminism.

Many of the themes in my laboratory's work came into clear focus between 1978 and 1981, pivotal years in the study of proviruses and their genetic potential (reviewed in reference 9). During those years, proviruses were first closely mapped, molecularly cloned, and sequenced. The discoveries of long terminal repeats and the regular features of host-virus junctions brought home what was previously conjectured and is now powerfully documented: that retroviral proviruses are dramatically similar in structure and function to the transposable elements and prophages of bacteria. The structural similarities were, of course, most readily appreciated; proviruses, like transposons such as Tn9, contain genes flanked by long terminal repeats that conclude with short inverted repeats, and both kinds of elements are flanked by copies of a short sequence of host DNA that was duplicated during integration. However, the functional analogies have been more provocative and wide-ranging: insertion of proviral DNA after infection can inactivate preexisting genes; integration of proviral DNA appears to be the first step in the process of transduction of the host genetic information that evolves into a retroviral oncogene; infection of germ cells or their precursors establishes an endogenous, genetically transmitted provirus; retroviruses can be redesigned to serve as genetic vectors for both experimental and therapeutic purposes; and suitable placement of a provirus can activate expression of nearby cellular genes. If the adjacent gene is a proto-oncogene, the insertion can initiate oncogenic change.

The idea that insertional activation of proto-oncogenes might constitute the first step in a cancerous process initiated by retroviruses lacking viral oncogenes took flight with the demonstration that avian leukosis viruses augment expression of the c-*myc* proto-oncogene by nearby integration of its provirus (3). These findings stimulated similar work with the mouse mammary tumor virus (MMTV), resulting in the discovery of the *Wnt-1* gene (5), the first cloned member of a large gene family known to govern early developmental events in diverse organisms (6).

The most compelling evidence that insertional activation of *Wnt-1* actually contributes to the induction of mammary cancers by MMTV emerged from studies of transgenic mice (8). Mice programmed to express the *Wnt-1* gene in the mammary gland under the influence of an MMTV long terminal repeat undergo mammary hyperplasia, and all of the females and some of the males develop one or more mammary carcinomas in the first 4 to 12 months of life. The kinetics of appearance of these tumors argue that additional events, not just the inheritance of the *Wnt-1* transgene, are required for full oncogenic transformation in mammary epithelial cells, as has been proposed for most human cancers. Genetic crosses with other transgenic mice (4) and experiments in which *Wnt-1* transgenic mice are infected with MMTV (7) both indicate that enhanced expression of certain fibroblast growth factor (*FGF*) genes can contribute to tumorigenesis, apparently through a direct or indirect collaboration in this model between *FGF* genes and *Wnt* genes.

More recently, *Wnt-1* transgenic mice were crossed to mice carrying genetically engineered mutations in the *p53* gene, a tumor suppressor gene that is often mutant in human breast cancers. Although *p53*-null mice rarely develop mammary cancers in the absence of the *Wnt-1* transgene (2), p53 deficiency has a strong contributory effect on tumorigenesis in the *Wnt-1* transgenic model (1). This can be seen at least three ways. First, mammary carcinomas arise much earlier in *Wnt-1* transgenic–*p53*-null animals, especially males, than in *Wnt-1* transgenic–*p53* wild-type animals. Second, in *Wnt-1* transgenic animals heterozygous for the *p53* mutation, tumors often show loss of the normal *p53* allele, implying selection for cells lacking normal *p53*. Third, tumors deficient in *p53* exhibit several differences from those that retain wild-type *p53:* they are histologically different, with much less fibrotic tissue and more anaplastic cells; they display a much greater tendency to aneuploidy, especially subtetraploidy; and they show a much higher frequency of gross genetic lesions (increases or decreases in DNA copy number, discerned by comparative genomic hybridization).

In the studies summarized in the preceding paragraphs, our group has moved beyond the study of MMTV proviruses as insertional mutagens to consider more broadly the multistep nature of mammary carcinogenesis in hopes of understanding the genetic and physiological events that produce premalignant (e.g., hyperplastic) lesions, primary cancers, and metastases in human beings as well as mice. Sadly, at about the same time, Howard also began to think about cancer from a new perspective, that of the cancer patient. The coincidence of events brought into focus for many of us a contrast of special poignancy: the distance that the field of oncogenetics had come, in no small part under Howard's influence, and the distance it still had to travel to offer something powerful to him, and to us, in a moment of need. It is now left to those who learned to do science through Howard's example to teach that example to others, a legacy to cancer research that may prove to be as great as the provirus itself.

REFERENCES

1. **Donehower, L. A., L. A. Godley, C. M. Aldaz, R. Pyle, Y.-P. Shi, D. Pinkel, J. Gray, A. Bradley, D. Medina, and H. E. Varmus.** Deficiency of *p53* accelerates mammary tumorigenesis in *Wnt-1* transgenic mice and promotes chromosomal instability. *Genes Dev.,* in press.
2. **Donehower, L. A., M. Harvey, B. L. Slagle, M. J. McArthur, C. A. Montgomery, Jr., J. S. Butel, and A. Bradley.** 1992. Mice deficient for p53 are developmentally normal but susceptible to spontaneous tumours. *Nature* (London) **356:**215–221.
3. **Hayward, W. S., B. G. Neel, and S. M. Astrin.** 1981. Activation of a cellular *onc* gene by promoter insertion in ALV-induced lymphoid leukosis. *Nature* (London) **209:**475–479.
4. **Kwan, H., V. Pecenka, A. Tsukamoto, T. G. Parslow, R. Guzman, T.-P. Lin, W. J. Muller, F. S. Lee, P. Leder, and H. E. Varmus.** 1992. Transgenes expressing the *Wnt-1* and *int-2* proto-oncogenes cooperate during mammary carcinogenesis in doubly transgenic mice. *Mol. Cell. Biol.* **12:**147–154.

5. **Nusse, R., and H. E. Varmus.** 1982. Many tumors induced by the mouse mammary tumor virus contain a provirus integrated in the same region of the host genome. *Cell* **31:**99.

6. **Nusse, R., and H. E. Varmus.** 1992. *Wnt* genes. *Cell* **69:**1073–1088.

7. **Shackleford, G. M., C. A. MacArthur, H. C. Kwan, and H. E. Varmus.** 1993. Mouse mammary tumor virus infection accelerates mammary carcinogenesis in *Wnt*-1 transgenic mice by insertional activation of *int-2/Fgf*-3 and *hst/Fgf*-4. *Proc. Natl. Acad. Sci. USA* **90:**740–744.

8. **Tsukamoto, A. S., R. Grosschedl, R. C. Guzman, T. Parslow, and H. E. Varmus.** 1988. Expression of the INT-1 gene in transgenic mice is associated with mammary gland hyperplasia and adenocarcinomas in male and female mice. *Cell* **55:**619–625.

9. **Varmus, H. E.** 1982. Form and function of retroviral proviruses. *Science* **216:**812.

The DNA Provirus: Howard Temin's Scientific Legacy
Edited by G. M. Cooper, R. Greenberg Temin, and B. Sugden
© 1995 American Society for Microbiology, Washington, DC 20005

Chapter 14

Herpes Simplex Virus DNA Replication and Genome Maturation

Sandra K. Weller

I sense an unusually strong feeling for Howard Temin in the scientific community. People want to remember him and talk about him. And they are moved when they do so. This is more than honoring an outstanding colleague. There is a feeling of warmth and a feeling of loss. Why? Perhaps it is right to say that Howard differed from most of us in the way he lived, worked and died. There is a wholeness about him. A wholeness that has departed. What were the components of the wholeness? I can see three: logic, courage and social consciousness. You might add intellectual honesty, but, then, it is part of all three. Actually all three follow from it and it follows from all three.

— George Klein, excerpted from an introduction to the Howard Temin Memorial Lecture Series at the Annual Meeting sponsored by the Laboratory of Tumor Cell Biology
(Rockville, Maryland, August 1994)

I derive great pleasure from the realization that Howard and I both became virologists at the California Institute of Technology (Caltech) in Pasadena, he as a graduate student and I as a high school student. Howard went to Caltech in the early 1950s as a graduate student, originally to study developmental biology. However, after a year and a half, he was inspired by the observation made by Manaker and Groupe (81) that RNA tumor viruses, as they were known then, could cause morphological changes in infected cells. He realized that animal viruses could provide a system with which to study development. Although he worked on Rous sarcoma virus, Howard was greatly influenced by the phage group at Caltech and by the elegant genetic systems being developed at the time. I, too, was influenced by the phage group at Caltech and had the good fortune as a high school student to be able to work in the laboratory of Robert Sinsheimer. In that laboratory, I was exposed to many aspects of phage biology and became fascinated by how one could study elementary aspects of the phage life cycle by using genetic and physical methodologies. My interest in retroviruses began the summer after high school; I had the opportunity to work in the laboratory of

Sandra K. Weller • Department of Microbiology, University of Connecticut Health Center, 263 Farmington Avenue, Farmington, Connecticut 06030.

Robert McAllister at the University of Southern California. In that laboratory, I became fascinated by the life cycles of mammalian retroviruses and by genetic variation between different strains of viruses.

Although as an undergraduate student at Stanford University I worked with other systems, my interest in phage and viruses persisted. I became fascinated by the concept that the study of phage would provide insights into the mode of animal viral reproduction. Twenty years later, I am even more fascinated by the remarkable similarities. As an undergraduate, I wrote a student paper that compared and contrasted the oncogene theory espoused by Huebner and colleagues with Howard's protovirus theory. My early interest in Howard's work and in his 1970 report of an RNA-dependent DNA polymerase in virions of Rous sarcoma virus (136) influenced my choice for graduate study. I was very pleased to get a phone call from Howard in the spring of 1974, inviting me to join his laboratory. I immediately accepted and joined the laboratory in the fall of 1974.

I was Howard's first female graduate student. In 1974, I had very little awareness of the pitfalls and challenges that lay ahead of me as a woman in science. I came to realize much later that Howard was concerned about his ability to be a mentor to a female student, and he had given the topic a lot of thought. For instance, when it came to picking my advisory committee, I was surprised that the names on his list were all women. At the time, I thought that was a little unusual and unnecessary, but I now realize that he recognized the need for me to be exposed to successful women scientists as role models. As the director of the graduate program here at the University of Connecticut Health Center, I am now very aware of the need of students to have appropriate role models and mentors. Howard took his job as mentor and teacher very seriously. I am proud to have had the benefit of his wisdom and direction. He was a tough mentor, demanding accuracy and precision at all times. Howard was very dedicated to encouraging and challenging us to become as rigorous and clear thinking as we could be. His concern for his former students and fellows was quite evident and continued until his death.

In Howard's laboratory, I always found myself drawn to questions of genetic variation within the avian leukosis and sarcoma viruses. The major focus of my thesis was the discovery that certain subgroups of avian leukosis viruses could kill chicken embryo fibroblasts; cell killing correlated well with the accumulation of unintegrated viral DNA (135, 145, 149). Subgroup B, D, and F viruses were cytopathic, whereas subgroup A, C, and E viruses were not. The implication was that the relationship between the envelope glycoprotein and its receptor was responsible for the patterns of cytopathicity, since the various subgroups of leukosis viruses were known to differ primarily in the envelope glycoprotein. Cell killing correlated well with spread and superinfection by the cytopathic subgroup viruses. We speculated that some inherent differences in the kinetics of establishment of superinfection immunity were responsible for the differences in cytopathicity of certain subgroups. The receptor for the A subgroup of avian leukosis viruses has recently been cloned and characterized (9, 50). Further characterization of this receptor and cloning and characterization of the receptors for other avian viruses of different subgroups may elucidate the molecular basis for the differences in cytopathicity. The mechanism of cell killing by retroviruses is even more relevant today, since many strains of human immunodeficiency virus are known to be cytopathic, and depletion of T lymphocytes plays an important role in immunodeficiency disease (43, 101). We may find that T-lymphocyte killing occurs by a combination of mechanisms, including massive superinfection, syncytium formation, apoptosis, or an as yet poorly defined signaling mechanism.

When it came time to choose a research area for my postdoctoral work, I made the decision to leave retroviruses for two major reasons, both of which show some lack of prescience on my part. First, I wanted to study a virus that had a larger genome, since most retroviruses studied at that time encoded only three or four gene products. Second, it seemed clear in 1979 that retroviruses did not directly cause any relevant human diseases. I was of course proved wrong on both counts by the discovery of human retroviruses that, in addition to being much more genetically complex than simple retroviruses, clearly are major human pathogens. But I have not regretted my decision to study herpesviruses, which have proved to be rich in genetic complexity and an excellent experimental system.

In 1981, as a postdoctoral fellow, I started working on herpes simplex virus type 1 (HSV-1), which has a double-stranded DNA genome and encodes approximately 80 viral gene products. HSV is amenable to genetic and biochemical analyses and in many ways bears a striking resemblance to the T-even bacteriophages. The rest of this chapter gives an overview of the work done by my laboratory during the last 10 years on the replication and maturation of the HSV-1 genome, drawing parallels with phage systems where appropriate.

GENERAL ASPECTS OF THE MECHANISM OF DNA REPLICATION AND GENOME MATURATION

The basic principles of DNA replication are conserved in all bacterial and animal viruses and generally involve semiconservative replication and initiation at specific sequences, termed origins of replication. However, the molecular mechanisms of DNA replication vary widely from virus to virus. Circular duplex DNAs are generally thought to replicate in a theta structure (24) or through a rolling-circle mechanism (51). The small eukaryotic papovaviruses such as simian virus 40 replicate entirely by theta replication, leading to interlocked circles that must be separated prior to genome encapsidation. In contrast, bacteriophage lambda replicates initially as a theta form and switches to rolling-circle replication later in infection.

The DNA bacteriophage T4 replicates as a linear DNA molecule, and initiation occurs by two markedly different strategies (79). Because of similarities in genome size and the large number of viral gene products required for DNA replication, T4 serves as an excellent model system for the study of DNA replication in HSV. The initial phase of T4 DNA replication occurs at specific origins of replication (73) and is dependent on the RNA polymerase of the host cell to synthesize primers that are utilized in the initial replication forks. Later during infection, concatemers are formed by a mechanism that does not use specific origin sequences (79, 94). Late DNA replication occurs by a recombination-dependent strategy in which DNA synthesis is primed from the ends of DNA strands that have invaded homologous segments of other chromosomes (Fig. 1A). Single-strand invasion provides 3' hydroxyl primers for initiation of further DNA replication (Fig. 1B). Thus, these recombination intermediates can be converted into replication forks. Repeated rounds of strand invasion followed by elongation result in the generation of branched DNA molecules (94) (Fig. 1C). Branch migration and subsequent resolution of these recombination intermediates are required for efficient late viral DNA synthesis and packaging in T4.

The 152-kb HSV-1 genome is composed of two components termed L and S. The L component consists of unique sequences (U_L) flanked by the inverted repeated sequences *ab* and *b'a'*, whereas the S component consists of unique sequences (U_S) flanked by inverted repeated sequences *ac* and *c'a'* (Fig. 2) (60, 118). During the course of viral DNA replication, the two unique regions invert relative to each other (60). The mechanism of DNA replication in HSV is not well understood, nor is the relationship between replication and recombination known. Several lines of evidence suggest that the two processes are closely linked (3, 142). Not only is recombination between two coinfecting viral genomes a frequent event, but, as just mentioned, intramolecular recombination leading to genome isomerization also occurs. Although it was initially proposed that the inversion events occur via a site-specific recombination event, it now appears more likely

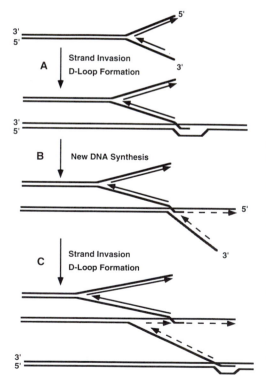

FIGURE 1. Model of T4 DNA replication and recombination. Recombination-dependent late DNA replication of T4 is depicted here. The topmost sketch represents a linear molecule that has replicated at an origin of replication. At the end of the molecule, a free 3' end that cannot be replicated is generated. (A) The 3' end of the parental DNA strand invades a homologous segment of another molecule. (B) Leading-strand synthesis can be primed from the 3' end of the invading DNA strand. Lagging-strand synthesis would require the activity of primase. (C) This process can be repeated with a new 3' end invading another molecule. Heavy solid lines represent parental DNA molecules. Thin solid lines represent DNA synthesized following initiation at an origin of replication. Thin dashed lines represent DNA synthesized following recombination events. This figure was based on the model of T4 DNA replication described by Mosig (93).

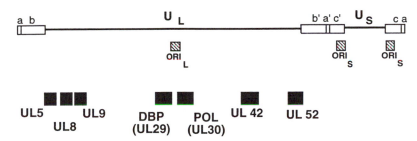

FIGURE 2. Locations of HSV origins and genes encoding DNA synthetic functions. The sequence arrangement of the HSV-1 genome is shown on the top line. Locations of the origins of DNA replication are depicted on the second line. On the third line are map locations of genes that encode functions involved in DNA synthesis. Genes and origins are not drawn to scale.

that HSV genome inversion can result from generalized recombination stimulated by double-stranded breaks (115, 123). In this sense, HSV may resemble a large number of organisms that use double-stranded break repair for recombination (134). An overview of what is known about HSV DNA replication and recombination follows.

Formation of Circular DNA Intermediates

Analysis of parental HSV DNA by restriction enzyme and sedimentation analyses suggests that most viral DNA molecules lose their free ends shortly after infection, most likely as a result of circularization (47, 67, 68, 103), which probably occurs by direct ligation of the ends. Fusion of the termini occurs rapidly after the virus enters the cell by a process that does not require de novo protein synthesis, suggesting that ligation can be carried out by a host cell enzyme or by a viral protein brought into the cells during infection (103).

Formation of Greater-Than-Unit-Length Replication Intermediates

The sedimentation behavior of replicating DNA suggests that greater-than-unit-length molecules of HSV DNA appear in cells during infection (14, 63, 67, 68). Electron microscopic examination of DNA from infected cells has revealed that viral DNA comprises complex networks, many of which are larger than unit size (12–14, 68). In addition, analysis of newly replicated DNA by restriction enzyme digestion suggests that such DNA does not possess detectable termini (67, 69). Available data suggest that DNA replication results in the formation of large head-to-tail concatemers consisting of tandem repeats of the viral genome.

It is not clear how large concatemeric molecules are generated from the initial circular molecule; however, the model most often invoked is a rolling-circle mechanism (67). Although there is evidence that the products of DNA replication are of high molecular weight and can be chased into low-molecular-weight DNA (68, 69, 97), direct evidence for rolling-circle DNA replication has been lacking. One model to explain how a rolling circle could be formed from a circular intermediate was presented by Morgan and Severini (92). Alternatively, concatemers could be formed by a recombination mechanism similar to that used during T4 bacteriophage DNA replication (Fig. 1). In this case, DNA synthesis could be primed from the ends of DNA strands that have invaded homologous segments of other chromosomes (Fig. 1A). Unlike that of T4, the HSV-1 genome is circular; however, free ends could be generated as a result of the cleavage and packaging process (discussed later in this chapter) or as a result of double-stranded breaks that arise during DNA replication across the inverted repeats of the genome, as suggested by Sarisky and Weber (115). The free end could invade either a homologous segment on another molecule (as shown in Fig. 1A) or a homologous region within the same molecule.

This model is consistent with several recent observations. (i) Severini and his colleagues at the University of Alberta have shown that replication intermediates in HSV-1-infected cells are present in a nonlinear structure that cannot enter a pulsed-field gel even after being digested with a restriction enzyme that has a single recognition site within the HSV genome (117). (ii) Analysis of newly replicated DNA by pulsed-field gel electrophoresis indicates that inversion has occurred at the earliest times that replicated DNA can be detected (8, 84a, 153). Taken together, these results suggest that recombination may play an obligatory role in the generation of replication intermediates. The model suggests that "branched DNA" is formed during DNA replication by a mechanism involving recombination. The resolution of branched DNA (see below) would lead to the generation of concatemeric DNA.

Resolution of Branched Recombination Intermediates

The branched structures that are created during late-stage DNA replication of the bacteriophage T4 are resolved by a phage protein endonuclease VII (product of gene *49*), which can process Holliday junctions to produce unconnected DNA duplexes (89). Resolution of the branched structures is required before genomic DNA is packaged into phage proheads (71, 72). If HSV DNA replication proceeds via recombination intermediates, we predict that branched molecules that will need to be resolved prior to cleavage and packaging will be generated. We propose that the viral enzyme alkaline nuclease may be involved in the resolution of replication-recombination intermediates. Evidence for the possible involvement of alkaline nuclease in this process is presented later in this chapter.

Cleavage of Concatemers into Unit-Length Virion DNA and Packaging of Unit-Length Virion DNA into Capsids

Regardless of how they are formed, the tandemly repeated HSV molecules generated during replication must be processed and packaged into capsids. Empty capsids (containing no DNA) assemble first in the nucleus, and these empty capsids apparently are the precursors of DNA-containing nucleocapsids. Electron microscopy reveals several forms of capsids in the nuclei of infected cells, ranging from "empty" forms (capsids that lack viral DNA) to "filled" or dense forms (capsids that contain packaged viral DNA) (45). Pulse-label experiments demonstrate that endless replicative intermediates are indeed the precursors to virion DNA (69, 97). Ladin et al. (77, 78) described a series of *ts* mutants of pseudorabies virus (a closely related porcine herpesvirus) that are defective in the cleavage of concatemeric DNA and the formation of nucleocapsids at the nonpermissive temperature. The existence of these mutants has been taken as evidence that maturation of viral DNA is closely correlated with the formation of full capsids. Similarly, HSV-1 mutants that are also defective in both cleavage and packaging have been described (1, 2, 5, 86, 105, 109, 112, 119, 120). The cleavage and packaging mutants define at least six complementation groups, which suggests that cleavage and packaging are likely to be complex processes.

The processes of capsid formation and encapsidation also have many parallels in phage systems. Most double-stranded DNA bacteriophages assemble preformed capsids (procapsids) that are competent to take up viral DNA (reviewed in references 11, 15, and 16). One peculiar structure present in bacteriophage procapsids is the portal vertex, a dodecameric ring of 12 proteins that forms a structure at one unique vertex of the capsid through which the DNA enters and exits. The portal vertex is the docking site for the packaging proteins, including terminase, which is responsible for cleavage of monomeric units from concatemeric DNA. DNA is believed to be taken up into capsids by the putative translocase by a poorly understood mechanism that requires ATP hydrolysis. In herpesviruses, it is likely that analogous proteins will be required for cleavage and packaging, including a portal protein, terminase, and a translocase (discussed below).

CIS- AND *TRANS*-ACTING ELEMENTS INVOLVED IN VIRAL DNA SYNTHESIS

Despite the fact that little is known about the mechanism of DNA replication in HSV, considerable information has been collected with regard to the *cis-* and *trans*-acting elements involved. These topics have been reviewed previously (99, 144) and are only briefly updated here.

SIGNALS NEEDED IN *CIS* FOR VIRAL DNA SYNTHESIS

The HSV-1 genome has three internal origins of replication: one is in the middle of U_L (ori_L), and a diploid origin is in the short repeated region (ori_S) (Fig. 2) (126, 130, 148). Both ori_S and ori_L lie between two divergently transcribed genes; ori_S is located between *ICP4* and either *ICP22* or *ICP47,* and ori_L is situated between the replication genes *UL29* and *UL30*. The *cis*-acting signals required for ori_S activity reside within a 90-bp segment that contains an almost perfect palindromic sequence of 45 bp with a central 18-bp A/T-rich region (130, 132). Wong and Schaffer demonstrated that regions that flank ori_S and contain transcriptional regulatory elements stimulate origin function as much as 80-fold (151). Precise localization and characterization of ori_L have been severely hampered by our inability to clone in bacteria sequences that contain ori_L in an undeleted form (125). We reported the successful cloning of an HSV fragment containing the deletion-prone region in a yeast cloning vector (148). This fragment contains a perfect 144-bp palindrome with striking homology to ori_S. Deletions that removed the center of symmetry of the palindrome were no longer active. The smallest fragment reported to exhibit origin activity is a 136-bp fragment within the 144-bp palindrome (144). The effects of transcriptional regulatory signals flanking ori_L have not been examined in detail. Both ori_S and ori_L contain binding sites for UL9, one of the *trans*-acting HSV proteins required for viral DNA replication (see below).

Although both origins can serve to direct the amplification of plasmids in transfected cells, little is known about the functional significance of origin sequences in the mechanism of DNA replication, nor is it clear why HSV has three origins (one copy of ori_L and two copies of ori_S). Viruses with a deletion of ori_L (104) or a deletion of both copies of ori_S are viable (66). Thus, it appears that none of the HSV origins of replication is individually required for viral DNA synthesis in an infected cell and that one copy of either ori_S or ori_L can suffice.

TRANS-ACTING PROTEINS INVOLVED IN DNA REPLICATION

In addition to structural elements required for the initiation of DNA synthesis, the HSV genome encodes many *trans*-acting proteins involved in DNA synthesis. Of the 77 or so proteins predicted to be encoded by HSV-1, at least 12 are involved in DNA synthesis directly or indirectly. These proteins can be divided into two classes: (i) those involved in nucleotide metabolism and (ii) those directly involved in DNA synthesis.

Viral Enzymes Involved in Nucleotide Metabolism

Several HSV enzymes involved in nucleotide metabolism were first identified through enzymatic assay of infected cell extracts, and many of these, including thymidine kinase, ribonucleotide reductase, dUTPase, and uracil-DNA glycosylase, have now been shown to be encoded by the virus (reviewed in reference 144). In general, the enzymes involved in nucleotide metabolism are dispensable for viral growth in exponentially growing cells in culture, although many of these enzymes may be required in growth-arrested cells or during in vivo infection of animal hosts.

Viral Enzymes Involved Directly in DNA Synthesis

A key approach to the identification of gene products essential for viral DNA replication in infected cells has been to isolate mutants that exhibit alterations in

DNA synthesis. The analysis of DNA-negative *ts* and null mutants has led to the identification of seven distinct complementation groups encoding seven viral gene products that are absolutely essential for viral DNA replication in infected cells (reviewed in reference 144). A transient transfection assay in which cloned fragments of HSV DNA were tested for their abilities to support the amplification of a cotransfected HSV-origin-containing plasmid also indicated that the same seven viral genes are necessary and sufficient for origin-dependent plasmid amplification (24, 121). The seven genes encode a two-subunit DNA polymerase (UL30 and UL42); a single-stranded DNA-binding protein, ICP8 (UL29); a three-protein complex with helicase-primase activities (UL5, UL8, and UL52); and an HSV-origin-specific DNA-binding protein (UL9) (reviewed in references 99 and 144). This review concentrates on the helicase-primase (UL5, UL8, and UL52) and the origin-binding protein (UL9), since these are the most intensively studied in my laboratory.

Two-subunit DNA polymerase (UL30 and UL42)

The HSV DNA polymerase is known to comprise two distinct subunits, a 130-kDa catalytic subunit encoded by the *UL30* gene and a 60-kDa associated protein encoded by the *UL42* gene (59, 61). Sequence analysis of the *UL30* gene has revealed several regions of striking sequence similarity with a variety of animal and bacterial virus DNA polymerases, including the mammalian replicative-DNA polymerase α (49, 111, 152). The *UL30* gene product is catalytically active on its own; moreover, it contains intrinsic 3'-5' exonuclease proofreading activity and 5'-3' exonuclease activity that can function as an RNase H activity (31, 76, 82, 98). The *UL42* gene product has a high affinity for double-stranded DNA, and it has been proposed to increase the processivity of the HSV DNA polymerase by acting as a sliding clamp (58). A combined pharmacological and genetic analysis of the *UL30* and *UL42* genes has provided a powerful approach for investigating functional domains responsible for catalytic activities, interactions with antiviral drugs, and subunit interactions (reviewed in references 29 and 85).

Major single-stranded DNA-binding protein (ICP8, UL29)

HSV-infected cells contain a DNA-binding protein (ICP8) that binds preferentially to single-stranded DNA with no detectable sequence specificity (10, 107, 108) and is essential for viral DNA synthesis (30, 146). ICP8 binds single-stranded DNA in a cooperative fashion that is not dependent on sequence (113, 114). Several properties have been ascribed to ICP8, indicating that it may play multiple roles in the viral life cycle. These include helix destabilization of duplex DNA (19), stimulation of HSV polymerase activity (61, 114), specific interaction with the origin-binding protein UL9 (17), distribution of viral proteins into replication compartments (23), promotion of homologous pairing and strand transfer (21), and regulation of viral gene expression (46, 52, 53). The function of ICP8 may be similar to that of other single-stranded binding proteins: it binds single-stranded

DNA and causes destabilization of duplex DNA during unwinding and movement of the replication fork. It may interact directly or indirectly with the HSV polymerase to stimulate the activity of this enzyme. In addition, it may play other roles; for instance, specific interaction with UL9 may facilitate the initial binding at an origin of replication. Its abilities to redistribute viral and perhaps cellular proteins within the nucleus, promote homologous pairing and strand transfer, and influence gene expression are also of considerable interest.

Helicase and primase (UL5, UL8, and UL52)

Helicase activity is essential for DNA replication in many organisms. For instance, bacteriophage T4 encodes at least two distinct DNA helicases, the gene *41*-encoded helicase and the DNA-dependent ATPase/helicase (dda) helicase (48, 96). The product of gene *41* in association with primase (gene *61*) is able to couple ATP hydrolysis to the unwinding of duplex DNA and is required for viral DNA synthesis. In HSV, helicase and primase activities are also associated with one another in a three-protein complex consisting of the products of the *UL5, UL8,* and *UL52* genes (32). The UL5-UL8-UL52 complex displays DNA-dependent ATPase, primase, and helicase activities. A subcomplex consisting of UL5 and UL52 displays similar activities (26, 41). UL52 contains a motif conserved in many primases that, when altered, abolishes primase activity but not ATPase and helicase activities (74). This suggests that UL52 is probably the primase of the complex. Sequence analysis indicates that UL5 contains six motifs found in a large superfamily of known and putative helicases (55, 56) and thus is likely to be the helicase of the complex. The importance of the helicase motifs to UL5 function has been demonstrated by the isolation and characterization of point mutations within each motif (see below). The function of UL8 in this complex is not clear. In vitro, helicase and primase activities are not absolutely dependent on UL8; however, the ATPase and primase activities of the UL5-UL52 subcomplex can be stimulated in vitro by the addition of UL8 (121, 137). Nevertheless, UL8 is absolutely essential for viral DNA replication in vivo (28). UL8 may play a role in the proper localization of the UL5-UL8-UL52 complex to the nucleus (25). A recent report indicates that UL8 may interact specifically with the origin-binding protein, UL9 (88). Of interest is the recent finding that the bacteriophage T4 gene *59* product appears to stimulate helicase and primase activities of the 41 (helicase) and 61 (primase) proteins and may be a helicase assembly factor that can stimulate the binding of the 41 helicase to DNase onto replication forks (48, 124).

Structure-function analysis of the *UL5* gene has indicated that all six helicase motifs are essential in vivo for replication (Fig. 3) (154, 155). A series of single-amino-acid substitution mutations in the most highly conserved amino acids in each motif were constructed, and an in vivo complementation test was used to study the effect of each mutation on the function of the UL5 protein in viral DNA replication. In this assay, a mutant UL5 protein expressed from a transfected plasmid is used to complement a replication-deficient null mutant with a mutation in the *UL5* gene for the amplification of HSV-origin-containing plasmids. Motifs

FIGURE 3. Conserved helicase motifs and motif mutations in *UL5*. The *UL5* gene is shown with six black boxes depicting each of the motifs shared within a superfamily of helicases. Below are shown the mutations introduced into conserved residues within each motif (55, 56). aas, amino acids.

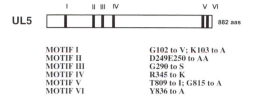

MOTIF I	G102 to V; K103 to A
MOTIF II	D249E250 to AA
MOTIF III	G290 to S
MOTIF IV	R345 to K
MOTIF V	T809 to I; G815 to A
MOTIF VI	Y836 to A

I and II are typical of a consensus nucleotide-binding site (141); mutations in these motifs fail to complement a UL5 null mutant in the in vivo complementation assay. The functional significance of the four other well-conserved motifs (Fig. 3, III through VI) is not known for any helicase. The biochemical analysis of purified wild-type and mutant UL5 proteins expressed from recombinant baculoviruses is expected to reveal the function of motifs III through VI. For instance, preliminary results indicate that motif V may be involved in DNA binding, since DNA-dependent ATPase activity is severely reduced in a purified preparation of UL5-UL52 expressed in cells doubly infected with mutant UL5 and wild-type UL52 (59a). Furthermore, extensive deletion analysis is expected to reveal the domains of each protein involved in protein-protein interactions.

UL9

Viral mutations in the *UL9* gene indicate that this gene is absolutely essential for viral DNA synthesis in vivo (27, 80). Purified recombinant UL9 protein has the following properties: cooperative origin-specific DNA-binding activity; DNA-dependent nucleoside 5'-triphosphatase; DNA helicase on partially double-stranded substrates; ability to form dimers in solution; and specific interaction with at least two other HSV replication proteins, ICP8 and UL8 (17, 18, 20, 22, 42, 44, 88, 143). The ability to bind specifically to origin DNA has been localized to the carboxy-terminal one-third (residues 564 to 832) of UL9 (4, 37, 83, 143). Cooperative DNA binding requires, in addition, the N-terminal two-thirds of UL9 (42, 128).

The N-terminal region of UL9 (residues 150 to 171) contains a putative leucine zipper motif. Mutagenic analysis indicates that the leucine zipper is required for UL9 function but is not involved in dimerization (80a). Interestingly, the region of UL9 required for interaction with UL8 has been localized to the N-terminal two-thirds of the molecule (88). The region required for interaction with ICP8 lies within the C-terminal 317 amino acids. UL9, when expressed by itself in a transient transfection, is capable of localizing to the nucleus in the absence of any other viral proteins, indicating that it contains its own nuclear localization signal. This signal on UL9 has recently been mapped to the C-terminal 105 amino acid residues (80a).

Although it is assumed that UL9 plays a role in the initiation of viral DNA replication, its mechanism of action is far from clear. UL9 shares several properties with the initiator protein of simian virus 40, large T antigen, which binds the origin of DNA replication in a sequence-specific manner and unwinds the duplex in

MOTIF I   K87 to R, K87 to A
MOTIF II   D175 to A
MOTIF III   A213 toS, T214 to S, T214 to A
MOTIF IV   F303W
MOTIF V   G354 to A
MOTIF VI   R387 to K

FIGURE 4. Conserved helicase motifs and motif mutations in *UL9*. The *UL9* gene is shown with six black boxes depicting each of the motifs shared within a superfamily of helicases. The putative leucine zipper within the N-terminal portion of *UL9* is represented by a hatched box overlapping motif II. The DNA-binding C-terminal domain is depicted by a stippled box. Below are shown the mutations introduced into conserved residues within each motif (84). aas, amino acids.

an ATP-dependent fashion (35, 36, 40, 129). However, despite numerous attempts, origin unwinding activity has never been demonstrated for UL9 (18, 44).

Genetic analysis has provided evidence that unwinding activity may be important for UL9 function in vivo. Protein sequence analysis indicates that UL9 is a member of a superfamily of established or putative helicases (56, 57, 64). Six motifs that are shared among these proteins have been identified (Fig. 4). We examined the functional significance of these six motifs for the UL9 protein through the introduction of site-specific mutations resulting in single amino acid substitutions of the most highly conserved residues within each motif (Fig. 4). Mutations in five of the six conserved motifs inactivated the function of the UL9 protein in the in vivo complementation assay described above. These results provide direct evidence for the importance of the conserved motifs (84).

The study of transdominant mutations has proven to be a powerful tool for characterizing specific regions of multifunctional proteins (62, 116). In principle, it is possible to isolate variant forms of a multifunctional protein that retain a subset of wild-type functions and are able to inhibit wild-type activity. The DNA-binding domain of UL9 by itself is a transdominant inhibitor of wild-type UL9 function (102, 131). We have investigated the abilities of mutations in the helicase motifs described above to inhibit wild-type UL9 in an infectious DNA assay. In these assays, wild-type UL9 is somewhat inhibitory to plaque formation by intact infectious viral DNA. Mutations in motifs I, II, and VI were strongly transdominant in this assay. Several possible mechanisms of transdominance have been considered. First, it is possible that motif mutations are transdominant because they can still dimerize, and mixed dimers are unable to initiate viral DNA replication. Alternatively, it is possible that motif mutations are transdominant because they bind viral DNA origins, resulting in titration of available origins. A third possibility is that titration of a viral or cellular factor(s) may be responsible for inhibition. We have begun testing these models and have found that addition of an origin-binding mutation to either wild-type UL9 or to the motif mutation diminishes the transdominant phenotype (80a). This result indicates that transdominance involves binding to the origin of replication. It is still conceivable that the ability to form dimers or interactions with other viral or cellular proteins also contributes to transdominance, and these possibilities are currently being investigated.

Several lines of evidence indicate that overexpression of the wild-type UL9 protein can be inhibitory to viral replication. First, as described above, in trans-dominance assays, plasmids that express wild-type UL9 are somewhat inhibitory to plaque formation by wild-type virus. Second, we previously reported that complementing cell lines that contain high copy numbers (over 100 per haploid genome) of a UL9 expression plasmid do not efficiently support wild-type HSV-1 infection (80). The third line of evidence comes from experiments in which the HSV replication proteins are expressed in insect cells from recombinant baculoviruses. In this assay, UL9 inhibits DNA synthesis from plasmids that contain a viral origin of replication (122). The inhibition of viral DNA synthesis by overexpression of wild-type UL9 could be explained by several models. (i) Overexpression of UL9 may be inhibitory if it titrates some viral factor into a form that is not productive. For instance, UL9 and UL8 have been shown by coimmunoprecipitation experiments to interact with each other (88). One possible role for this interaction would be that UL9, once bound at an origin of replication, would then recruit the three members of the helicase-primase to the complex. It is possible that overexpression of UL9 would result in the titration of helicase-primase in a nonproductive state unable to carry out DNA synthesis. (ii) Another possibility is that UL9 titrates a cellular factor required for efficient initiation. A corollary to this possibility is that UL9 is required to direct the initiation complex to a specific location within the nucleus, for instance, to an attachment on the nuclear matrix. Overexpression of UL9 may titrate cellular sites or factors, resulting in the inability of a functional initiation complex to form. (iii) Another model posits that DNA replication occurs in two stages, similar to T4 DNA bacteriophage replication. The first stage would be origin dependent and require at least one HSV origin and the UL9 protein. By analogy to T4, the second stage may be recombination dependent and may not require specific initiation at an origin of replication. We propose that UL9 would not be required for the second stage of DNA replication and that it may in fact be inhibitory. Further experiments will be required to distinguish between these possible models, which are not mutually exclusive.

Intracellular localization of replication proteins

HSV DNA synthesis occurs within defined regions of the nucleus that are termed replication compartments (38, 75). These globular structures were initially identified on the basis of ICP8 staining patterns (110); however, other replication proteins, including the polymerase/UL42 complex and UL9, the origin-binding protein, colocalize with ICP8 (23, 54, 100). Earlier studies with members of the helicase-primase complex (UL5, UL8, and UL52) suggested that these proteins are nuclear; however, the staining was too faint to determine whether they colocalize in replication compartments (100). Using epitope-tagged UL5, UL8, and UL52, we have now demonstrated that they colocalize with ICP8 in replication compartments (79b). Interestingly, upon HSV-1-infection, several cellular proteins also redistribute into these complexes. In particular, proteins involved in cellular DNA replication move to replication compartments (150). Since it has not been possible to reconstitute origin-dependent viral DNA synthesis with purified viral components, the role of cellular proteins in viral DNA replication cannot be assessed.

If DNA synthesis is blocked, either by the addition of viral DNA polymerase inhibitors or by infection with polymerase or polymerase-associated protein mutants, ICP8 localizes in a punctate pattern to sites termed prereplicative sites (38, 75). Some controversy exists over whether polymerase and UL42 colocalize with ICP8 in a punctate pattern (54), but at least one group claims that all three proteins colocalize in these structures (23). We recently showed that UL9 and all three members of the helicase-primase complex colocalize to prereplicative sites when viral DNA synthesis is blocked with phosphonoacetic acid or acyclovir (79a). It is not clear whether prereplicative sites are functional intermediates in the formation of replication compartments or whether they form only under conditions in which the viral polymerase is inhibited.

RESOLUTION OF REPLICATION INTERMEDIATES

Although the actual mechanism of DNA replication in HSV is unknown, available evidence indicates that mature viral genomes are processed from complex nonlinear, probably concatemeric DNA structures that accumulate during DNA replication (84a, 117). We propose that by analogy with bacteriophage T4, these structures represent branched DNA that results from recombination. Furthermore, we suggest that the viral enzyme alkaline nuclease may be involved in the resolution of these replication-recombination intermediates. Some of the evidence in support of this hypothesis is outlined below.

Alkaline Nuclease

Infection of mammalian cells with HSV-1 or HSV-2 results in the induction of a novel activity, the alkaline nuclease (UL12) (65, 70). HSV alkaline nuclease is a relatively abundant phosphoprotein in infected cells (6, 7); however, its precise role in the viral life cycle is unknown. Viruses carrying a null mutation (AN-1) or a frameshift mutation in the HSV-1 alkaline nuclease gene were isolated in our laboratory (84a, 117a, 147). Both *nuc* mutants form tiny plaques on Vero cells, and mutant virus yields in nonpermissive cells are approximately 0.1 to 1% that seen for wild-type virus (117a, 147).

Phenotype of *nuc* Mutants

To determine the nature of the defect in *nuc* mutants, several stages in the viral life cycle have been analyzed, including viral DNA and protein synthesis, cleavage, DNA structure, encapsidation, capsid formation, and production of mature virions. The results can be summarized as follows. Viral DNA synthesis, late protein synthesis, and cleavage into unit-sized monomers occur at or near wild-type levels in cells infected with *nuc* mutants. The two major differences observed between wild-type and *nuc* mutants were in encapsidation and in DNA structure.

Encapsidation of Viral DNA in Wild-Type and Mutant-Infected Cells

Encapsidation was analyzed by measuring the amount of DNase I-resistant DNA in infected cells. A slight reduction in encapsidation of *nuc* mutant DNA compared with that of wild-type (two- to threefold at most) was observed; however, this decrease is not sufficient to explain the 100- to 1,000-fold decrease in virus production (117a). Further analysis of encapsidation was carried out by digesting nuclear and cytoplasmic extracts from infected cells with staphylococcal nuclease (SN). In *nuc* mutant-infected Vero cells, significant amounts of SN-resistant DNA are observed in the nucleus, but very little SN-resistant DNA is present in the cytoplasm (117a). In contrast, in wild-type-infected cells, considerable amounts of protected viral DNA are present in both the nucleus and the cytoplasm. These results indicate that in *nuc* mutant-infected Vero cells, viral DNA is encapsidated but fails to egress efficiently into the cytoplasm. We suggest that the process of encapsidation in the nucleus is aberrant in some way, leading to the formation of capsids that are not competent to leave the nucleus.

Structure of Replicating DNA in Wild-Type and Mutant-Infected Cells

Pulsed-field gel electrophoresis was used to characterize the DNA that accumulates in AN-1-infected cells in an attempt to understand the nature of the defect. Infected cells contain two major species of viral DNA: one band that does not enter the gel (well DNA) and a band that corresponds to free monomers (152 kb) (84a). Treatment of total DNA with SN results in the total digestion of well DNA, leaving only the protected monomeric virion DNA. DNA that does not enter a pulsed-field gel is generally assumed either to be circular or to have an unusual structure; a simple linear structure, even if it were very large, would be expected to leave the well. To better understand the nature of the well DNA, this material was isolated from a gel and digested to completion with a restriction enzyme (*Spe*I) that cleaves once in the viral genome. This treatment would be expected to produce linear monomers from either circular or concatemeric viral DNA. If genomic inversion has occurred within the concatemeric DNA, three bands are expected because of genomic inversion. At 24 or 48 h postinfection, the three predicted fragments can be released from well DNA from wild-type-virus-infected cells, although a considerable amount of viral DNA remains in the well. In contrast, no discrete fragments are released from the *nuc* mutant DNA even at late times. One explanation for DNA fragments that cannot enter a pulsed-field gel is that the DNA is branched. Experiments using two-dimensional gel electrophoresis and electron microscopy are in progress to confirm whether the nonlinear DNA is indeed branched.

Our conclusions are that (i) both wild-type and mutant DNAs exist in a complex nonlinear form (possibly branched) during replication and (ii) discrete monomer-length DNA cannot be released from *nuc* DNA by a single cutting enzyme (84a). These results are consistent with the model that branched DNA accumulates during DNA replication (117). We propose that one possible function of alkaline nuclease is to resolve such branched structures that may arise as a direct contribution of recombination to viral DNA replication. In phage T4, the recombination-dependent replication mode results in the formation of branched concatemers that are resolved through the action of a T4-encoded endonuclease VII, the product of gene *49* (71, 72).

The second major defect in *nuc* mutants appears to be the failure of DNA-containing capsids to egress into the cytoplasm, as described above. We are considering two possible reasons for this failure. (i) The DNA that accumulates in the nuclei of mutant-infected cells is protected from SN digestion but retains some unresolved structure (i.e., a small branch) that prevents a "locking step" necessary for egress. (ii) The alkaline nuclease plays a role in the conformational or maturational changes required for egress of DNA-containing capsids that is completely separate from its role in the resolution of recombination intermediates. The isolation of mutants that are affected in one function but not the other, i.e., in resolution of intermediates but not in maturation of capsids, will be necessary to distinguish between these two possibilities.

CLEAVAGE AND ENCAPSIDATION OF VIRAL GENOMES

cis-Acting Elements Required for Cleavage and Packaging

Several lines of evidence suggest that viral genome maturation involves site-specific cleavage of viral DNA concatemers. The *cis*-acting sequence required for cleavage is located within the *a* sequence (91, 127, 133, 139). The *a* sequence is present as a direct repeat at both molecular termini and in inverted orientation at the L-S junction (Fig. 2) (34, 140). The *a* sequence is present in a single copy at the S terminus of the viral genome and in one to several tandem copies at the L terminus and at the L-S junction (90). Two separate *cis*-acting signals within the *a* sequence (called *pac*-1 and *pac*-2) appear to be essential for the cleavage and packaging processes (39, 138). The signal for cleavage of HSV-1 has been defined further to a 179-bp fragment across an *a-a* junction (95).

trans-Acting Proteins Involved in Cleavage and Packaging

Given the existence of *cis*-acting signals required for cleavage and maturation, it follows that there must be *trans*-acting factors that presumably recognize these sites and act to carry out cleavage and facilitate packaging. As described above, six complementation groups of HSV mutants that make but fail to process viral DNA under nonpermissive conditions have been isolated. These mutants have been mapped to six distinct viral genes: *UL6, UL15, UL25, UL28, UL32*, and *UL33* (1, 2, 5, 86, 105, 109, 112, 119, 120). Many of these viral genes are extremely well conserved within the herpesvirus family, indicating that they probably play important roles in the viral life cycle. Little is known about the functions of the products of these six viral genes. It is likely that they include proteins that bind and cleave viral DNA at the *a* sequences (terminases), proteins that initiate the uptake of DNA into the capsid (portal proteins), and proteins responsible for the translocation of DNA into the capsid (translocases). *UL15* is highly conserved among the herpesvirus family and is reported to have homology with the T4 terminase gene, gene *17* (33, 106). Like known terminases, the predicted open reading frame of UL15 also contains a predicted nucleotide triphosphate-binding site (87). Preliminary evidence indicates that UL6 is found in A and B capsids as well as in mature virions, suggesting that it may be a minor virion protein (78a).

SUMMARY

In summary, seven HSV replication factors have been identified and studied in vitro. Structure-function analysis of several of these proteins revealed multiple motifs important for their function, and it is anticipated that this information will be useful in designing improved antiviral therapies. Experiments designed to elucidate the functions of proteins involved in maturation, cleavage, and encapsidation of viral genomes are well on their way and may provide additional targets for antiviral strategies. Genetic analysis has provided important insights not only into the identification of viral proteins required in DNA replication and genome maturation but also into their roles in these complex processes. For instance, although purified UL9 has not been shown to unwind origin-containing plasmids in vitro, the importance of the helicase activity of this protein can be inferred by genetic analysis: single-amino-acid substitution mutations in helicase motifs of UL9 abolish its in vivo activity. Likewise, it is hoped that further analysis of mutations in the alkaline nuclease will continue to provide insight into the role of this unusual enzyme. For these analyses, the phage systems, especially T4, have guided our thinking at many steps.

Despite recent progress in identification of individual *cis*- and *trans*-acting elements required for viral DNA replication and processing, several outstanding questions remain. What is the role of cellular replication proteins in these processes? Are there specific sites in the infected-cell nucleus at which DNA replication and maturation occur? What is the mode of DNA replication? Do theta structures form initially at early times, to be followed by rolling-circle replication? Do greater-than-unit-length concatemers form solely by rolling-circle replication, or do inter- and intramolecular recombinations play a role? Are there different modes of viral DNA synthesis that differ in their requirements for viral proteins such as UL9? How do the proteins required for cleavage and packaging work together to accomplish their functions? We are looking forward to using genetic and biochemical approaches to begin to answer these questions.

I wish I could sit down with Howard Temin as I have done many times in the past to discuss questions about the biology of viruses. As a scientist, Howard possessed an amazing intuition about biology. He provided inspiration and confidence to pursue what I perceive to be the important biological questions. He was a great teacher. I close this chapter with a story about taking my son William, then age 4, to visit Howard during the summer before Howard's death. Before we arrived at the house, I told William that Dr. Temin was my teacher, and William immediately understood that that was an important relationship. We had a very nice visit with Howard in his backyard. I remember that Howard and William chased a squirrel. As we drove away, William was very worried about Howard's illness; he asked me if Dr. Temin ate fruit and vegetables. He wanted to send Dr. Temin a big box of broccoli. I could not figure out the reason for these questions until I finally remembered that in my efforts to get William to eat his vegetables, I had told him that broccoli can keep you from getting sick.

But I still do not have an answer to his most recent question: "Mama, who will be your teacher now?"

ACKNOWLEDGMENTS. I am grateful to members of my laboratory for helpful comments. Many thanks to Josh Goldstein for discussions and preparation of Fig. 1.

This investigation was supported by Public Health Service grant AI21747 and American Cancer Society grant VM-9. S.K.W. is the recipient of an American Heart Association-Genentech Established Investigator Award.

REFERENCES

1. **Addison, C., F. J. Rixon, and V. G. Preston.** 1990. Herpes simplex virus type 1 UL28 gene product is important for the formation of mature capsids. *J. Gen. Virol.* **71:**2377–2384.
2. **Al-Kobaisi, M. F., F. J. Rixon, I. McDougall, and V. G. Preston.** 1991. The herpes simplex virus UL33 gene product is required for the assembly of full capsids. *Virology* **180:**380–388.
3. **Amundsen, S. K., and D. S. Parris.** 1984. Detection of herpes simplex virus intertypic recombinant genomes in infected cell DNA. *J. Virol. Methods* **8:**19–25.
4. **Arbuckle, M. I., and N. D. Stow.** 1993. A mutational analysis of the DNA-binding domain of the herpes simplex virus type 1 UL9 protein. *J. Gen. Virol.* **74:**1349–1355.
5. **Baines, J. D., A. P. W. Poon, and B. Roizman.** 1994. The herpes simplex virus 1 UL 15 gene encodes two proteins and is required for cleavage of genomic viral DNA. *J. Virol.* **68:**8118–8124.
6. **Banks, L., D. J. Purifoy, P. F. Hurst, R. A. Killington, and K. L. Powell.** 1983. Herpes simplex virus non-structural proteins. IV. Purification of the virus-induced deoxyribonuclease and characterization of the enzyme using monoclonal antibodies. *J. Gen. Virol.* **64:**2249–2260.
7. **Banks, L. M., I. W. Halliburton, D. J. M. Purifoy, R. A. Killington, and K. L. Powell.** 1985. Studies on the herpes simplex virus alkaline nuclease: detection of type-common and type-specific epitopes on the enzyme. *J. Gen. Virol.* **66:**1–14.
8. **Bataille, D., and A. Epstein.** 1994. Herpes simplex virus replicative concatemers contain L components in inverted orientation. *Virology* **203:**384–388.
9. **Bates, P., J. A. Young, and H. E. Varmus.** 1993. A receptor for subgroup A Rous sarcoma virus is related to the low density lipoprotein receptor. *Cell* **74:**1043–1051.
10. **Bayliss, G. J., H. S. Marsden, and J. Hay.** 1975. Herpes simplex virus proteins: DNA-binding proteins in infected cells and the virus structure. *Virology* **68:**124–134.
11. **Bazinet, C., and J. King.** 1985. The DNA translocating vertex of DSDNA bacteriophage. *Annu. Rev. Microbiol.* **39:**109–129.
12. **Ben-Porat, T., and F. J. Rixon.** 1979. Replication of herpesvirus DNA. IV. Analysis of concatemers. *Virology* **94:**61–70.
13. **Ben-Porat, T., F. J. Rixon, and M. L. Blankenship.** 1979. Analysis of the structure of the genome of pseudorabies virus. *Virology* **95:**285–294.
14. **Ben-Porat, T., and S. A. Tokazewski.** 1977. Replication of herpesvirus DNA. II. Sedimentation characteristics of newly synthesized DNA. *Virology* **79:**292–301.
15. **Black, L.** 1989. DNA packaging in ds DNA bacteriophages. *Annu. Rev. Microbiol.* **43:**267–292.
16. **Black, L. W., M. K. Showe, and A. C. Steven.** 1995. Morphogenesis of the T4 head, p. 219–245. *In* C. K. Mathews, E. M. Kutter, G. Mosig, and P. B. Berget (ed.), *Bacteriophage T4.* American Society for Microbiology, Washington, D.C.

17. **Boehmer, P. E., M. C. Craigie, N. D. Stow, and I. R. Lehman.** 1994. Association of origin binding protein and single strand DNA-binding protein, ICP8, during herpes simplex virus type 1 DNA replication in vivo. *J. Biol. Chem.* **269:**29329–29334.

18. **Boehmer, P. E., M. S. Dodson, and I. R. Lehman.** 1993. The herpes simplex virus type-1 origin binding protein. DNA helicase activity. *J. Biol. Chem.* **268:**1220–1225.

19. **Boehmer, P. E., and I. R. Lehman.** 1993. Herpes simplex virus type 1 ICP8: helix-destabilizing properties. *J. Virol.* **67:**711–715.

20. **Boehmer, P. E., and I. R. Lehman.** 1993. Physical interaction between the herpes simplex virus 1 origin-binding protein and single-stranded DNA-binding protein ICP8. *Proc. Natl. Acad. Sci. USA* **90:**8444–8448.

21. **Bortner, C., T. R. Hernandez, I. R. Lehman, and J. Griffith.** 1993. Herpes simplex virus 1 single-strand DNA-binding protein (ICP8) will promote homologous pairing and strand transfer. *J. Mol. Biol.* **231:**241–250.

22. **Bruckner, R. C., J. J. Crute, M. S. Dodson, and I. R. Lehman.** 1991. The herpes simplex virus 1 origin binding protein: a DNA helicase. *J. Biol. Chem.* **266:**2669–2674.

23. **Bush, M., D. R. Yager, M. Gao, K. Weisshart, A. I. Marcy, D. M. Coen, and D. M. Knipe.** 1991. Correct intranuclear localization of herpes simplex virus DNA polymerase requires the viral ICP8 DNA-binding protein. *J. Virol.* **65:**1082–1089.

24. **Cairns, J.** 1963. The chromosome of Escherichia coli. *Cold Spring Harbor Symp. Quant. Biol.* **28:**43–46.

25. **Calder, J. M., E. C. Stow, and N. D. Stow.** 1992. On the cellular localization of the components of the herpes simplex virus type 1 helicase-primase complex and the viral origin-binding protein. *J. Gen. Virol.* **73:**531–538.

26. **Calder, J. M., and N. D. Stow.** 1990. Herpes simplex virus helicase-primase: the UL8 protein is not required for DNA-dependent ATPase and DNA helicase activities. *Nucleic Acids Res.* **18:** 3573–3578.

27. **Carmichael, E. P., M. J. Kosovsky, and S. K. Weller.** 1988. Isolation and characterization of herpes simplex virus type 1 host range mutants defective in viral DNA synthesis. *J. Virol.* **62:** 91–99.

28. **Carmichael, E. P., and S. K. Weller.** 1989. Herpes simplex virus type 1 DNA synthesis requires the product of the UL8 gene: isolation and characterization of an *ICP6::lacZ* insertion mutation. *J. Virol.* **63:**591–599.

29. **Coen, D. M.** 1992. Molecular aspects of anti-herpesvirus drugs. *Semin. Virol.* **3:**3–12.

30. **Conley, A. J., D. M. Knipe, P. C. Jones, and B. Roizman.** 1981. Molecular genetics of herpes simplex virus. VII. Characterization of a temperature-sensitive mutant produced by in vitro mutagenesis and defective in DNA synthesis and accumulation of γ polypeptides. *J. Virol.* **37:** 191–206.

31. **Crute, J. J., and I. R. Lehman.** 1989. Herpes simplex-1 DNA polymerase: identification of an intrinsic 5'-3' exonuclease with ribonuclease H activity. *J. Biol. Chem.* **264:**19266–19270.

32. **Crute, J. J., T. Tsurumi, L. Zhu, S. K. Weller, P. D. Olivo, M. D. Challberg, E. S. Mocarski, and I. R. Lehman.** 1989. Herpes simplex virus 1 helicase-primase: a complex of three herpes-encoded gene products. *Proc. Natl. Acad. Sci. USA* **86:**2186–2189.

33. **Davison, A. J.** 1992. Channel catfish virus: a new type of herpesvirus. *Virology* **186:**9–14.

34. **Davison, A. J., and N. M. Wilkie.** 1981. Nucleotide sequences of the joint between the L and S segments of herpes simplex virus types 1 and 2. *J. Gen. Virol.* **55:**315–331.

35. **Dean, F. B., P. Bullock, Y. Murakami, C. R. Wobbe, L. Weissbach, and J. Hurwitz.** 1987. Simian virus 40 (SV40) DNA replication: SV40 large T antigen unwinds DNA containing the SV40 origin of replication. *Proc. Natl. Acad. Sci. USA* **84:**16–20.

36. **Dean, F. B., M. Dodson, H. Echols, and J. Hurwitz.** 1987. ATP-dependent formation of a specialized nucleoprotein structure by simian virus 40 (SV40) large tumor antigen at the SV40 replication origin. *Proc. Natl. Acad. Sci. USA* **84:**8981–8985.

37. **Deb, S., and S. P. Deb.** 1991. A 269-amino-acid segment with a pseudo-leucine zipper and a helix-turn-helix motif codes for the sequence-specific DNA-binding domain of herpes simplex virus type 1 origin-binding protein. *J. Virol.* **65:**2829–2838.

38. **de Bruyn Kops, A., and D. M. Knipe.** 1988. Formation of DNA replication structures in herpes virus-infected cells requires a viral DNA binding protein. *Cell* **55:**857–868.

39. **Deiss, L. P., J. Chou, and N. Frenkel.** 1986. Functional domains within the *a* sequence involved in the cleavage-packaging of herpes simplex virus DNA. *J. Virol.* **59:**605–618.

40. **Dodson, M., F. B. Dean, P. Bullock, H. Echols, and J. Hurwitz.** 1987. Unwinding of duplex DNA from the SV40 origin of replication by T antigen. *Science* **238:**964–967.

41. **Dodson, M. S., J. J. Crute, R. C. Bruckner, and I. R. Lehman.** 1989. Overexpression and assembly of the herpes simplex virus type 1 helicase-primase in insect cells. *J. Biol. Chem.* **264:** 20835–20838.

42. **Elias, P., C. M. Gustafsson, O. Hammarsten, and N. D. Stow.** 1992. Structural elements required for the cooperative binding of the herpes simplex virus origin binding protein to oriS reside in the N-terminal part of the protein. *J. Biol. Chem.* **267:**17424–17429.

43. **Fauci, A. S.** 1988. The human immunodeficiency virus: infectivity and mechanisms of pathogenesis. *Science* **239:**617–622.

44. **Fierer, D. S., and M. D. Challberg.** 1992. Purification and characterization of UL9, the herpes simplex virus type 1 origin-binding protein. *J. Virol.* **66:**3986–3995.

45. **Friedmann, A., J. E. Coward, H. S. Rosenkranz, and C. Morgan.** 1975. Electron microscopic studies on assembly of herpes simplex virus upon removal of hydroxyurea block. *J. Gen. Virol.* **26:**171–181.

46. **Gao, M., and D. M. Knipe.** 1991. Potential role for herpes simplex virus ICP8 DNA replication protein in stimulation of late gene expression. *J. Virol.* **65:**2666–2675.

47. **Garber, D. A., S. M. Beverley, and D. M. Coen.** 1993. Demonstration of circularization of herpes simplex virus DNA following infection using pulsed field gel electrophoresis. *Virology* **197:** 459–462.

48. **Gauss, P., K. Park, T. E. Spencer, and K. J. Hacker.** 1994. DNA helicase requirements for DNA replication during bacteriophage T4 infection. *J. Bacteriol.* **176:**1667–1672.

49. **Gibbs, J. S., H. C. Chiou, J. D. Hall, D. W. Mount, M. J. Retondo, S. K. Weller, and D. M. Coen.** 1985. Sequence and mapping analyses of the herpes simplex virus DNA polymerase gene predict a C-terminal substrate binding domain. *Proc. Natl. Acad. Sci. USA* **82:**7969–7973.

50. **Gilbert, J. M., P. Bates, H. E. Varmus, and J. M. White.** 1994. The receptor for the subgroup A avian leukosis-sarcoma viruses binds to subgroup A but not to subgroup C envelope glycoprotein. *J. Virol.* **68:**5623–5628.

51. **Gilbert, W., and D. Dressler.** 1968. DNA replication: the rolling circle model. *Cold Spring Harbor Symp. Quant. Biol.* **33:**473–484.

52. **Godowski, P. J., and D. M. Knipe.** 1985. Identification of a herpes simplex virus function that represses late gene expression from parental viral genomes. *J. Virol.* **55:**357–365.

53. **Godowski, P. J., and D. M. Knipe.** 1986. Transcriptional control of herpesvirus gene expression: gene functions required for positive and negative regulation. *Proc. Natl. Acad. Sci. USA* **83:** 256–260.

54. **Goodrich, L. D., P. A. Schaffer, D. I. Dorsky, C. S. Crumpacker, and D. S. Parris.** 1990. Localization of the herpes simplex virus type 1 65-kilodalton DNA-binding protein and DNA polymerase in the presence and absence of viral DNA synthesis. *J. Virol.* **64:**5738–5749.

55. **Gorbalenya, A. E., E. V. Koonin, A. P. Donchenko, and V. M. Blinov.** 1988. A conserved NTP-motif in putative helicases. *Nature* (London) **333:**22–23.

56. **Gorbalenya, A. E., E. V. Koonin, A. P. Donchenko, and V. M. Blinov.** 1988. A novel superfamily of nucleoside triphosphate-binding motif containing proteins which are probably involved in duplex unwinding in DNA and RNA replication and recombination. *FEBS Lett.* **235:**16–24.

57. **Gorbalenya, A. E., E. V. Koonin, A. P. Donchenko, and V. M. Blinov.** 1989. Two related superfamilies of putative helicases involved in replication, recombination, repair and expression of DNA and RNA genomes. *Nucleic Acids Res.* **17:**4713–4730.

58. **Gottlieb, J., and M. D. Challberg.** 1994. Interaction of herpes simplex virus type 1 DNA polymerase and the UL42 accessory protein with a model primer template. *J. Virol.* **68:**4937–4945.

59. **Gottlieb, J., A. I. Marcy, D. M. Coen, and M. D. Challberg.** 1990. The herpes simplex virus type 1 UL42 gene product: a subunit of DNA polymerase that functions to increase processivity. *J. Virol.* **64:**5976–5987.

59a. **Graves, K. L., and S. K. Weller.** Unpublished data.

60. **Hayward, G. S., R. J. Jacob, S. C. Wadsworth, and B. Roizman.** 1975. Anatomy of herpes simplex virus DNA: evidence for four populations of molecules that differ in the relative orientations of their long and short components. *Proc. Natl. Acad. Sci. USA* **72:**4243–4247.

61. **Hernandez, T. R., and I. R. Lehman.** 1990. Functional interaction between the herpes simplex-1 DNA polymerase and UL42 protein. *J. Biol. Chem.* **265:**11227–11232.

62. **Herskowitz, I.** 1987. Functional inactivation of genes by dominant negative mutations. *Nature* (London) **329:**219–222.

63. **Hirsch, I., G. Cabral, M. Patterson, and N. Biswal.** 1977. Studies on intracellular replicating DNA of herpes simplex virus type 1. *Virology* **81:**48–61.

64. **Hodgman, T. C.** 1988. A new superfamily of replicative proteins. *Nature* (London) **333:**22–23.

65. **Hoffmann, P. J., and Y.-C. Cheng.** 1978. The deoxyribonuclease induced after infection of KB cells by herpes simplex virus type 1 or type 2. I. Purification and characterization of the enzyme. *J. Biol. Chem.* **253:**3557–3562.

66. **Igarashi, K., R. Fawl, R. J. Roller, and B. Roizman.** 1993. Construction and properties of a recombinant herpes simplex virus 1 lacking both S-component origins of DNA synthesis. *J. Virol.* **67:**2123–2132.

67. **Jacob, R. J., L. S. Morse, and B. Roizman.** 1979. Anatomy of herpes simplex virus DNA. XII. Accumulation of head to tail concatemers in the nuclei of infected cells and their role in the generation of four isomeric arrangements of viral DNA. *J. Virol.* **29:**448–457.

68. **Jacob, R. J., and B. Roizman.** 1977. Anatomy of herpes simplex virus DNA. VIII. Properties of the replicating DNA. *J. Virol.* **23:**394–411.

69. **Jongeneel, C. V., and S. L. Bachenheimer.** 1981. Structure of replicating herpes simplex virus DNA. *J. Virol.* **39:**656–660.

70. **Keir, H. M., and E. Gold.** 1963. Deoxyribonucleic acid nucleotidyltransferase and deoxyribonuclease from cultured cells infected with herpes simplex virus. *Biochim. Biophys. Acta* **72:**263–276.

71. **Kemper, B., and D. T. Brown.** 1976. Function of gene 49 of bacteriophage T4. II. Analysis of intracellular development and the structure of very fast-sedimenting DNA. *J. Virol.* **18:**1000–1015.

72. **Kemper, B., M. Garabett, and U. Courage.** 1981. Studies on the function of gene 49 controlled endonuclease of phage T4 (endonuclease VII). *Prog. Clin. Biol. Res.* **64:**151–166.

73. **King, G. J., and W. M. Huang.** 1982. Identification of the origins of T4 DNA replication. *Proc. Natl. Acad. Sci. USA* **79:**7248–7252.

74. **Klinedinst, D. K., and M. D. Challberg.** 1994. Helicase-primase complex of herpes simplex virus type 1: a mutation in the UL52 subunit abolishes primase activity. *J. Virol.* **68:**3693–3701.

75. **Knipe, D. M.** 1989. The role of viral and cellular nuclear proteins in herpes simplex virus replication. *Adv. Virus Res.* **37:**85–123.

76. **Knopf, K. W.** 1979. Properties of herpes simplex virus DNA polymerase and characterization of its associated exonuclease activity. *Eur. J. Biochem.* **98:**231–244.

77. **Ladin, B. F., M. L. Blankenship, and T. Ben-Porat.** 1980. Replication of herpesvirus DNA. V. Maturation of concatemeric DNA of pseudorabies virus to genome length is related to capsid formation. *J. Virol.* **33:**1151–1164.

78. **Ladin, B. F., S. Ihara, H. Hampl, and T. Ben-Porat.** 1982. Pathway of assembly of herpesvirus capsids: an analysis using DNA + temperature-sensitive mutants of pseudorabies virus. *Virology* **116:**544–561.

78a. **Lamberti, C., and S. K. Weller.** Unpublished data.

79. **Luder, A., and G. Mosig.** 1982. Two alternative mechanisms for initiation of DNA replication forks in bacteriophage T4: priming by RNA polymerase and by recombination. *Proc. Natl. Acad. Sci. USA* **79:**1101–1105.

79a. **Lukonis, C. J., A. K. Malik, and S. K. Weller.** Unpublished data.

79b. **Lukonis, C. J., and S. K. Weller.** Unpublished data.

80. **Malik, A. K., R. Martinez, L. Muncy, E. P. Carmichael, and S. K. Weller.** 1992. Genetic analysis of mutations in the HSV-1 UL9 origin specific DNA binding protein: isolation of an ICP6::lacZ insertion mutant. *Virology* **190:**702–715.

80a. **Malik, A. K., and S. K. Weller.** Unpublished data.

81. **Manaker, R. A., and V. Groupe.** 1956. Discrete foci of altered chicken embryo cell associated with Rous sarcoma virus in tissue culture. *Virology* **2:**838–840.

82. **Marcy, A. I., P. D. Olivo, M. D. Challberg, and D. M. Coen.** 1990. Enzymatic activities of overexpressed herpes simplex virus DNA polymerase purified from recombinant baculovirus-infected insect cells. *Nucleic Acids Res.* **18:**1207–1215.

83. **Martin, D. W., R. M. Munoz, D. Oliver, M. A. Subler, and S. Deb.** 1994. Analysis of the DNA-binding domain of the HSV-1 origin-binding protein. *Virology* **198:**71–80.

84. **Martinez, R., L. Shao, and S. K. Weller.** 1992. The conserved helicase motifs of the herpes simplex virus type 1 origin-binding protein UL9 are important for function. *J. Virol.* **66:**6735–6746.

84a. **Martinez, R., and S. K. Weller.** Unpublished data.

85. **Matthews, J. T., B. J. Terry, and A. K. Field.** 1993. The structure and function of the HSV DNA replication proteins: defining novel antiviral targets. *Antiviral Res.* **20:**89–114.

86. **Matz, B., S. J. H. Subak, and V. G. Preston.** 1983. Physical mapping of temperature-sensitive mutations of herpes simplex virus type 1 using cloned restriction endonuclease fragments. *J. Gen. Virol.* **64:**2261–2270.

87. **McGeoch, D. J., M. A. Dalrymple, A. J. Davison, A. Dolan, M. C. Frame, D. McNab, L. J. Perry, J. E. Scott, and P. Taylor.** 1988. The complete DNA sequence of the long unique region in the genome of herpes simplex virus type 1. *J. Gen. Virol.* **69:**1531–1574.

88. **McLean, G. W., A. P. Abbotts, M. E. Parry, H. S. Marsden, and N. D. Stow.** 1994. The herpes simplex virus type 1 origin-binding protein interacts specifically with the viral UL8 protein. *J. Gen. Virol.* **75:**2699–2706.

89. **Mizuuchi, K., B. Kemper, H. Hays, and R. A. Wiesberg.** 1988. T4 endonuclease VII cleaves Holliday structures. *Cell* **29:**357–365.

90. **Mocarski, E. S., and B. Roizman.** 1982. Herpesvirus-dependent amplification and inversion of cell-associated viral thymidine kinase gene flanked by viral a sequences and linked to an origin of viral DNA replication. *Proc. Natl. Acad. Sci. USA* **79:**5626–5630.

91. **Mocarski, E. S., and B. Roizman.** 1982. Structure and role of the herpes simplex virus DNA termini in inversion, circularization and generation of virion DNA. *Cell* **31:**89–97.

92. **Morgan, A. R., and A. Severini.** 1990. Interconversion of replication and recombination structures: implications for terminal repeats and concatemers. *J. Theor. Biol.* **144:**195–202.

93. **Mosig, G.,** 1983. Relationship of T4 DNA replication and recombination, p. 120–130. *In* C. K. Mathews, E. M. Kutter, G. Mosig, and P. B. Berget (ed.), *Bacteriophage T4.* American Society for Microbiology, Washington, D.C.

94. **Mosig, G.** 1987. The essential role of recombination in phage T4 growth. *Annu. Rev. Genet.* **21:**347–371.

95. **Nasseri, M., and E. S. Mocarski.** 1988. The cleavage recognition signal is contained within sequences surrounding an a-a junction in herpes simplex virus DNA. *Virology* **167:**25–30.

96. **Nossal, N. G.** 1992. Protein-protein interactions at a DNA replication fork: bacteriophage T4 as a model. *FASEB J.* **6:**871–878.

97. **O'Callaghan, D. J., M. C. Kemp, and C. C. Randall.** 1977. Properties of nucleocapsid species isolated from an in vivo herpesvirus infection. *J. Gen. Virol.* **37:**585–594.

98. **O'Donnell, M. E., P. Elias, and I. R. Lehman.** 1987. Processive replication of single-stranded DNA templates by the herpes simplex virus-induced DNA polymerase. *J. Biol. Chem.* **262:**4252–4259.

99. **Olivo, P. D., and M. D. Challberg,** 1990. Functional analysis of the herpes simplex virus gene products involved in DNA replication, p. 137–150. *In* E. Wagner (ed.), *Herpesvirus Transcription and Its Regulation.* CRC Press, Inc., Boca Raton, Fla.

100. **Olivo, P. D., N. J. Nelson, and M. D. Challberg.** 1989. Herpes simplex virus type 1 gene products required for DNA replication: identification and overexpression. *J. Virol.* **63:**196–204.

101. **Pauza, C. D., J. E. Galindo, and D. D. Richman.** 1990. Reinfection results in accumulation of unintegrated viral DNA in cytopathic and persistent human immunodeficiency virus type 1 infection of CEM cells. *J. Exp. Med.* **172:**1035–1042.

102. **Perry, H. C., D. J. Hazuda, and W. L. McClements.** 1993. The DNA binding domain of herpes simplex virus type 1 origin binding protein is a transdominant inhibitor of virus replication. *Virology* **193:**73–79.

103. **Poffenberger, K. L., and B. Roizman.** 1985. A noninverting genome of a viable herpes simplex virus 1: presence of head-to-tail linkages in packaged genomes and requirements for circularization after infection. *J. Virol.* **53:**587–595.

104. **Polvino, B. M., P. K. Orberg, and P. A. Schaffer.** 1987. Herpes simplex virus type 1 *oriL* is not required for virus replication or for the establishment and reactivation of latent infection in mice. *J. Virol.* **61:**3528–3535.

105. **Poon, A. P., and B. Roizman.** 1993. Characterization of a temperature-sensitive mutant of the UL15 open reading frame of herpes simplex virus 1. *J. Virol.* **67:**4497–4503.

106. **Powell, D., J. Franklin, F. Arisaka, and G. Mosig.** 1990. Bacteriophage T4 DNA packaging genes 16 and 17. *Nucleic Acids Res.* **18:**4005.

107. **Powell, K., and D. J. M. Purifoy.** 1976. DNA-binding proteins of cells infected by herpes simplex virus type 1 and 2. *Intervirology* **7:**225–239.

108. **Powell, K. L., E. Littler, and D. J. Purifoy.** 1981. Nonstructural proteins of herpes simplex virus. II. Major virus-specific DNa-binding protein. *J. Virol.* **39:**894–902.

109. **Preston, V. G., J. A. Coates, and F. J. Rixon.** 1983. Identification and characterization of a herpes simplex virus gene product required for encapsidation of virus DNA. *J. Virol.* **45:**1056–1064.

110. **Quinlan, M. P., L. B. Chen, and D. M. Knipe.** 1984. The intranuclear location of a herpes simplex virus DNA-binding protein is determined by the status of viral DNA replication. *Cell* **36:**857–868.

111. **Quinn, J. P., and D. J. McGeoch.** 1985. DNA sequence of the region in the genome of herpes simplex virus type 1 containing the genes for DNA polymerase and the major DNA binding protein. *Nucleic Acids Res.* **13:**8143–8163.

112. **Rixon, F. J., A. M. Cross, C. Addison, and V. G. Preston.** 1988. The products of herpes simplex virus type 1 gene UL26 which are involved in DNA packaging are strongly associated with empty but not with full capsids. *J. Gen. Virol.* **69:**2879–2891.

113. **Ruyechan, W. T.** 1983. The major herpes simplex virus DNA-binding protein holds single-stranded DNA in an extended configuration. *J. Virol.* **46:**661–666.

114. **Ruyechan, W. T., and A. C. Weir.** 1984. Interaction with nucleic acids and stimulation of the viral DNA polymerase by the herpes simplex virus type 1 major DNA-binding protein. *J. Virol.* **52:**727–733.

115. **Sarisky, R. T., and P. C. Weber.** 1994. Requirement for double-strand breaks but not for specific DNA sequences in herpes simplex virus type 1 genome isomerization events. *J. Virol.* **68:**34–47.

116. **Schimmel, P.** 1990. Hazards and their exploitation in the applications of molecular biology of structure-function relationships. *Biochemistry* **29:**9495–9502.

117. **Severini, A., A. R. Morgan, D. R. Tovell, and L. J. Tyrrel.** 1994. Study of the structure of replicative intermediates of HSV-1 DNA by pulsed-field gel electrophoresis. *Virology* **200:**428–435.

117a.**Shao, L., L. M. Rapp, and S. K. Weller.** 1993. Herpes simplex virus 1 alkaline nuclease is required for efficient egress of capsids from the nucleus. *Virology* **196:**146–162.

118. **Sheldrick, P., and N. Berthelot.** 1975. Inverted repetitions in the chromosome of herpes simplex virus. *Cold Spring Harbor Symp. Quant. Biol.* **2:**667–678.

119. **Sherman, G., and S. Bachenheimer.** 1987. DNA processing in temperature-sensitive morphogenic mutants of HSV-1. *Virology* **158:**427–430.

120. **Sherman, G., and S. L. Bachenheimer.** 1988. Characterization of intranuclear capsids made by ts morphogenic mutants of HSV-1. *Virology* **163:**471–480.

121. **Sherman, G., J. Gottlieb, and M. D. Challberg.** 1992. The UL8 subunit of the herpes simplex virus helicase-primase complex is required for efficient primer utilization. *J. Virol.* **66:**4884–4892.

122. **Skaliter, R., and I. R. Lehman.** 1994. Rolling circle DNA replication in vitro by a complex of herpes simplex virus type 1-encoded enzymes. *Proc. Natl. Acad. Sci. USA* **91:**10665–10669.

123. **Smiley, J. R., J. Duncan, and M. Howes.** 1990. Sequence requirements for DNA rearrangements induced by the terminal repeat of herpes simplex virus type 1 KOS DNA. *J. Virol.* **64:**5036–5050.

124. **Spacciapoli, P., and N. G. Nossal.** 1994. Interaction of DNA polymerase and DNA helicase within the bacteriophage T4 DNA replication complex. Leading strand synthesis by the T4 DNA polymerase mutant A737V (tsL141) requires the T4 gene 59 helicase assembly protein. *J. Biol. Chem.* **269:**447–455.

125. **Spaete, R. R., and N. Frenkel.** 1982. The herpes simplex virus amplicon: a new eucaryotic defective-virus cloning-amplifying vector. *Cell* **30:**295–304.

126. **Spaete, R. R., and N. Frenkel.** 1985. The herpes simplex virus amplicon: analyses of cis-acting replication functions. *Proc. Natl. Acad. Sci. USA* **82:**694–698.

127. **Spaete, R. R., and E. S. Mocarski.** 1985. The a sequence of the cytomegalovirus genome functions as a cleavage/packaging signal for herpes simplex virus defective genomes. *J. Virol.* **54:**817–824.

128. **Stabell, E. C., and P. D. Olivo.** 1993. A truncated herpes simplex virus origin binding protein which contains the carboxyl terminal origin binding domain binds to the origin of replication but does not alter its conformation. *Nucleic Acids Res.* **21:**5203–5211.

129. **Stahl, H., P. Droege, and R. Knippers.** 1986. DNA helicase activity of SV40 large tumor antigen. *EMBO J.* **5:**1939–1944.

130. **Stow, N. D.** 1982. Localization of an origin of DNA replication within the TRS/IRS repeated region of the herpes simplex virus type 1 genome. *EMBO J.* **1:**863–867.

131. **Stow, N. D., O. Hammarsten, M. I. Arbuckle, and P. Elias.** 1993. Inhibition of herpes simplex virus type 1 DNA replication by mutant forms of the origin-binding protein. *Virology* **196:**413–418.

132. **Stow, N. D., and E. C. McMonagle.** 1983. Characterization of the TRS/IRS origin of DNA replication of herpes simplex virus type 1. *Virology* **130:**427–438.

133. **Stow, N. D., E. C. McMonagle, and A. J. Davison.** 1983. Fragments from both termini of the herpes simplex virus type 1 genome contain signals required for the encapsidation of viral DNA. *Nucleic Acids Res.* **11:**8205–8220.

134. **Szostak, J. W., T. L. Orr-Weaver, and R. J. Rothstein.** 1983. The double-strand-break repair model for recombination. *Cell* **33:**25–35.

135. **Temin, H. M., E. Keshet, and S. K. Weller.** 1980. Correlation of transient accumulation of linear unintegrated viral DNA and transient cell killing by avian leukosis and reticuloendotheliosis viruses. *Cold Spring Harbor Symp. Quant. Biol.* **2:**773–778.

136. **Temin, H. M., and S. Mizutani.** 1970. RNA-dependent DNA polymerase in virions of Rous sarcoma virus. *Nature* (London) **226:**1211–1213.

137. **Tenney, D. J., W. W. Hurlburt, P. A. Micheletti, M. Bifano, and R. K. Hamatake.** 1994. The UL8 component of the herpes simplex virus helicase-primase complex stimulates primer synthesis by a subassembly of the UL5, and UL52 components. *J. Biol. Chem.* **269:**5030–5035.

138. **Varmuza, S. L., and J. R. Smiley.** 1985. Signals for site-specific cleavage of HSV DNA: maturation involves two separate cleavage events at sites distal to the recognition sequences. *Cell* **41:**793–802.

139. **Vlazny, D. A., A. Kwong, and N. Frenkel.** 1982. Site-specific cleavage/packaging of herpes simplex virus DNA and the selective maturation of nucleocapsids containing full-length viral DNA. *Proc. Natl. Acad. Sci. USA* **79:**1423–1427.

140. **Wagner, M. J., and W. C. Summers.** 1978. Structure of the joint region and the termini of the DNA of herpes simplex virus type 1. *J. Virol.* **27:**374–384.

141. **Walker, J. E., M. Saraste, M. J. Runswick, and N. J. Gay.** 1982. Distantly related sequences in the α and β-subunits of ATP synthase, myosin, kinases and other ATP-requiring enzymes and a common nucleotide binding fold. *EMBO J.* **1:**945–951.

142. **Weber, P. C., M. D. Challberg, N. J. Nelson, M. Levine, and J. C. Glorioso.** 1988. Inversion events in the HSV-1 genome are directly mediated by the viral DNA replication machinery and lack sequence specificity. *Cell* **54:**369–381.

143. **Weir, H. M., J. M. Calder, and N. D. Stow.** 1989. Binding of the herpes simplex virus type 1 UL9 gene product to an origin of viral DNA replication. *Nucleic Acids Res.* **17:**1409–1425.

144. **Weller, S. K.** 1990. Genetic analysis of HSV genes required for genome replication, p. 105–135. *In* E. Wagner (ed.), *Herpesvirus Transcription and Its Regulation.* CRC Press, Inc., Boca Raton, Fla.

145. **Weller, S. K., A. E. Joy, and H. M. Temin.** 1980. Correlation between cell killing and massive second-round superinfection by members of some subgroups of avian leukosis virus. *J. Virol.* **33:**494–506.

146. **Weller, S. K., K. J. Lee, D. J. Sabourin, and P. A. Schaffer.** 1983. Genetic analysis of temperature-sensitive mutants which define the gene for the major herpes simplex virus type 1 DNA-binding protein. *J. Virol.* **45:**354–366.

147. **Weller, S. K., R. M. Seghatoleslami, L. Shao, D. Rowse, and E. P. Carmichael.** 1990. The herpes simplex virus type 1 alkaline nuclease is not essential for viral DNA synthesis: isolation and characterization of a 1acZ insertion mutant. *J. Gen. Virol.* **71:**2941–2952.

148. **Weller, S. K., A. Spadaro, J. E. Schaffer, A. W. Murray, A. M. Maxam, and P. A. Schaffer.** 1985. Cloning, sequencing, and functional analysis of *oriL,* a herpes simplex virus type 1 origin of DNA synthesis. *Mol. Cell. Biol.* **5:**930–942.

149. **Weller, S. K., and H. M. Temin.** 1981. Cell killing by avian leukosis viruses. *J. Virol.* **39:**713–721.

150. **Wilcock, D., and L. D. P.** 1991. Localization of p53, retinoblastoma and host replication proteins at sites of viral replication in herpes-infected cells. *Nature* (London) **349:**429–431.

151. **Wong, S. W., and P. A. Schaffer.** 1991. Elements in the transcriptional regulatory region flanking herpes simplex virus type 1 *oriS* stimulate origin function. *J. Virol.* **65:**2601–2611.

152. **Wong, S. W., A. F. Wahl, P. M. Yuan, N. Arai, B. E. Pearson, K. Arai, D. Korn, M. W. Hunkapiller, and T. S. Wang.** 1988. Human DNA polymerase alpha gene expression is cell proliferation dependent and its primary structure is similar to both prokaryotic and eukaryotic replicative DNA polymerases. *EMBO J.* **7:**37–47.

153. **Zhang, X., S. Efstathiou, and A. Simmons.** 1994. Identification of novel herpes simplex virus replicative intermediates by field inversion gel electrophoresis: implications for viral DNA amplification strategies. *Virology* **202:**530–539.

154. **Zhu, L., and S. K. Weller.** 1992. The six conserved helicase motifs of the UL5 gene product, a component of the herpes simplex virus type 1 helicase-primase, are essential for its function. *J. Virol.* **66:**469–479.

155. **Zhu, L., and S. K. Weller.** 1992. The UL5 gene of the herpes simplex virus type 1: isolation of a *lacZ* insertion mutant and association of the UL5 gene product with other members of the helicase-primase complex. *J. Virol.* **66:**458–468.

Chapter 15

Reprint of Temin's 1993 Paper on the Inherent Contributions of Reverse Transcription to Retroviral Variation

For two decades, Howard Temin analyzed retroviral replication in detail in order to elucidate the high rate of genetic variation inherent to this process. In "Retrovirus Variation and Reverse Transcription: Abnormal Strand Transfers Result in Retrovirus Genetic Variation," he summarized his understanding of the kinds of retroviral variation and proposed a mechanism by which several of them might occur.

Virologists have appreciated for many years that viruses with RNA genomes vary rapidly and that retroviruses do, too. One contribution to this high rate is thought to be inherent in retroviral RNA replicases: they lack the editing activities of DNA replicases, which serve to limit some types of errors potentially made by nucleic acid replicases. What additional sources might contribute to retroviral variation could not be guessed in the 1970s. In fact, the actual rates of variation were not known, because no assays were available to accurately count the mutations introduced per round of viral replication. Most experiments performed to analyze genetic variation in animal RNA viruses in the 1970s and early 1980s measured the frequency of accumulated mutations in progeny viruses after serial propagation of a parent wild-type virus. The number of rounds of replication the viruses underwent was unknown, and thus ascertaining the rate of formation of mutant derivatives was impossible.

One elegant contribution to the resolution of the problem of multiple rounds of viral replication was the development and application of helper cells by Howard Temin and his colleagues. They developed cells that constitutively express the structural genes of spleen necrosis virus, a simple retrovirus. They could then introduce into these cells the DNA vectors that contained all the *cis*-acting genetic elements required by spleen necrosis virus for its replication. Transcription of the DNA vectors by cellular RNA polymerase yields RNAs that can be packaged as viral genomes and released from the helper cells as infectious virus particles. The virus particles contain reverse transcriptase and other viral replicative enzymes synthesized in the helper cells but do not contain the genes for these enzymes. They can therefore undergo only one round of replication. Howard, his students, and his postdoctoral researchers developed several protocols for selecting and enumerating the frequency of mutations generated in these viral vectors after one round of their replication. The identities of the mutants were revealed by sequencing their DNA after its rescue from the infected cells. This general approach under-lies more than half of the work cited in this review.

Howard and his colleagues devoted much effort to identifying the nucleic acid sequences of the mutant viral genomes and reconstructing from their sequences possible mechanisms for their formation. This effort led to his hypothesis that reverse transcriptase is "error prone" because retroviral replication requires it to tolerate the transfer of itself bound to its newly synthesized DNA product from one site on a nucleic acid template to another at least twice per viral replicative cycle. This inherent lack of processiveness of reverse transcriptase, for example, fosters mutations generated when a growing strand of newly synthesized DNA transfers its 3' nucleotide from its template to another homologous site on the same or a different template. Such transfers can cause insertions and deletions in the completed product DNA.

Howard's hypothesis is appealing because it provides a single explanation for a variety of both simple and complex mutations known to arise in retroviruses. It also helps explain how this family of viruses can in general acquire cellular genes and in particular transduce oncogenes.

Bill Sugden

Proc. Natl. Acad. Sci. USA
Vol. 90, pp. 6900–6903, August 1993

Review

Retrovirus variation and reverse transcription: Abnormal strand transfers result in retrovirus genetic variation

Howard M. Temin

McArdle Laboratory, University of Wisconsin, 1400 University Avenue, Madison, WI 53706

Contributed by Howard M. Temin, May 28, 1993

ABSTRACT **Human immunodeficiency virus variation is extensive and is based on numerous mistakes in reverse transcription. All retrovirus replication requires two strand transfers (growing point jumps) to synthesize the complete provirus. I propose that the numerous mistakes in reverse transcription are the result of this requirement for the two strand transfers needed to form the provirus.**

Retroviruses vary at a notoriously high rate. For example, antibody- and drug-resistant human immunodeficiency virus type 1 (HIV-1) strains rapidly appear in infected and treated persons, and it is estimated that the HIV-1 sequences (*env* gene) in an infected person change ≈1% per year (1, 2). Retroviruses recombine frequently, and simpler retroviruses often contain captured cellular protooncogenes (3, 4). [Simpler retroviruses contain only genes for virion proteins—*gag*, *pol*, and *env*. More complex retroviruses, like HIV-1, encode additional genes involved in regulation (5, 6).]

I propose that this high rate of retrovirus variation is a direct consequence of the requirement for transfer of the nascent strand at the reverse transcriptase growing point during retrovirus DNA synthesis. [A similar suggestion was made by Bebenek *et al.* (7) on the basis of studies with purified HIV-1 reverse transcriptase.] Of course, selection and other processes will finally determine the effects of this variation (8, 9). However, the high rate of genetic change in each replication cycle ensures that there is a wide field for selection and other processes. Reverse transcriptase is coded for by the retroviral *pol* gene and has associated RNase H activity, which may be required for one of the primer transfers (10–12). Because of its multiple roles—RNA-directed DNA synthesis, DNA-directed DNA synthesis, digesting RNA·DNA hybrid molecules, and strand transfers—reverse transcriptase must be quite flex-

ible in structure and action (H. Buc, personal communication). I shall also consider in this paper the hypotheses (*i*) that misincorporation promotes strand transfers (13) and (*ii*) that misincorporation accompanies strand transfers (14).

Retroviruses are a family of animal viruses that alternate their genetic material between RNA in the virion and DNA in the infected cell (15, 16). In addition, all retrovirus virions contain two identical molecules of virion RNA—the dimer RNA. The DNA form of a retrovirus, the provirus, is larger than the viral RNA form (Fig. 1). During reverse transcription, promoter/enhancer sequences found at the 3′ end of viral RNA within the unique 3′ RNA sequences (u3) are duplicated at the 5′ end of viral DNA to form the U3 DNA sequences (capital U indicates DNA rather than RNA), and downstream polyadenylylation sequences at the 5′ end of viral RNA within the unique 5′ RNA sequences (u5) are duplicated at the 3′ end of viral DNA to form the U5 DNA sequences. These duplications result in the formation of long terminal repeats (LTRs) at both ends of the proviral DNA and provide autonomy in the cis-acting sequences needed for transcription and replication, which are the same for retroviruses. This autonomy results from the virus U3 sequences containing promoter/enhancer elements that are recognizable by cellular transcription factors and 3′ LTR sequences that are recognizable by cellular polyadenylylation factors.

The duplications in the LTRs are a result of two jumps, switches, or transfers of the reverse transcriptase growing point from one end of each template to the other end during replication (17) (Fig. 2). (In this article, I use the term strand transfers for these processes.)

Retrovirus genetic variation consists of base-pair substitutions, frameshifts, deletions, deletions with insertions, homologous recombination, and nonhomologous recombination. I shall discuss, in relation to the strand-transfer hypothesis, minus-strand and plus-strand DNA primer transfers and each of these types of genetic variation. All of these processes with the exception of some deletions with insertions and the two types of recombination involve only one molecule of the retrovi-

rus dimer RNA (J. S. Jones, R. W. Allan, and H.M.T., unpublished data).

Another way to state the strand-transfer hypothesis is that, instead of steady processive polymerization, the reverse transcriptase growing point frequently pauses and enters a metastable state, leaving this metastable state to continue polymerization either at the next base or at another base at a different location. This transfer can be a result of the growing point moving or of another portion of the template displacing the template at the growing point. Polymerization at locations other than the next base gives rise to all of these types of variation except some base-pair substitutions that require misincorporation before continuing polymerization. Other base-pair substitutions involve dislocation (18).

Primer Transfers

To synthesize the LTR and then have a primer for copying the bulk of the viral genome, retroviruses start minus-strand DNA synthesis near the 5′ end of viral RNA using a base-paired cellular tRNA as a primer. This primer is annealed to the primer binding site (pbs) in viral RNA. After copying of the u5 and repeat (r) regions, the nascent minus-strand DNA transfers to the r sequences at the 3′ end of the same molecule of viral RNA, next to the poly(A) sequence. RNase H activity, associated with the reverse transcriptase molecule, may be involved in this transfer, removing the RNA r and u5 sequences (10–12). [Because a retrovirus virion contains two molecules of viral RNA the minus-strand primer DNA could theoretically transfer to the same molecule or to the other one (19). Recent work has clearly established that, in the absence of breaks, the minus-strand primer DNA always transfers from the 5′ to the 3′ end of the same RNA molecule, designated intramolecular minus-strand

Abbreviations: HIV-1, human immunodeficiency virus type 1; LTR, long terminal repeat; u3 (U3), unique 3′ RNA (DNA); u5 (U5), unique 5′ RNA (DNA); pbs (PBS), primer binding site in RNA (DNA); ppt (PPT), polypurine tract in RNA (DNA); r (R), repeat region in RNA (DNA).

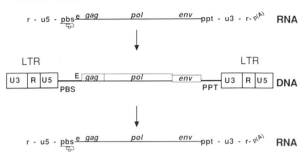

FIG. 1. RNA and DNA genomes of a simpler retrovirus. Infecting viral RNA is reverse transcribed to viral DNA, which is transcribed to form progeny viral RNA. *gag, pol,* and *env* are genes for virion proteins. The other named genes act in cis and are described in the text. p(A) is a polyadenylate tail, which is not reverse transcribed. The tRNA primer is shown annealed to pbs. In the virion, there are two copies of the genomic RNA.

DNA primer transfer (J. S. Jones, R. W. Allan, and H.M.T., unpublished data).]

Other recent work indicates that the minus-strand and plus-strand DNA primer transfers take place during the elongation or synthesis phase of DNA synthesis rather than at the end of the template (refs. 13, 17, 20, and 44; G. Pulsinelli and H.M.T., unpublished data; J. Zhang and H.M.T., unpublished data). Therefore, minus- and plus-strand strong stop DNAs, as they are traditionally termed, are not the usual intermediates for the reverse transcriptase growing point primer transfers.

The transferred minus-strand DNA can then be used as a primer to copy the viral RNA up to the end of the remaining 5′ RNA sequences, thus generating most of the minus-strand DNA.

After the minus-strand DNA is elongated through the R U3 regions, a reverse transcriptase RNase H activity cleaves the RNA template near its 3′ end just after a polypurine tract (ppt). The 3′ end of the viral RNA ppt forms a primer for plus-strand DNA synthesis. Elongation from this point occurs. At some point during the copying of U5 and the tRNA primer, the reverse transcriptase growing point transfers to the 3′ end of the minus-strand DNA molecule, annealing to the complementary PBS sequences. [In ≈20% of cases the strand transfer happens upon reaching the end of the pbs sequences in the minus-strand tRNA primer (G.

Pulsinelli and H.M.T., unpublished data). Much more rarely, the transfer is to DNA copied from the other molecule in the virion, designated intermolecular plus-strand primer transfer (J. S. Jones, R. W. Allan, and H.M.T., unpublished data). The low rate of intermolecular transfers may reflect the low rate of synthesis of complete minus-strand DNA from both RNA molecules in one virion.]

Base-Pair Substitutions

Base-pair substitution mutations involve dislocation mutagenesis or misincorporation by reverse transcriptase at the growing point, followed by polymerization beyond the misincorporation. There are definite hot spots for substitution mutations by reverse transcriptase as there are with other DNA polymerases (7, 18, 21–24). The retroviral reverse transcriptase does not have any error-correcting function (25, 26), perhaps because it lacks necessary accessory proteins and nuclease activities (27).

I propose that after misincorporation, the surrounding sequence determines whether or not there is polymerization at the base adjacent to the mismatch, thereby maintaining the reading frame, or transfer to another position on the template, forming a deletion, insertion, or recombinant. Reverse transcriptases appear to differ from other DNA polymerases more by the frequency of extension from a misincorporation than from the frequency of misincorporation itself (18, 22–24). This observation indicates that the reverse transcriptase can add some base-paired nucleotides relatively efficiently to a nucleotide that is not base-paired.

Frameshifts

Frameshifts, the additions or subtractions of 1 base, commonly occur during retrovirus replication, as in all other replication, within runs of a single nucleotide, and their frequency increases as the runs become longer (21, 28). Thus, with spleen necrosis virus, a simpler avian retrovirus, runs of 9 or 10 thymines or of 9 or 10 adenines result in frameshifts in 20–40% of replications (21, D. P. W. Burns and H.M.T., unpublished data). Frameshifts usually add or delete 1 base from the run itself. Dislocation mutagenesis (7, 18), where the frameshift involves a base-pair substitution incorporating the nucleotide next to the run, is a good illustration of the process.

Deletions

Deletions in retrovirus replication, as in many other systems, usually involve removal of nucleotides between small direct repeats (Fig. 3) (29–31). In addition, misincorporation can lead to deletions

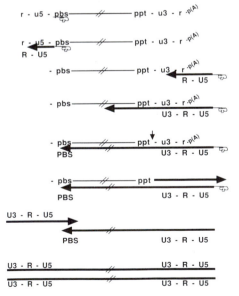

FIG. 2. Synthesis of retrovirus DNA. Thin lines, RNA; thick lines, DNA. Complete description is given in text.

Proc. Natl. Acad. Sci. USA 90 (1993)

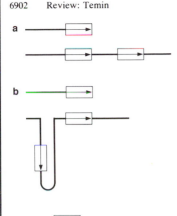

FIG. 3.   Simple deletion. Misalignment to an identical sequence (box with arrow) on the template can result in deletion as illustrated.

when the reverse transcriptase growing point scans downstream for an identical sequence rather than polymerizing through the misincorporation (13). However, the high rate, almost 100%, of deletions of long tandem repeats makes it unlikely that misincorporation is required for all deletions (32).

Deletions with Insertions

In retrovirus replication, it is not uncommon to find extra nucleotides inserted in the deletion, substituting for the deleted bases (31). Analysis of these inserted sequences reveals that they result from the reverse transcriptase growing point transferring to a small region of sequence identity on another template molecule in the virion (refs. 13, 31, 37, and 38; L. M. Mansky and H.M.T., unpublished data). This new template is the RNase H-digested plus-strand virion RNA or another RNA that is encapsidated in the virion. A second reverse transcriptase growing point transfer is then required to return to the original template. Often several abnormal strand transfers are required before returning to the original template.

Dimer RNA

All of the processes discussed above—primer transfers, base-pair substitutions, frameshifts, deletions, and deletions with insertions—involve only one RNA template molecule. The second viral RNA molecule does not seem to be required for normal reverse transcription (J. S. Jones, R. W. Allan, and H.M.T., unpublished data). However, recombination between the two strands of RNA provides a strong positive selective advantage for retroviruses, allowing them to repair breaks in the RNA and to exchange nucleic acid

sequences. Retroviruses do not have a pool of replicative intermediates or other molecules that can recombine (15). Thus, they have evolved dimer virion RNA to provide substrates for recombination (35).

Homologous Recombination

Homologous recombination during retrovirus replication almost always occurs during the original minus-strand DNA synthesis (36, 37). Homologous recombination results from the reverse transcriptase growing point transferring to an identical sequence on the other RNA molecule of the dimer RNA. Homologous recombination can be the result of usual reverse transcriptase growing point transfer, called copy-choice, or the result of an RNA break that forces the reverse transcriptase growing point to transfer, called forced copy-choice (38, 39). It has also been proposed that misincorporation is necessary for recombination (14). This hypothesis is based on experiments with purified HIV-1 reverse transcriptase, which showed that when the reverse transcription growing point transfers from RNA to RNA at a blunt-ended RNA·DNA hybrid molecule there is addition of an untemplated nucleotide. Since such reverse transcription growing point transfers do not usually occur at a blunt end, except possibly during forced copy-choice recombination, the hypothesis is unlikely to apply generally. In fact, when a modification of the system described by Zhang and Temin (34) was used, direct sequencing of recombinants in a region of sequence identity in the midst of nonidentical sequences showed no base-pair substitutions in 22 of 22 recombinants (J. Zhang and H.M.T., unpublished data).

Thus, an earlier misincorporation is not necessarily involved in homologous recombination. This is not surprising, since the rate of recombination is so high that the rate of base-pair substitution would be too high for viability if misincorporation were a necessary precursor for homologous recombination (refs. 40 and 41; J. S. Jones, R. W. Allan, and H.M.T., unpublished data).

Nonhomologous Recombination

When the retrovirus virion contains nonviral RNA sequences, the reverse transcriptase growing point can transfer to this RNA. When the nonviral RNA sequences are in a chimeric RNA molecule, a single reverse transcriptase growing point transfer will result in formation of a virus capable of replication with helper virus or helper cell proteins. The chimeric RNA usually results from read-through of transcription past the normal retrovirus polyadenylylation sequences. The transfer is usually to a short region of

sequence identity in the otherwise nonidentical sequence (33, 34). This process has given rise to naturally occurring highly oncogenic retroviruses, which contain an insertion of cellular protoon-cogene sequences (34, 42).

Increasing the size of the region of sequence identity in the midst of an otherwise nonidentical sequence increases the rate of such nonhomologous recombination (J. Zhang and H.M.T., unpublished data). At its maximum, however, the rate of such nonhomologous recombination is 1000 times less than that of homologous recombination. This result, together with other evidence that the relative location of the regions of sequence identity in the midst of otherwise nonidentical sequences affects the recombination rate (J. Zhang and H.M.T., unpublished data), indicates a higher order of virion organization that is not yet described and that can influence the reverse transcriptase growing point transfers.

Is One Property of Reverse Transcriptase Responsible for All of These Processes?

In this article, I have suggested that the necessity for the reverse transcriptase growing point to transfer from one place on the template to another place on the template, in order to form the primer molecules for much of the DNA synthesis and LTRs, underlies all of these processes of genetic variation. Some evidence in favor of one underlying process comes from a comparison of rates of mutations and types of mutations in two different viruses. As mentioned earlier, spleen necrosis virus is a simpler avian retrovirus, similar to murine leukemia viruses. Bovine leukemia virus is a more complex retrovirus, similar to human T-cell leukemia viruses. The overall rate of forward mutations in bovine leukemia virus replication is significantly less than the rate for spleen necrosis virus (L. M. Mansky and H.M.T., unpublished data). However, the distribution of different types of mutations is the same for both bovine leukemia and spleen necrosis viruses (L. M. Mansky and H.M.T., unpublished data). Thus, the bovine leukemia virus reverse transcriptase growing point seems to have a lower propensity to transfer during normal viral replication than the spleen necrosis virus reverse transcriptase growing point, but the results of the transfers are similar.

Attempts to Measure Kinetic Parameters

Numerous attempts have been made to model these processes in cell-free systems with purified reverse transcriptase and defined templates (for a recent review, see ref. 18). The results are similar to those found in experiments that analyze a single cycle of retrovirus replication (but see ref. 43).

All experiments with defined templates run into the inescapable problem of local sequence effects, which I have already indicated are an important feature of the reverse transcriptase growing point transfers. Thus, any experimental rates are the average for a particular template. [It should be noted that the comparisons of spleen necrosis and bovine leukemia viruses discussed above was done with the exact same template but in the opposite orientation (L. M. Mansky and H.M.T., unpublished data).]

Given this problem, we have measured the rates of each of these steps in a single cycle of replication of a simpler avian retrovirus. The rates are expressed as mutations per base pair per replication cycle and are as follows: base-pair substitutions, 1×10^{-5}; frameshifts, 1×10^{-6}; deletions, 2×10^{-6}; deletions with insertions, 1×10^{-6}; homologous recombination, 2×10^{-4}; nonhomologous recombination, 5×10^{-8}; recombination of a limited sequence identity in the midst of otherwise nonidentical sequence, 6×10^{-6} (refs. 21, 31, and 41; J. S. Jones, R. W. Allan, and H.M.T., unpublished data; J. Zhang and H.M.T., unpublished data).

In terms of the strand-transfer hypothesis, the most informative rates are perhaps the rates of frameshifts. In the formation of a frameshift within a run of 10 thymines or 10 adenines, the sum of the rates of formation of the metastable state and the probability of continuing misincorporation is $\approx 20\%$ (ref. 21; D. P. W. Burns and H.M.T., unpublished data). A simple interpretation of this result would be that there is a 40% probability of the reverse transcriptase growing point entering the metastable state for each 10 thymines or adenines incorporated and a 50% probability of slippage within the run. (I assume that the probability of a mistaken polymerization is <50%.) The lower rates of the genetic processes other than frameshifts discussed in this article would reflect the lower probability that the reverse transcriptase growing point would make an inappropriate transfer to resolve the metastable state in the absence of a nearby run of the same nucleotide.

In contrast, the rate of base-pair substitution would first include misincorporation, which would induce the metastable state of the reverse transcriptase growing point, and then resolution of the metastable state by readthrough or transfer controlled by the local and nearby sequences.

Summary

Retroviruses developed reverse transcriptase growing point transfers to form a provirus that is autonomous with respect to cis-acting sequences for transcription; that is, the enhancer/promoter sequences in viral DNA are copied from the viral RNA genome. This strand-transfer process can occur during polymerization of internal sequences as well as during primer synthesis. The rate of transfer and the genetic effects of the transfers depend on the local nucleotide sequence, distant or foreign sequences, and the sequence of the second RNA strand in the dimer. On the average, a simpler retrovirus seems to have for each round of replication at least one additional transfer that can have a genetic effect in addition to the two transfers required to make the LTRs and primers. Retroviruses have made a virtue of necessity by using the reverse transcriptase growing point transfer mechanism both in their replication and in their high rate of mutation. The AIDS epidemic is just one striking expression of this ability.

I thank K. Boris-Lawrie, S. Broder, H. Buc, D. Burns, J. Jones, T. Kunkel, D. Loeb, L. Mansky, G. Pulsinelli, B. Sugden, H. Varmus, S. Yang, and J. Zhang for useful comments on the manuscript. The research in H.M.T.'s laboratory is supported by U.S. Public Health Service Grants CA-22443 and CA-07175 from the National Cancer Institute. H.M.T. is an American Cancer Society Research Professor.

1. Albert, J., Abrahamsson, B., Nagy, K., Aurelius, E., Gaines, H., Nyström, G. & Fenyö, E. M. (1990) *AIDS* **4**, 107–112.
2. Myers, G. & Pavlakis, G. N. (1992) in *The Retroviridae*, ed. Levy, J. A. (Plenum, New York), Vol. 1, pp. 51–106.
3. Hu, W. S., Pathak, V. K. & Temin, H. M. (1993) in *Reverse Transcriptase*, eds. Skalka, A. M. & Goff, S. P. (Cold Spring Harbor Lab. Press, Plainview, NY), pp. 251–274.
4. Bishop, J. M. & Varmus, H. (1985) in *RNA Tumor Viruses: Molecular Biology of Tumor Viruses*, eds. Weiss, R., Teich, N., Varmus, H. & Coffin, J. (Cold Spring Harbor Lab. Press, Plainview, NY), 2nd Ed., Vol. 2, pp. 249–356.
5. Cullen, B. R. (1991) *J. Virol.* **65**, 1053–1056.
6. Temin, H. M. (1992) in *The Retroviridae*, ed. Levy, J. A. (Plenum, New York), Vol. 1, pp. 1–18.
7. Bebenek, K., Abbotts, J., Roberts, J. D., Wilson, S. H. & Kunkel, T. A. (1989) *J. Biol. Chem.* **264**, 16948–16956.
8. Coffin, J. (1993) in *Reverse Transcriptase*, eds. Skalka, A. M. & Goff, S. P. (Cold Spring Harbor Lab. Press, Plainview, NY), pp. 445–479.
9. Sánchez-Palomino, S., Rojas, J. M., Martínez, M. A., Fenyö, E. M., Nájera, R., Domingo, E. & López-Galíndez, C. L. (1993) *J. Virol.* **67**, 2938–2943.
10. DeStefano, J. J., Mallaber, L. M., Rodriguez-Rodriguez, L., Fay, P. J. & Bambara, R. A. (1992) *J. Virol.* **66**, 6370–6378.
11. Tanese, N., Telesnitsky, A. & Goff, S. P. (1991) *J. Virol.* **65**, 4387–4397.
12. Luo, G. & Taylor, J. (1990) *J. Virol.* **64**, 4321–4328.
13. Pulsinelli, G. & Temin, H. M. (1991) *J. Virol.* **65**, 4786–4797.
14. Peliska, J. A. & Benkovic, S. J. (1992) *Science* **258**, 1112–1118.
15. Weiss, R., Teich, N., Varmus, H. & Coffin, J. eds. (1985) *RNA Tumor Viruses: Molecular Biology of Tumor Viruses* (Cold Spring Harbor Lab. Press, Plainview, NY), 2nd Ed., Vol. 2.
16. Coffin, J. M. (1990) in *Virology*, eds. Fields, B. N., Knipe, D. M., Chanock, R. M., Hirsch, M. S., Melnick, J. L., Monath, T. P. & Roizman, B. (Raven, New York), Second Ed., Vol. 1, pp. 1437–1500.
17. Tekesnitsky, A. & Goff, S. P. (1993) in *Reverse Transcriptase*, eds. Skalka, A. M. & Goff, S. P. (Cold Spring Harbor Lab. Press, Plainview, NY), 1st Ed., Vol. 1, pp. 49–83.
18. Bebenek, K. & Kunkel, T. A. (1993) in *Reverse Transcriptase*, eds. Skalka, A. M. & Goff, S. P. (Cold Spring Harbor Lab. Press, Plainview, NY), pp. 85–102.
19. Panganiban, A. T. & Fiore, D. (1988) *Science* **241**, 1064–1069.
20. Lobel, L. I. & Goff, S. P. (1985) *J. Virol.* **53**, 447–455.
21. Pathak, V. & Temin, H. M. (1990) *Proc. Natl. Acad. Sci. USA* **87**, 6019–6023.
22. Preston, B. D., Poiesz, B. J. & Loeb, L. A. (1988) *Science* **242**, 1168–1171.
23. Mendelman, L. V., Petruska, J. & Goodman, M. F. (1990) *J. Biol. Chem.* **265**, 2338–2346.
24. Ricchetti, M. & Buc, H. (1990) *EMBO J.* **9**, 1583–1593.
25. Battula, N. & Loeb, L. (1976) *J. Biol. Chem.* **251**, 982–986.
26. Roberts, J. D., Bebenek, K. & Kunkel, T. A. (1988) *Science* **242**, 1171–1173.
27. Sancar, A. & Hearst, J. (1993) *Science* **259**, 1415–1420.
28. Ripley, L. S. (1990) *Annu. Rev. Genet.* **24**, 189–213.
29. Streisinger, G., Okada, Y., Emrich, J., Newton, J., Tsugita, A., Terzaghi, E. & Inouye, M. (1966) *Cold Spring Harbor Symp. Quant. Biol.* **31**, 77–84.
30. Streisinger, G. & Owen, J. (1985) *Genetics* **109**, 633–659.
31. Pathak, V. K. & Temin, H. M. (1990) *Proc. Natl. Acad. Sci. USA* **87**, 6024–6028.
32. Rhode, B. W., Emerman, M. & Temin, H. M. (1987) *J. Virol.* **61**, 925–927.
33. Zhang, J. & Temin, H. M. (1993) *Science* **259**, 234–238.
34. Zhang, J. & Temin, H. M. (1993) *J. Virol.* **67**, 1747–1751.
35. Temin, H. M. (1991) *Trends Genet.* **7**, 71–74.
36. Hu, W.-S. & Temin, H. M. (1992) *J. Virol.* **66**, 4457–4463.
37. Jones, J. S., Allan, R. W. & Temin, H. M. (1993) *J. Virol.* **67**, 3151–3158.
38. Coffin, J. M. (1979) *J. Gen. Virol.* **42**, 1–26.
39. Xu, H. & Boeke, J. D. (1987) *Proc. Natl. Acad. Sci. USA* **84**, 8553–8557.
40. Nowak, M. (1990) *Nature (London)* **347**, 522.
41. Hu, W.-S. & Temin, H. M. (1990) *Proc. Natl. Acad. Sci. USA* **87**, 1556–1560.
42. Swain, A. & Coffin, J. M. (1992) *Science* **255**, 841–845.
43. Varela-Echavarria, A., Garvey, N., Preston, B. D. & Dougherty, J. P. (1992) *J. Biol. Chem.* **267**, 24681–24688.
44. Ramsey, C. M. & Panganiban, A. T. (1993) *J. Virol.* **67**, 4114–4121.

The DNA Provirus: Howard Temin's Scientific Legacy
Edited by G. M. Cooper, R. Greenberg Temin, and B. Sugden
© 1995 American Society for Microbiology, Washington, DC 20005

Chapter 16

Retrovirus Variation and Evolution

John M. Coffin

In the 25 years since the simultaneous discovery of reverse transcriptase by Temin and Mizutani (95) and Baltimore (7), retroviruses have remained in the forefront of biomedical science. The study of them has provided the key insights into novel mechanisms of information transfer used by elements found in most or all eukaryotes (44, 92) and at least some bacteria (52); into the genetic and biochemical mechanisms underlying malignant and normal cell growth (10, 11); and into the causation of newly recognized diseases, including adult T-cell lymphoma and AIDS (60). The remarkable power of these simple elements to continually inform our understanding of important biological and pathological processes is due partly to their unique evolutionary association with the host and partly to the historical accident that brought retroviral diseases to epidemiological prominence in this country at a time when the medical community had already developed the tools necessary to figure them out.

That retroviruses have a unique mode of association with their host was Howard Temin's key insight, arrived at while he was still a student at the California Institute of Technology and based on inference from genetic experiments (88). In 1967, when I joined the laboratory as one of the first graduate students, the provirus hypothesis remained untested and generally unaccepted by the scientific community. Even among members of the Temin Laboratory, opinion in its favor was less than unanimous. Howard often returned from scientific conferences no less convinced of the correctness of his views but depressed and frustrated by his inability to perform the experiments that would prove them to the skeptical world. In fact, in the late 1960s, only a few members of the laboratory were studying retrovirus replication (see, for example, reference 12). Howard had directed his own experimental work toward the study of growth factors and the differential responses of normal and transformed cells to these factors (89). This work was also far ahead of its time but much less controversial.

The situation changed dramatically with the arrival of Satoshi Mizutani. In a short time, he found that infection with Rous sarcoma virus was insensitive to inhibitors of protein synthesis early after infection, implying that the necessary enzyme must preexist in the virion. With this information and the precedent set

John M. Coffin • Department of Molecular Biology and Microbiology, Tufts University School of Medicine, 136 Harrison Avenue, Boston, Massachusetts 02111.

by RNA polymerases in vaccinia and vesicular stomatitis viruses, it was relatively straightforward to assay virions for the predicted RNA-directed DNA polymerase.

The excitement surrounding this discovery was remarkable, especially when considered from an era in which we are, it seems, bombarded with media hype about every scientific "discovery," no matter how trivial or irreproducible, from certain laboratories. Howard first presented the observation at an International Cancer Congress in Houston. Shortly before, he asked the laboratory for advice on whether he should submit the results, then in a somewhat preliminary state, for immediate publication or wait to complete a more thorough study. Some others and I counseled waiting at least until the product of the reaction was well characterized as viral DNA. This was terrible advice, fortunately not followed. Indeed, the story is told that one well-known scientist left the congress immediately after Howard's presentation in Houston, flew home to New York, and repeated the whole study the next day. Also, *Nature,* which only a few weeks before had printed a rather sarcastic article describing Howard's views (5), rushed the paper into print with only a 12-day delay from receipt to publication and with the breathless banner "Central Dogma Reversed" (4).

The crush by other scientists to duplicate and extend the discovery was also unprecedented, at least in the experience of a young graduate student. Within a few months, many other laboratories had duplicated and extended the work, leading to another *Nature* banner: "Apres Temin, le Deluge" (3). Particularly memorable was the Cold Spring Harbor tumor virus meeting that year, where many well-known scientists fought vigorously over the biochemical details of reverse transcriptase (even to the extent of engaging in a food fight at the end of the meeting), while Howard, now secure that the real discovery was his, could view all the competition from a lofty viewpoint.

The laboratory was visited often by the media that year. I remember a particularly amusing dialogue between Howard and the photographer sent by *Newsweek* to shoot a cover, when Howard discovered that the photographer also worked for *Playboy.* I also recall a photo crew for a large-circulation German magazine who were far more interested in photographing the young female technicians than their famous boss.

The discovery of reverse transcriptase also provided me with a long-sought thesis project. My task was to use reverse transcriptase as a marker to study virus structure and to learn whether subviral structures could be detected and studied in infected cells by virtue of their endogenous reverse transcriptase activity. I was able to demonstrate such an activity but found it to be present in uninfected cells as well (28, 29). Although the project was continued for a few years by Young Kang (55, 56), it became clear, at least to us, that we could not distinguish a true reverse transcriptase reaction from a chance association of cellular DNA polymerase and RNA. Unfortunately, some other laboratories seemed not to be aware of our findings, since assays similar to the ones we used play a key role in several of the transiently famous "human rumor virus" discoveries of the mid-1970s (103).

After I left the Temin Laboratory in 1972 and moved to Switzerland to study retrovirus RNA structure, synthesis, and genetic organization with Charles

Weissmann, Howard and I maintained parallel interests in many areas of retrovirology. Over the ensuing decades, both our laboratories were involved in elucidating fundamental aspects of retroviruses, including genome organization (27, 42), the mechanism of viral DNA synthesis and integration (27, 58, 59, 67, 77), virus-receptor interaction and cell killing (33, 104, 105), endogenous viruses (8, 41, 57, 97), the mechanism of recombination (18, 50, 85, 109), and genetic variation (16, 18, 20, 21, 35, 68, 85). We met frequently at conferences during the years, and our conversations during these encounters, although brief, often formed the high point of the whole meeting in terms of the information, insight, and common-sense advice that I received. In this respect, I was the luckiest of Howard's trainees, since our joint interests often brought us together in this way.

Two of Howard's long-term interests were the origin and evolution of retroviruses, and he published many papers speculating on these topics (78, 90, 91–94). Although we still do not know where retroviruses came from, we do have the tools available to answer other important questions regarding retrovirus evolution. In this paper, I discuss two of these topics: the long-term association of virus and host as revealed by the study of endogenous proviruses and the short-term "evolution" that characterizes the association of some viruses, most notably human immunodeficiency virus (HIV), with their host.

ENDOGENOUS VIRUSES

Retroviruses are the only group of viruses known to have a fossil record, and we all carry around as part of our genetic makeup retroviruses in the form of endogenous proviruses that resulted from infection of the germ lines of our distant ancestors (19, 80). It is convenient to divide endogenous proviruses into two groups: ancient and modern.

Ancient endogenous proviruses were inserted into the germ line of a species prior to separation of that species from related species. In general, such proviruses are not polymorphic in location among members of a species and are shared among closely related species, allowing the time of their insertion to be estimated. Because of their great age and the apparent absence of selection for viability, they have suffered many inactivating mutations over time, and a certain amount of molecular paleontology, such as correction of reading frames, is necessary to infer their original genetic structure. Such proviruses are widely dispersed in mammals and birds and probably in other vertebrates and invertebrates as well. Thousands of endogenous proviruses inhabit the genomes of humans, mice, and other species (70, 106). These belong to a number of distantly related groups, some of which are very similar to modern retroviruses of several genera, including the avian and mammalian C-type virus and the B- and D-type viruses. Their presence is a clear indication of a time when ancestral species were subject to infection with exogenous viruses, a small residue of which have remained behind as a fossil record of these infections (24). Study of the relationship of these elements to modern viruses, to one another, and to the genome they occupy could be highly rewarding in uncovering the evolution of the host-retrovirus association.

At present, the analysis of the ancient proviruses is primitive at best. In contrast, in recent years we have learned a considerable amount about the more recent active proviruses. Like the ancient proviruses, they belong to a number of groups, each of which is quite closely related to exogenous viruses of the same species. Members of a group differ in their positions within the genome. They are, however, very closely related in sequence to one another and differ largely by simple deletions. Each group of active proviruses often has one or more members that can be expressed to yield infectious virus. Their insertion into the genome is recent and ongoing, as can be inferred both from the polymorphism of their location in one individual host (or inbred strain) compared to another and from their occasional appearance at new locations. It is the polymorphism in location that makes these proviruses more tractable than the ancient proviruses for genetic analysis. Unlike the ancient proviruses, which are widely dispersed among animal species, the recent proviruses are sporadically distributed and limited to a relatively small number of hosts, including many species of mice, cats, swine, baboons and a few other monkeys (but not humans or other higher primates), and a few birds, including chickens and some pheasants. The numbers of these proviruses also vary considerably, from one to a few in chickens to 60 or so in mice (19).

To understand the host-provirus relationship better, we chose as a model the proviruses of mice, in particular the C-type proviruses, which form the largest group related to known viruses. We started by cloning and analyzing as many proviruses as possible from one strain of inbred mice (HRS/J). This analysis, performed by Jonathan Stoye in my laboratory, provided the major insight that allowed us to develop the system further. Jonathan discovered that although they are very closely related, all the sequences could be divided into four groups, called ecotropic, xenotropic, polytropic, and modified polytropic. Each group shared a set of polymorphisms that distinguish it from all the other groups (81) (Fig. 1). Most usefully, a small sequence polymorphism in the SU portion of *env* allowed us to develop a set of oligonucleotide probes that unambiguously detected all

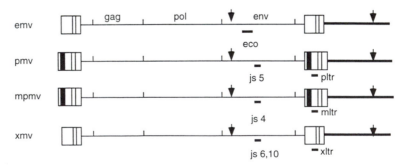

FIGURE 1. Polymorphism among endogenous murine leukemia viruses. The provirus maps show linked sequence differences useful for distinguishing the four groups. These include sequences in *env* and the long terminal repeat that generate useful oligonucleotide probes, restriction site polymorphisms, receptor utilization differences, and an insertion of a small transposable element into the long terminal repeat (81). emv, ecotropic murine virus; pmv, polytropic murine virus; mpmv, modified polytropic murine virus; xmv, xenotropic murine virus.

members of the three largest groups (collectively referred to as nonecotropic viruses) (82). The use of these probes enabled us to divide the proviruses in these groups into three subsets of about 20 each in each strain of inbred mice. The separation permitted ready detection and precise identification and enumeration of all proviruses by use of a simple Southern blot strategy (82). This strategy relies on identification of individual proviruses by their chromosomal locations,

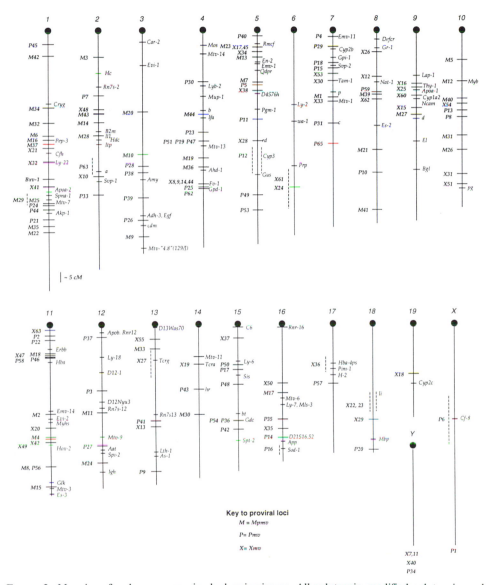

FIGURE 2. Mapping of endogenous murine leukemia viruses. All polytropic, modified polytropic, and xenotropic proviruses were mapped onto the mouse genome by using the proviral DNA distribution in recombinant inbred and intercross mice. The proviruses are shown on the left side of each chromosome, and standard markers are shown on the right side (39–41).

as indicated by the size of the restriction fragment that includes the region of viral DNA reactive with the probe joined to a 3' flanking sequence. This strategy was used to study several aspects of the association between endogenous proviruses and the murine host.

First, we found considerable polymorphism in the distribution of the endogenous proviruses among inbred strains of mice. Of approximately 150 proviruses analyzed, only 1 was common to all strains (41). With a few exceptions, about one-half of the proviruses in any one strain were shared with some other strain, and every strain examined displayed a unique and characteristic pattern of proviruses. In fact, the pattern of endogenous proviruses is the best and simplest way to unambiguously identify an inbred mouse strain, although it has not been widely applied for this purpose.

Given this polymorphism, it was a relatively straightforward task to map the proviruses to specific chromosomal sites by analysis of their distribution among sets of recombinant inbred strains derived by repeated inbreeding of a number of progeny from a single initial cross between two parental strains (87). Assignment of relative map locations is a simple matter of comparison of the strain distribution profile of a provirus with that of other known markers. In other cases, direct analysis of F_2 and backcross progeny of crosses between mice of two strains was used. In this way, we determined the chromosomal locations of about 150 different proviruses in about 20 inbred strains (41). The results of this study show that endogenous proviruses are widely distributed in the mouse genome with no significant bias in distribution or obvious clustering (Fig. 2).

The large number and widespread distribution of proviruses, together with their ease of identification, have made them quite useful as markers for identifying regions of the mouse genome, and they have been used in this way to map a number of genes, such as *mnd* (63), as well as a transgene (43), a set of deletion mutants (72), and a complex trait such as epilepsy (73). They have also been useful for identifying chromosomal alterations in tumors (96) and for positional cloning of a nearby locus (79a).

In addition to serving as markers for genes, a provirus inserted into an identifiable gene can lead at the same time to a mutation and the means of cloning the affected sequence. Among the 150 proviruses studied, we identified two such mutagenic insertions, one into the retinal degeneration (*rd*) loc (13) and the other into hairless (*hr*) (83). In the former case, the gene had been cloned in another way; in the latter, the provirus provided the initial entry for its ultimate cloning and sequencing (17).

In the course of these studies, we traced the distribution of a large number of proviruses over many generations. The data set we accumulated allowed us to estimate the rate of accumulation of new proviruses at about 1 per 3,500 generations and the rate of loss at about 1 per 250,000 generations (41). These rates suggest that the proviruses are gradually accumulating, at least in inbred mice. We would also have been able to detect gene conversion events (as two different types of proviruses in the same position in different strains) with similar sensitivity, but none were observed. Similarly, we saw no evidence for size or restriction site polymorphism in any provirus or flanking region during the course of inbreeding.

The distributions of endogenous proviruses and other genetic markers among inbred strains can also be used to draw inferences regarding the origin and "evolution" of these strains. Inbred mouse strains were derived from fancy mice, which themselves were derived from wild mice of Asian and European origin, with selection for traits (such as color) of interest to fanciers. While the general outlines of the pattern of inbreeding and the specific relationships of a few strains are known, there are large gaps in our knowledge of how most strains were derived and of the origins of the genetic polymorphisms that distinguish them (64). Were the endogenous proviruses derived during the course of inbreeding, or do they reflect genetic polymorphism already present in the ancestors and differentially fixed in the genomes during inbreeding? Because the insertion of a specific provirus at a particular site is a very rare event, highly unlikely to happen twice independently, and because its loss is also rare, proviruses can be considered a set of irreversible genetic markers, each of which indelibly tags a segment of DNA and all its direct ancestors, starting from the time of insertion. Simple inspection of the distribution of a sample of informative proviruses among strains reveals a pattern of inheritance inconsistent with an evolutionary model in that there is virtually no congruence among sets of inbred strains sharing common proviruses (Fig. 3). To address this issue further, we derived the optimum phylogenetic tree for the strains examined and compared it to a similar tree derived using biochemical markers (6). As shown in Fig. 4, although the two trees are equally consistent with the historical record (heavy lines), they are otherwise completely different. This result is not consistent with an evolutionary model for the derivation of the strains but is consistent with a model in which most or all of the informative proviruses (those found in more than one but less than all strains) were already present in the varied ancestral species, and subsets were randomly fixed during inbreeding.

If already present at the time of the initial crosses, then the individual proviruses might have been present for a considerable time prior to that point. Indeed, we have begun an examination of wild species and subspecies and have observed a number of proviruses yielding bands of sizes identical to those found for inbred

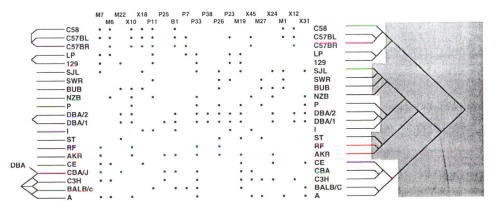

FIGURE 3. Distribution of some endogenous proviruses among common inbred strains of mice. The chart on the right depicts the relationships of some strains inferred from the historical record (64).

228   COFFIN

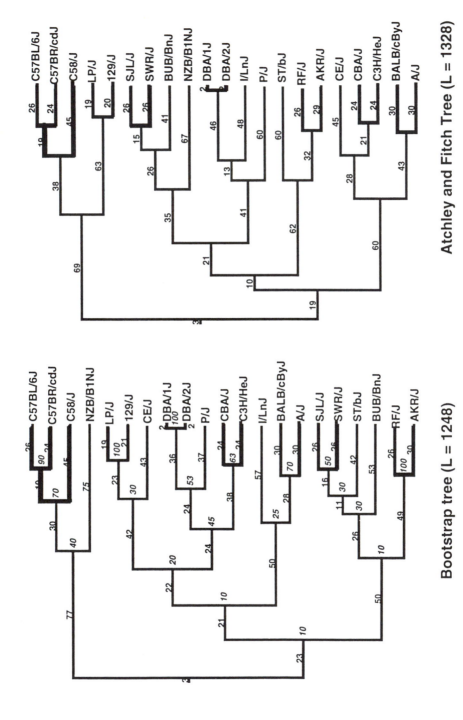

Bootstrap tree (L = 1248)

Atchley and Fitch Tree (L = 1328)

FIGURE 4. Phylogeny of inbred strains. (Left) Tree derived from the endogenous provirus distribution by using PAUP (86); (right) relationships of inbred strains derived by analysis of biochemical polymorphisms (6). Branches depicted by heavy lines are consistent with the historical record.

mice, suggesting that a number of the proviruses were fixed in the ancestral mice. We also observed that although the three classes of nonecotropic viruses are more or less equally present in inbred mice, each of the wild species tends to have only one. Finally, in some species of mice, the predominant endogenous proviruses are recombinant relative to those we had been studying in laboratory mice. That is, *env* and long-terminal-repeat markers that are always linked in a certain way in inbred mice are also linked in wild mice, but in a different way. Taken together, these results suggest that the different species were invaded by closely related but not identical viruses at a time roughly corresponding to their separation from one another. These viruses spread clonally within the germ line of each species, only to meet when the different species were mated to one another to form inbred strains. This meeting is not totally without pathogenic consequence. The expression of one or a few proviruses followed by recombination events involving the virus progeny of no fewer than three proviruses can give rise to pathogenic viruses and eventually lead to the death from lymphoma of all mice of strains of the correct genetic background, such as AKR (84).

Infection of the germ line is a rare event, yet endogenous proviruses have many times been independently fixed in the genomes of mice and other species. To account for the frequent fixation of rare events, a strong selective force is necessary. A likely clue to the selection mechanism is provided by the case of the mouse mammary tumor virus (MMTV). A few years ago, we (38) and others (2, 36, 66) observed that endogenous MMTV proviruses as well as exogenous viruses (62) contain a gene (now known as *sag*) that encodes a superantigen activity. A superantigen is a protein that stimulates T cells via interaction with their antigen receptor and with the major histocompatibility complex class 2 molecule on B cells. In this respect, it behaves just like a normal antigen. The difference is that it interacts nonspecifically with most class 2 molecules and all T cells belonging to one or a few specific subsets of Vβ specificity (1, 23, 53). As a consequence, a large fraction (typically, some 5%) of T cells are stimulated. If the superantigen is expressed from early in life (as with endogenous proviruses), then these cells are deleted. Among endogenous and exogenous MMTVs, there is considerable variation in Vβ recognition specificity, conferred by a short sequence at the C terminus of the Sag protein (9, 108). The presence of Sag activity is necessary for efficient infection of B cells and transmission of the virus from the site of infection (the gut) to the site of transmission (the mammary gland). To accomplish these processes, Sag protein expressed by infected B cells activates neighboring T cells, which in turn release cytokines to stimulate rapid expansion of the infected B cells (48). If the Sag-reactive T cells are deleted early in life by expression of identically reactive Sag proteins from an endogenous provirus, then the animal is protected and the transmission chain is broken (46). This protein in turn exerts a strong selective pressure on the virus to vary its Vβ reactivity, presumably leading to the observed polymorphism.

Although superantigen activity is unique to MMTV, similar opportunities exist for protection of the host by other types of proviruses via *env*-mediated interference with infection. At least two different endogenous proviruses of mice were originally identified as genetic loci protective against infection with exogenous virus of the same receptor specificity (14, 51). Similar phenomena have been seen in chickens, for which it has been suggested that even endogenous proviruses expressed at a very low level offer some protection against pathogenic effects of exogenous viruses by inducing tolerance and mitigating immune-mediated effects (19).

Taken together, these observations lead to a model for the origin of endogenous viruses in which a species is infected with an exogenous retrovirus that becomes endemic. Chance introduction of a few of these proviruses into the germ line confers on their hosts at least partial resistance to infection with the exogenous virus and therefore a strong selective advantage for them and their descendants. Once in the germ line, endogenous viruses continue to move around (largely or entirely by expression of virions and reinfection). At the same time, they evolve to a better fit within this different niche, for example, by becoming less pathogenic. Also at the same time, host mutations, such as polymorphism in receptors, are selected in response to pressure from the infecting virus. These effects combine to create the phenomenon of xenotropism, in which the endogenous virus can no longer infect its host because the host lacks an appropriate receptor. The combined

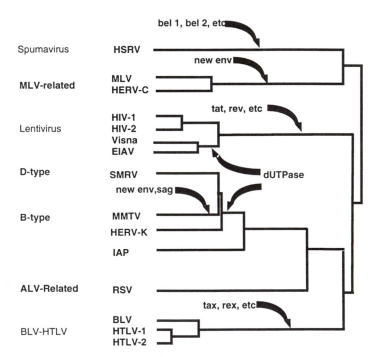

FIGURE 5. Phylogeny of retroviruses. The figure shows a phylogenetic tree based on comparison of reverse transcriptases and relating the presently defined genera (32) (listed on the left). Heavy curved arrows indicate probable recombination events. Retrovirus genera with endogenous members are shown in boldface. MLV, murine leukemia virus; ALV, avian leukemia virus; HSRV, human spuma-retrovirus; HERV-C, human endogenous retrovirus C; EIAV, equine infectious anemia virus; SMRV, squirrel monkey retrovirus; IAP, intracisternal A particle; RSV, Rous sarcoma virus; BLV, bovine leukemia virus; HTLV, human T-lymphotropic leukemia virus.

effects of endogenous proviruses and host resistance mutations eventually succeed in eradicating the exogenous virus infection and remove the selection pressure for active endogenous proviruses. In the absence of such selection, the endogenous proviruses accumulate inactivating mutations. Finally, no active members remain, and the remaining proviruses are fixed and fossilized in the germ line. This process has probably affected the ancestors of all mammals and birds many times in the distant past and continues today in a few species.

A final point concerning the evolution of endogenous viruses is an interesting correlation between virus lifestyle and endogenization. Retroviruses are currently classified into seven genera that are based on several criteria, including relationship among the reverse transcriptases (Fig. 5) (22). A useful distinction between the genera is based on simple versus complex lifestyle (Howard liked to call these ''more simple'' and ''more complex''). Simple retroviruses usually contain only genes encoding virion proteins and have a simple splicing pattern with only one, or occasionally two, subgenomic mRNAs. Complex viruses (the lentiviruses, spumaviruses, and human lymphotropic leukemia virus-bovine leukemia virus group) have multiple, partially spliced mRNAs and a number of additional genes encoding proteins that regulate transcription and processing of mRNA. The interesting point is that all four genera of simple viruses have endogenous relatives in some species; none of the complex viruses do. The basis for this correlation remains unknown. A possibility is that the complex viruses have evolved a lifestyle emphasizing infection of immunologically competent individuals; the simple viruses are generally restricted to vertical transmission into a fetus or newborn. It may be that germ line infection with all retroviruses occurs only during the stages of oogenesis occurring in newborn females, the way it does with some murine leukemia viruses (61).

GENETIC VARIATION IN VITRO AND IN VIVO

Retroviruses differ considerably in the amount of genetic variation observed among individuals within species. On the one hand, the avian C-type viruses show a very high level of sequence conservation: even isolates from different species and from the same species at very different times and on different continents are virtually identical in sequence over long stretches of *env*. Furthermore, the genetic variation that they do exhibit is limited to regions of obvious biological significance, such as the portions of *env* that contribute to diversity in receptor interaction (34). On the other hand, HIV and the other lentiviruses exhibit considerable diversity among isolates and even between genomes isolated from the same individual at the same time (47, 101). Divergence of up to 25% within the most variable regions (in *env*) has been observed, even when infection was initiated with known clonal virus (15, 54). These observations are consistent with the evolution of HIV in vivo into a complex quasispecies (37). It has been common practice to attribute these differences to a higher error rate of lentivirus reverse transcriptase, although the mutation rate of HIV remains to be measured, and the in vitro measurements of misincorporation rates that have been done (69, 74) are of doubtful significance.

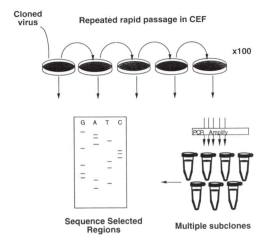

Cloned virus

Repeated rapid passage in CEF

x100

Sequence Selected Regions

PCR Amplify

Multiple subclones

FIGURE 6. Protocol for in vitro rapid passage of Rous sarcoma virus. A biologically cloned sample of the Prague strain of Rous sarcoma virus (subgroup B) was subjected to undiluted passage in cultures of chicken embryo fibroblasts (CEF) every 3 to 4 days. At selected passages, a sample was taken for analysis, initially by quantitative RNase T1 fingerprinting (30) and later by PCR amplification, cloning, and sequencing of a selected region on *env* (26).

The accumulation of mutations during replication of a population of virus is a function of three parameters: the mutation rate of a single replication cycle, the selective advantage or disadvantage of each specific change, and the number of replication cycles of the population since it was a single individual. To test the interplay of these factors in a retrovirus population, we subjected a biologically cloned population of Rous sarcoma virus to repeated rapid cycles of replication by passage every 3 to 4 days in chicken fibroblast cultures (Fig. 6). The rationale for this experiment was that this regime should represent a large change in the selective environment of the virus, which up to this point had been maintained as an oncogenic virus whose replication was incompatible with adverse effects

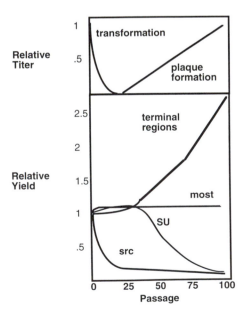

FIGURE 7. Selected changes in the Rous sarcoma virus genome as a function of passage. (Top) Biological properties: transformation (measured by focus assay) and cytopathicity (measured by plaque assay); (bottom) genome changes measured by quantitative fingerprinting, including loss of the *src* gene, overgrowth of a highly defective variant, and appearance of selected point mutations in the SU portion of *env* (30).

on cell viability. It seemed reasonable, therefore, that certain aspects of virus replication would be maintained at suboptimal rates. By rapidly passaging the virus, we relieved it of that restriction, selecting only for the virus most able to amplify itself in a short period without regard for the health of the infected cell. Changes in the genome over the course of 100 passages (probably 200 to 300 replication cycles) were monitored originally by fingerprinting the genomic RNA (30) and more recently by PCR amplification and sequencing of selected regions (26).

The results shown in Fig. 7 illustrate the changes in the biology and genome of the virus over the course of the experiment. Consistent with the rationale, the first event observed was the loss of the *src* gene by deletion, coincident with the expected loss in transforming ability. Also consistent was the increasing amount of cytopathic virus, as indicated by a simple plaque assay (30). Another effect of passaging was the appearance of a highly deleted genome, which we have characterized extensively (99, 100). Despite these changes, the majority of the genome was quite stable, with only a few point mutations observed. For the purposes of this discussion, the most interesting of these changes were a pair of nearby mutations in a highly conserved region of *env*. Fitting of the changes in these bases to theoretical curves (see below) implies that each imparts a selective advantage to the genome of about 4%. Sequence analysis of these two mutations (Fig. 8) was quite revealing. One of them was a nonconservative change in a very

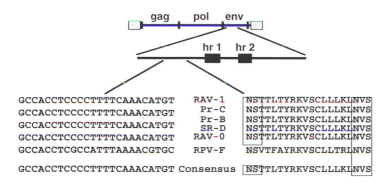

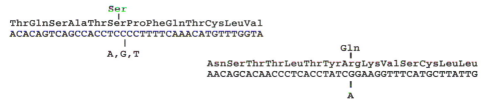

FIGURE 8. Selected point mutations in the Rous sarcoma virus *env* gene. The region of the SU portion of *env* shown in the top lines was amplified, cloned, and sequenced from selected passages of the experiment shown in Fig. 6 and 7. The two most prominent selected changes are indicated in the nucleotide and amino acid sequences shown at the bottom, and the same region from a number of avian leukemia virus isolates of chickens and pheasants is shown for comparison. RAV-1, Rous-associated virus-1; Pr-C, Prague strain RSV, subgroup C; SR-D, Schmidt-Ruppin RSV, subgroup D; RPV-F, ring-necked pheasant virus, subgroup F.

highly conserved region of the genome (even between viruses of chickens and pheasants). Thus, although this is one of the first changes to be seen, its appearance must be the consequence of an altered selective environment, not a hypervariable region or mutational hotspot. The other change is a noncoding change (in fact, all three possible changes at this position were observed), again in a highly conserved sequence.

Although we do not know the nature of the selective forces responsible for these changes, there are some important general points. First, even very highly conserved sequences are not immutable and can be maintained by relatively subtle selective forces and rapidly altered when these forces change. Second, synonymous mutations are not neutral mutations, and assumptions to the contrary (as in application of neutral theory to the problem of virus variation) are likely to lead to erroneous conclusions. Third, relatively subtle selective forces can contribute significantly to shaping the viral genome under conditions of mass replication.

The results given above suggest that the rapid generation of genetic diversity in a normally very stable region of a very stable genome was a consequence of subtle selection forces acting on a rapidly replicating population. To study these effects further, multiple clones of the region including these two sequences were selected and sequenced following amplification by PCR from virus at various passages. These sequences revealed that the region contains a number of selected mutations at other conserved bases that increase in frequency at different rates. Very little evidence for accumulation of neutral mutations could be observed, suggesting selection for conservation of sequence at most positions, even synonymous ones. Figure 9 shows the pattern of accumulation of changes in the region analyzed as a function of time in culture. As might be expected from the shape of the underlying curves, the curve obtained is clearly sigmoidal, not linear. Thus, it is impossible (and very misleading) to calculate a "rate" of variation from this kind of data, although many such calculations are found in the literature (15, 65, 79). This figure also shows data of the same type obtained for simian immunodeficiency virus (SIV) following inoculation of monkeys with molecularly cloned virus (15). Although the data are not as smooth (owing to the smaller number of sequences analyzed), both the extent of variation and the shape of the curve for the two viruses are very similar. Thus, under conditions of rapid replication, a virus

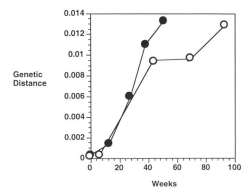

FIGURE 9. Genetic change as a function of time in Rous sarcoma virus passaged in vitro and SIV growing in vivo. The genetic distance of the passaged virus from the experiment shown in Fig. 6 was calculated from the sequences of at least 50 clones from each selected passage (closed circles). For comparison, the data of Burns and Desrosiers (15) for molecularly cloned SIV growing in monkeys are shown by open circles.

that displays relatively low levels of genetic variation is not distinguishable from a "highly variable" virus. This result strongly suggests that differences in lifestyle are more important than differences in underlying biology (such as error rate of reverse transcription) in determining the extent of genetic variation.

To see the effects of mutation and selection in such populations, we did some simple modeling, using a straightforward algorithm to estimate the frequency of a given mutation in a population as a function of the number of replication cycles, the mutation frequency μ, and its selective advantage or disadvantage s [defined as relative growth rate (mutant/wild type) -1]. Some curves derived in this way are shown in Fig. 10. These reveal several surprising results. First, under the conditions of this sort of experiment (large numbers of replication cycles of populations large enough that all possible mutations are likely to appear many times at each replication cycle), subtle positive selective advantages are much more

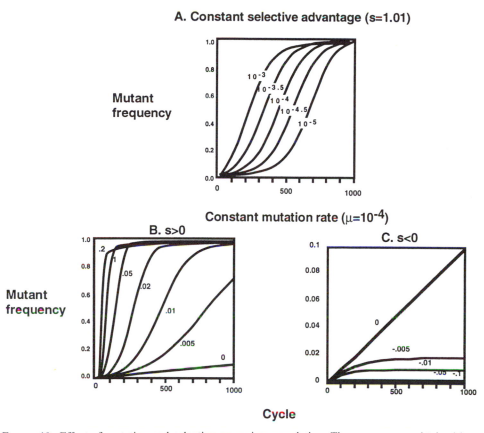

FIGURE 10. Effect of mutation and selection on a virus population. The curves were obtained by a simple computer simulation in which the frequency of a given mutation was incremented by the effects of forward mutation (μ) and selective advantage (s if positive), and decremented by reverse mutation (assumed to be equal to μ) and s (if negative). (A) Effect of varying μ from 10^{-3} to 10^{-5} while holding s constant at 10^{-2}; (B and C) effects of varying positive (B) and negative (C) s while holding μ constant at 10^{-4} (21, 25).

important than large differences in mutation rate in determining variation. For a mutation that confers a 1% selective advantage, a 10-fold difference in mutation rate makes less than a 2-fold difference in the pattern of its appearance (Fig. 10A). A twofold difference in selective advantage (corresponding to about a 1% difference in relative growth rate) has about the same effect (Fig. 10B). For mutations with small negative s (probably the majority in any virus population), the pattern is rather different. These mutations accumulate initially at a rate equal to the forward mutation rate but then level off at a value roughly equal to $\mu/-s$ (Fig. 10C). Thus, unlike the case with positive selection, the frequency of negatively selected mutants is equally dependent on μ and s. This relationship can have important consequences for therapeutic strategies against HIV, since drug resistance mutations (in protease or reverse transcriptase, for example) are expected to have a slight negative effect on virus replication in the absence of drug selection (25).

The powerful molding effect of subtle selective forces on a viral genome is exerted only at large numbers of replication cycles as a population. With most virus infections, the number of replication cycles in a single infected host is relatively small (probably less than 20), and genetic variation will be much more strongly determined by mutation rate and chance. Does HIV replicate sufficiently in vivo to be describable in this way? While the pattern of genetic variation shown in Fig. 9 is consistent with this idea, it does not prove it. Proof of the extensive replication of HIV in vivo was obtained recently by short-term studies looking at the amount of circulating virus as a function of time after treatment with newly developed reverse transcriptase or protease inhibitors (49, 76, 102). The results of these studies (Fig. 11) show that the amount of virus in circulation declines rapidly after initiation of treatment, dropping to 1% of its initial value or less after 1 to 2 weeks, and then rises again as resistant mutants take over the virus population. Since the decline of circulating virus must follow the death of infected cells, the mean lifetime of an infected cell must be less than 1 to 2 days, and the average replication cycle time of the virus must be still less (25). Furthermore, the same studies imply that the population of the replicating virus pool is quite large: 10^9 or more cells are being infected, producing virus, and dying every day. Thus, the virus is replicating at the rate of some 300 generations a year, so that 1,000 generations are accomplished in somewhat more than 3 years. There is plenty of time

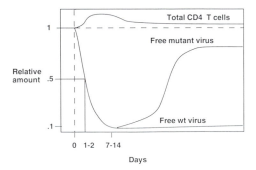

FIGURE 11. Kinetics of HIV infection after inhibitor treatment in vivo. Shown is a compilation of data from a number of studies (49, 71, 76, 102) in which the concentration of virus (as genome RNA) and the CD4 cell number were monitored after administration of one of a number of (nucleoside or nonnucleoside) reverse transcriptase or protease inhibitors (25). wt, wild type.

and more than enough infected cells for the population to be molded by the sorts of subtle selection pressures illustrated in Fig. 10.

These considerations also imply that HIV infection can be described by a simple steady-state model in which the majority population of productively infected cells is turning over very rapidly but at a rate that closely balances the rate of replacement, so that only a nearly imperceptible change in cell number is occurring. Furthermore, the number of infected cells and the concentration of virions in blood remain constant, implying that the rate of their clearance is equal to the rate of release of new virus into the circulation. Within such a population, three types of infected cells are (at least theoretically) present (Fig. 12). The first are the productively infected cells whose mean lifetimes are less than 2 days and that produce the vast majority (>99%) of the replicating virus. The second are latently infected cells that remain in a nonproductive state for much longer periods but can be stimulated in some way (by T-cell activation signals, for example) to release infectious virus. The third (at least in theory) are chronically producing cells that release virus for long periods before dying. The replication kinetics

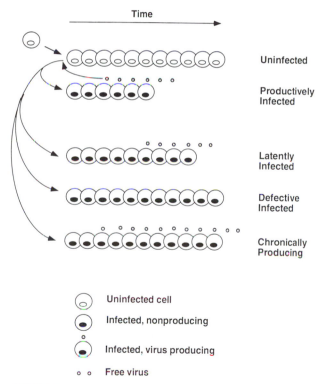

FIGURE 12. Types of HIV-infected cells in vivo. At steady state, the rate of creation of each cell type by infection matches its rate of death, and the overall rate of replenishment matches the death rates of all types combined. The horizontal axis is labeled "time" to indicate that at steady state, the number of cells in each state (nonproducing or virus producing) is proportional to the time spent in that state (25).

inferred from the drug experiments imply that the contribution to the virus pool of the last two cell types must be less than 1% at any given time and, conversely, that 30 to 50% of the virus in circulation at any one time comes from cells infected the previous day.

The final type of infected cell is the one that, because of errors during reverse transcription, acquires a provirus defective for replication and unable either to kill a cell directly or to allow it to be recognized by the immune system. Even if only a small fraction of the proviruses made in one replication cycle are of this type, they will rapidly accumulate to become a large fraction of the total as the cells that acquire intact proviruses die off. That these cells do form a large fraction of the proviruses to be found in cell DNA is indicated by the very high frequency of isolation of defective sequences from blood cell DNA in HIV- and SIV-infected individuals (31, 54, 98, 107) and by the much slower rate of loss of wild-type genomes from viral DNA than from circulating viral RNA after drug treatment (102). Unfortunately, this aspect of viral quasispecies evolution has been largely ignored by workers in the field, and there has been too little recognition of the unreliability of viral DNA as a marker for the current extent of virus replication or the genetic structure of the virus population.

The high rate of replication of HIV during the entire course of the infection has important consequences for understanding genetic variation and its role in evolution, pathogenesis, and therapy.

First, given the very large numbers of replication cycles involved and the consequent extreme sensitivity of the genetic structure of the virus population to subtle selective influences, very few mutations are likely to be truly neutral. Thus, blind application of the tenets of neutral theory (45) are likely to give very misleading results. Such misapplication includes assuming that there is a constant rate of variation and using it to estimate the time of divergence of the lentivirus groups as well as assuming that synonymous mutations are neutral and using them to standardize genetic distance measurements. The rapid development of the HIV quasispecies is not evolution. It is quite likely that the process of evolution of these viruses is one of very slow adaptation to a particular host species and gradual coevolution with it, punctuated by periods of rapid change upon colonization of a new species by the virus.

Second, given the ability of the virus quasispecies to rapidly adapt by expansion of mutant populations that are always present, it is highly unlikely that the pathogenic progression typical of AIDS is due to the slow generation of more "virulent" variants. Rather, the appearance, late in infection, of viruses with greater cytopathicity in cell cultures is much more likely a consequence of altered selective conditions (such as the declining immune response) that favor the growth of such variants.

Third, the kinetics of accumulation of mutants with small selective disadvantages must also apply to mutants that confer resistance to replication inhibitors used as therapeutic agents. Since mutants resistant to all agents tested to date have been found following treatment of HIV-infected individuals (75), it is important to realize that these mutations (as well as many combinations of them) will already be present in the patient at the time of treatment. Furthermore, these mutants will rapidly take over the population after treatment. Thus, the ideas that treatments that sufficiently block replication of wild-type virus will block the occurrence of resistant mutants, that multiply resistant mutants will be so rare as not to occur in the virus population, and that the drug that most strongly inhibits wild-type virus replication will be the most effective in vivo are all fallacious. Rather, it must be recognized that there are almost certain to be mutants resistant to any therapy based on inhibition of HIV replication and that any such mutant is likely to take over the replicating virus population within a few weeks of the onset of treatment. Therefore, the drug is seeing the virus against which it is aimed for only a very short time, and further treatment is likely to be of little additional help.

The news is not entirely bad, however. The good news is twofold. First, the fact that many drugs can so rapidly and completely block the replication of the virus in vivo with a corresponding increase in the CD4 cell number means that the pathogenic effects of the virus are directly related to its rapid replication. Even a relatively modest sustained reduction in the number of productively infected cells at steady state should give at least a proportional extension of the period of clinical latency. Second, mutations that confer resistance to replication inhibitors are predicted to confer reduced replication efficiency in the absence of inhibitors. This effect should lead to a slightly reduced steady-state level of any resistant mutant relative to that of its wild-type parent. It is likely that there are some drugs or drug combinations for which the "best" mutant that the virus can easily find is still substantially deficient in replication. For this reason, it is imperative that an assessment of the mutations that confer resistance and their cost to the virus be performed early in the drug development process and that only those drugs that meet this criterion be selected for further development. In this way, we might finally be able to convert some of our hard-won knowledge of the in vivo dynamics of this resilient pathogen into real clinical benefit.

ACKNOWLEDGMENTS. I thank S. Wolinsky, R. Taplitz, N. Rosenberg, C. Boucher, D. Ho, and G. Shaw for helpful discussion.

Work in my laboratory is supported by grant R35-CA-44385 from the National Cancer Institute. I am the recipient of a Research Professorship from the American Cancer Society.

REFERENCES

1. **Acha-Orbea, H., and E. Palmer.** 1991. Mls—a retrovirus exploits the immune system. *Immunol. Today* **12:**356–361.
2. **Acha-Orbea, H., A. N. Shakhov, L. Scarpellino, E. Kolb, V. Muller, A. Vessaz-Shaw, R. Fuchs, K. Blochlinger, P. Rollini, J. Billotte, M. Sarafidou, H. R. MacDonald, and H. Diggelmann.** 1991. Clonal deletion of Vβ14-bearing T cells in mice transgenic for mammary tumour virus. *Nature* (London) **250:**207–211.
3. **Anonymous.** 1970. Apres Temin, le deluge. *Nature* (London) **227:**998.
4. **Anonymous.** 1970. Central dogma reversed. *Nature* (London) **226:**1198–1199.
5. **Anonymous.** 1970. DNA from RNA template. *Nature* (London) **226:**1003.
6. **Atchley, W. R., and W. M. Fitch.** 1991. Gene trees and the origins of inbred strains of mice. *Science* **254:**554–558.
7. **Baltimore, D.** 1970. RNA-dependent DNA polymerase in virions of RNA tumour viruses. *Nature* (London) **226:**1209–1211.

8. **Bauer, G., and H. M. Temin.** 1979. RNA-directed DNA polymerase from particles released by normal goose cells. *J. Virol.* **29:**1006–1013.

9. **Beutner, U., W. N. Frankel, M. S. Cote, J. M. Coffin, and B. T. Huber.** 1992. Mls-1 is encoded by the long terminal repeat in the open reading frame of the mammary tumor provirus Mtv-7. *Proc. Natl. Acad. Sci. USA* **89:**5432–5436.

10. **Bishop, J. M.** 1983. Cellular oncogenes and retroviruses. *Annu. Rev. Biochem.* **52:**301–354.

11. **Bishop, J. M., and H. E. Varmus.** 1982. Functions and origins of retroviral transforming genes, p. 999–1108. *In* R. Weiss, N. Teich, H. Varmus, and J. Coffin (ed.), *RNA Tumor Viruses.* Cold Spring Harbor Laboratory, Cold Spring Harbor, N.Y.

12. **Boettiger, D., and H. M. Temin.** 1970. Light inactivation of focus formation by chicken embryo fibroblasts infected with avian sarcoma virus in the presence of 5-bromodeoxyuridine. *Nature* (London) **228:**622–624.

13. **Bowes, C., T. Li, W. N. Frankel, M. Danciger, J. M. Coffin, M. L. Applebury, and D. B. Farber.** 1993. Localization of a retroviral element within the *rd* gene coding for the β subunit of cGMP phosphodiesterase. *Proc. Natl. Acad. Sci. USA* **90:**2955–2959.

14. **Buller, R. S., A. Ahmed, and J. L. Portis.** 1987. Identification of two forms of an endogenous murine retroviral *env* gene linked to the Rmcf locus. *J. Virol.* **61:**29–34.

15. **Burns, D. P. W., and R. C. Desrosiers.** 1991. Selection of genetic variants of simian immunodeficiency virus in persistently infected rhesus monkeys. *J. Virol.* **65:**1843–1854.

16. **Burns, D. P. W., and H. M. Temin.** 1994. High rates of frameshift mutations within homo-oligomeric runs during a single cycle of retroviral replication. *J. Virol.* **68:**4196–4203.

17. **Cachon-Gonzales, M. B., S. Fenner, J. M. Coffin, C. Moran, S. Best, and J. P. Stoye.** 1994. Structure and expression of the hairless gene of mice. *Proc. Natl. Acad. Sci. USA* **91:**7717–7721.

18. **Coffin, J. M.** 1979. Structure, replication, and recombination of retrovirus genomes: some unifying hypotheses. *J. Gen. Virol.* **42:**1–26.

19. **Coffin, J. M.** 1982. Endogenous viruses, p. 1109–1204. *In* R. Weiss, N. Teich, H. Varmus, and J. Coffin (ed.), *RNA Tumor Viruses.* Cold Spring Harbor Laboratory, Cold Spring Harbor, N.Y.

20. **Coffin, J. M.** 1986. Genetic variation in AIDS viruses. *Cell* **46:**1–4.

21. **Coffin, J. M.** 1992. Genetic diversity and evolution of retroviruses. *Curr. Top. Microbiol. Immunol.* **176:**143–164.

22. **Coffin, J. M.** 1992. Structure and classification of retroviruses, p. 19–50. *In* J. A. Levy (ed.), *The Retroviridae.* Plenum Press, New York.

23. **Coffin, J. M.** 1992. Superantigens and endogenous retroviruses: a confluence of puzzles. *Science* **255:**411–413.

24. **Coffin, J. M.** 1993. Reverse transcription and evolution, p. 445–479. *In* S. Goff and A. M. Skalka (ed.), *Reverse Transcriptase.* Cold Spring Harbor Laboratory Press, Cold Spring Harbor, N.Y.

25. **Coffin, J. M.** 1995. HIV replication dynamics in vivo: implications for genetic variation, pathogenesis, and therapy. *Science* **267:**483–488.

26. **Coffin, J. M., C. Barker, D. Fiore, and C. Naugle.** Unpublished data.

27. **Coffin, J. M., and W. A. Haseltine.** 1977. Terminal redundancy and the origin of replication of Rous sarcoma virus RNA. *Proc. Natl. Acad. Sci. USA* **74:**1908–1912.

28. **Coffin, J. M., and H. M. Temin.** 1971. Comparison of Rous sarcoma virus-specific deoxyribonucleic acid polymerases in virions of Rous sarcoma virus and in Rous sarcoma virus-infected chicken cells. *J. Virol.* **7:**625–634.

29. **Coffin, J. M., and H. M. Temin.** 1971. Ribonuclease-sensitive deoxyribonucleic acid polymerase activity in uninfected rat cells and rat cells infected with Rous sarcoma virus. *J. Virol.* **8:**630–642.

30. **Coffin, J. M., P. N. Tsichlis, C. S. Barker, and S. Voynow.** 1980. Variation in avian retrovirus genomes. *Ann. N.Y. Acad. Sci.* **354:**410–425.

31. **Delassus, S., R. Cheynier, and S. Wain-Hobson.** 1991. Evolution of human immunodeficiency virus type 1 nef and long terminal repeat sequences over 4 years in vivo and in vitro. *J. Virol.* **65:**225–231.

32. **Doolittle, R. F., D. F. Feng, M. A. McClure, and M. S. Johnson.** 1990. Retrovirus phylogeny and evolution, p. 1–18. *In* R. Swanstrom and P. K. Vogt (ed.), *Retroviruses. Strategies of Replication.* Springer-Verlag, New York.

33. **Dorner, A. J., and J. M. Coffin.** 1986. Determinants for receptor interaction and cell killing on the avian retrovirus glycoprotein gp85. *Cell* **45**:365–374.

34. **Dorner, A. J., J. P. Stoye, and J. M. Coffin.** 1985. Molecular basis of host range variation in avian retroviruses. *J. Virol.* **53**:32–39.

35. **Dougherty, J. P., and H. M. Temin.** 1988. Determination of the rate of base-pair substitution and insertion mutations in retrovirus replication. *J. Virol.* **62**:2817–2822.

36. **Dyson, P. J., A. M. Knight, S. Fairchild, E. Simpson, and K. Tomonari.** 1991. Genes encoding ligands for deletion of Vβ11 T cells cosegregate with mammary tumour virus genomes. *Nature* (London) **349**:531–532.

37. **Eigen, M., and C. K. Biebricher.** 1988. Sequence space and quasispecies distribution, p. 3–22. *In* E. Domingo, J. J. Holland, and P. Ahlquist (ed.), *RNA Genetics,* vol. III. CRC Press, Inc., Boca Raton, Fla.

38. **Frankel, W. N., C. Rudy, J. M. Coffin, and B. T. Huber.** 1991. Linkage of Mls genes to endogenous mammary tumour viruses of inbred mice. *Nature* (London) **349**:526–528.

39. **Frankel, W. N., J. P. Stoye, B. A. Taylor, and J. M. Coffin.** 1989. Genetic analysis of endogenous xenotropic murine leukemia viruses: association with two common mouse mutations and the viral restriction locus Fv-1. *J. Virol.* **63**:1763–1774.

40. **Frankel, W. N., J. P. Stoye, B. A. Taylor, and J. M. Coffin.** 1989. Genetic identification of endogenous polytropic proviruses by using recombinant inbred mice. *J. Virol.* **63**:3810–3821.

41. **Frankel, W. N., J. P. Stoye, B. A. Taylor, and J. M. Coffin.** 1990. A linkage map of endogenous murine leukemia proviruses. *Genetics* **124**:221–236.

42. **Fritsch, E., and H. M. Temin.** 1977. Formation and structure of infectious DNA of spleen necrosis virus. *J. Virol.* **21**:119–130.

43. **Gerstein, R. M., W. N. Frankel, C.-L. Hsieh, J. M. Durdik, S. Rath, J. M. Coffin, A. Nisonoff, and E. Selsing.** 1990. Isotype switching of an immunoglobulin heavy chain transgene occurs by DNA recombination between different chromosomes. *Cell* **63**:537–548.

44. **Goff, S. J., and A. M. Skalka.** 1993. *Reverse Transcriptase.* Cold Spring Harbor Laboratory Press, Cold Spring Harbor, N.Y.

45. **Gojobori, T., E. N. Moriyama, and M. Kimura.** 1990. Molecular clock of viral evolution, and the neutral theory. *Proc. Natl. Acad. Sci. USA* **87**:10015–10018.

46. **Golovkina, T. V., A. Chervonsky, J. P. Dudley, and S. R. Ross.** 1992. Transgenic mouse mammary tumor virus superantigen expression prevents viral infection. *Cell* **69**:637–645.

47. **Groenink, M., A. C. Andeweg, R. A. M. Fouchier, S. Broersen, R. C. M. van der Jagt, H. Schuitemaker, R. E. Y. de Goede, M. L. Bosch, H. G. Huisman, and M. Tersmette.** 1992. Phenotype-associated *env* gene variation among eight related human immunodeficiency virus type 1 clones: evidence for in vivo recombination and determinants of cytotropism outside the V3 domain. *J. Virol.* **66**:6175–6180.

48. **Held, W., G. A. Waanders, A. N. Shakhov, L. Scarpellino, H. Acha-Orbea, and H. R. MacDonald.** 1993. Superantigen-induced immune stimulation amplifies mouse mammary tumor virus infection and allows virus transmission. *Cell* **74**:529–540.

49. **Ho, D. D., A. U. Neumann, A. S. Perelson, W. Chen, J. M. Leonard, and M. Markowitz.** 1995. Rapid turnover of plasma virions and CD4 lymphocytes in HIV infection. *Nature* (London) **373**:123–126.

50. **Hu, W.-S., and H. M. Temin.** 1990. Retroviral recombination and reverse transcription. *Science* **250**:1227–1233.

51. **Ikeda, H., and H. Sugimura.** 1989. Fv-4 resistance gene: a truncated endogenous murine leukemia virus with ecotropic interference properties. *J. Virol.* **63**:5405–5412.

52. **Inouye, S., M.-Y. Hsu, S. Eagle, and M. Inouye.** 1989. Reverse transcriptase associated with the biosynthesis of the branched RNA-linked msDNA in Myxococcus xanthus. *Cell* **56**:709–717.

53. **Janeway, C.** 1991. Mls: makes a little sense. *Nature* (London) **349**:459–461.

54. **Johnson, P. J., T. E. Hamm, S. Goldstein, S. Kitov, and V. M. Hirsch.** 1991. The genetic fate of molecular cloned simian immunodeficiency virus in experimentally infected macaques. *Virology* **185**:217–228.

55. **Kang, C. Y., and H. M. Temin.** 1972. Endogenous RNA-directed DNA polymerase activity in uninfected chicken embryos. *Proc. Natl. Acad. Sci. USA* **69**:1550–1554.

56. **Kang, C. Y., and H. M. Temin.** 1973. RNA-directed DNA synthesis in viruses and normal cells: a possible mechanism in differentiation, p. 339–348. *In* M. C. Niu and S. J. Segal (ed.), *The Role of RNA in Reproduction and Development.* North Holland, Amsterdam.

57. **Keshet, E., and H. M. Temin.** 1977. Nucleotide sequences derived from pheasant DNA in the genome of recombinant avian leukosis viruses with subgroup F specificity. *J. Virol.* **24:**505–513.

58. **Kitamura, Y., Y. M. H. Lee, and J. M. Coffin.** 1992. Nonrandom integration of retroviral DNA *in vitro:* effect of CpG methylation. *Proc. Natl. Acad. Sci. USA* **89:**5532–5536.

59. **Lee, Y. M. H., and J. M. Coffin.** 1991. Relationship of avian retrovirus DNA synthesis to integration in vitro. *Mol. Cell. Biol.* **11:**1419–1430.

60. **Levy, J. A.** 1993. Pathogenesis of human immunodeficiency virus infection. *Microbiol. Rev.* **57:** 183–289.

61. **Lock, L. F., E. Keshet, D. J. Gilbert, N. A. Jenkins, and N. G. Copeland.** 1988. Studies of the mechanism of spontaneous germline ecotropic provirus acquisition in mice. *EMBO J.* **7:** 4169–4177.

62. **Marrack, P., E. Kushnir, and J. Kappler.** 1991. A maternally inherited superantigen encoded by a mammary tumor virus. *Nature* (London) **349:**524–526.

63. **Messer, A., J. Plummer, P. Maskin, J. Coffin, and W. Frankel.** 1992. Mapping of the motor neuron degeneration (*Mnd*) gene, a mouse model of amyotrophic lateral sclerosis (ALS). *Genomics* **13:** 797–802.

64. **Morse, H.** 1981. The laboratory mouse: a historical perspective, p. 1–16. *In* J. D. Small, H. L. Foster, and J. G. Fox (ed.), *The Mouse in Biomedical Research.* Academic Press, Inc., New York.

65. **Myers, G., and G. N. Pavlakis.** 1992. Evolutionary potential of complex retroviruses, p. 51–106. *In* J. Levy (ed.), *The Retroviridae.* Plenum Press, New York.

66. **Palmer, E.** 1991. Infectious origin of superantigens. *Curr. Biol.* **1:**74–76.

67. **Panganiban, A. T., and H. M. Temin.** 1983. The terminal nucleotides of retrovirus DNA are required for integration but not virus production. *Nature* (London) **306:**155–160.

68. **Pathak, V. K., and H. M. Temin.** 1990. Broad spectrum of in vivo forward mutations, hypermutations, and mutational hotspots in a retroviral shuttle vector after a single replication cycle: substitutions, frameshifts and hypermutations. *Proc. Natl. Acad. Sci. USA* **87:**6019–6023.

69. **Preston, B. D., B. J. Poiesz, and L. A. Loeb.** 1988. Fidelity of HIV-1 reverse transcriptase. *Science* **242:**1168–1171.

70. **Resnick, R. M., M. T. Boyce-Jacino, Q. Fu, and A. J. Faras.** 1990. Phylogenetic distribution of the novel avian endogenous provirus family EAV-0. *J. Virol.* **64:**4640–4653.

71. **Richman, D. D.** Personal communication.

72. **Rinchik, E. M., J. P. Stoye, W. N. Frankel, J. M. Coffin, B. S. Kwon, and L. B. Russell.** 1993. Molecular analysis of viable spontaneous and radiation-induced albino (c)-locus mutations in the mouse. *Mutat. Res.* **286:**199–207.

73. **Rise, M. L., W. N. Frankel, J. M. Coffin, and T. N. Seyfried.** 1991. Genes for epilepsy mapped in the mouse. *Science* **253:**669–673.

74. **Roberts, J. D., K. Bebenek, and T. A. Kunkel.** 1988. The accuracy of reverse transcriptase from HIV-1. *Science* **242:**1171–1173.

75. **Schinazi, R., B. Larder, and J. Mellors.** 1994. Mutations in HIV-1 reverse transcriptase and protease associated with drug resistance. *Int. Antiviral News* **2:**72–75.

76. **Schurmann, R., M. Nijhuis, R. van Leeuwen, P. Schipper, P. Collis, S. Danner, J. Miulder, C. Loveday, C. Christopherson, S. Kwok, J. Sninsky, and C. Boucher.** Changes in HIV-1 RNA load and appearance of drug-resistant mutations in individuals treated with 3TC (lamuvidine). *J. Infect. Dis.,* in press.

77. **Shimotohno, K., S. Mizutani, and H. M. Temin.** 1980. Sequence of retrovirus provirus resembles that of bacterial transposable elements. *Nature* (London) **285:**550–554.

78. **Shimotohno, K., and H. M. Temin.** 1981. Evolution of retroviruses from cellular movable genetic elements. *Cold Spring Harbor Symp. Quant. Biol.* **2:**719–730.

79. **Smith, T. F., A. Srinivasan, G. Schochetman, M. Marcus, and G. Myers.** 1988. The phylogenetic history of immunodeficiency viruses. *Nature* (London) **333:**573–575.

79a. **Stoye, J.** Personal communication.

80. **Stoye, J. P., and J. M. Coffin.** 1985. Endogenous viruses, p. 357–404. *In* R. Weiss, N. Teich, H. Varmus, and J. Coffin (ed.), *RNA Tumor Viruses*. Cold Spring Harbor Laboratory, Cold Spring Harbor, N.Y.

81. **Stoye, J. P., and J. M. Coffin.** 1987. The four classes of endogenous murine leukemia virus: structural relationships and potential for recombination. *J. Virol.* **61:**2659–2669.

82. **Stoye, J. P., and J. M. Coffin.** 1988. Polymorphism of murine endogenous proviruses revealed by using virus class-specific oligonucleotide probes. *J. Virol.* **62:**168–175.

83. **Stoye, J. P., S. Fenner, G. E. Greenoak, C. Moran, and J. M. Coffin.** 1988. Role of endogenous retroviruses as mutagens: the hairless mutation of mice. *Cell* **54:**383–391.

84. **Stoye, J. P., C. Moroni, and J. Coffin.** 1991. Virological events leading to spontaneous AKR thymomas. *J. Virol.* **65:**1273–1285.

85. **Swain, A., and J. M. Coffin.** 1989. Polyadenylation at correct sites in genome RNA is not required for retrovirus replication or genome encapsidation. *J. Virol.* **63:**3301–3306.

86. **Swofford, D. L.** 1991. PAUP: phylogenetic analysis using parsimony, version 3.0s. Illinois Natural History Survey, Champaign.

87. **Taylor, B. A.** 1978. Recombinant inbred strains: use in gene mapping, p. 423–438. *In* H. C. Morse III (ed.), *Origins of Inbred Mice*. Academic Press, Inc., New York.

88. **Temin, H. M.** 1964. Nature of the provirus of Rous sarcoma. *Natl. Cancer. Inst. Monogr.* **17:**557–570.

89. **Temin, H. M.** 1969. Control of cell multiplication in uninfected chicken cells and chicken cells converted by avian sarcoma viruses. *J. Cell. Physiol.* **74:**9–16.

90. **Temin, H. M.** 1982. Viruses, protoviruses, development, and evolution. *J. Cell. Biochem.* **19:**105–118.

91. **Temin, H. M.** 1983. Evolution of RNA tumor viruses: analogy for nonviral carcinogenesis. *Prog. Nucleic Acid Res. Mol. Biol.* **29:**7–16.

92. **Temin, H. M.** 1985. Reverse transcription in the eukaryotic genome: retroviruses, pararetroviruses, retrotransposons, and retrotranscripts. *Mol. Biol. Evol.* **6:**455–468.

93. **Temin, H. M.** 1986. Retroviruses and evolution. *Cell Biophys.* **9:**9–16.

94. **Temin, H. M.** 1989. Retrovirus variation and evolution. *Genome* **31:**17–22. (Review.)

95. **Temin, H. M., and S. Mizutani.** 1970. RNA-dependent DNA polymerase in virions of Rous sarcoma virus. *Nature* (London) **226:**1211–1213.

96. **Thome, K., and N. Rosenberg.** Personal communication.

97. **Tsichlis, P. N., and J. M. Coffin.** 1980. Recombinants between endogenous and exogenous avian tumor viruses: role of the c region and other portions of the genome in the control of replication and transformation. *J. Virol.* **33:**238–249.

98. **Vartanian, J.-P., A. Meyerhans, B. Asjo, and S. Wain-Hobson.** 1991. Selection, recombination, and G—A hypermutation of human immunodeficiency virus type 1 genomes. *J. Virol.* **65:**1779–1788.

99. **Voynow, S. L., and J. M. Coffin.** 1985. Evolutionary variants of Rous sarcoma virus: large deletion mutants do not result from homologous recombination. *J. Virol.* **55:**67–78.

100. **Voynow, S. L., and J. M. Coffin.** 1985. Truncated *gag*-related proteins are produced by large deletion mutants of Rous sarcoma virus and form virus particles. *J. Virol.* **55:**79–85.

101. **Wain-Hobson, S.** 1992. Human immunodeficiency virus quasispecies in vivo and ex vivo. *Curr. Top. Microbiol. Immunol.* **176:**181–194.

102. **Wei, X., S. Ghosh, M. E. Taylor, V. A. Johnson, E. A. Emini, P. Deutsch, J. D. Lifson, S. Bonhoeffer, M. A. Nowak, B. H. Hahn, M. S. Saag, and G. M. Shaw.** 1995. Viral dynamics in human immunodeficiency virus type 1 infection. *Nature* (London) **373:**117–122.

103. **Weiss, R. A.** 1982. The search for human RNA tumor viruses, p. 1205–1281. *In* R. Weiss, N. Teich, H. Varmus, and J. Coffin (ed.), *RNA Tumor Viruses*. Cold Spring Harbor Laboratory, Cold Spring Harbor, N.Y.

104. **Weller, S. K., A. E. Joy, and H. M. Temin.** 1980. Correlation between cell killing and massive second-round superinfection by members of some subgroups of avian leukosis virus. *J. Virol.* **33:**494–506.

105. **Weller, S. K., and H. M. Temin.** 1981. Cell killing by avian leukosis viruses. *J. Virol.* **39:**713–721.

106. **Wilkenson, D. A., D. Mager, and J.-A. Leong.** 1994. Endogenous human retroviruses, p. 465–536. *In* J. A. Levy (ed.), *The Retroviridae.* Plenum Press, New York.
107. **Wolinsky, S. M.** Personal communication.
108. **Yazdanbakhsh, K., C. G. Park, G. M. Winslow, and Y. Choi.** 1993. Direct evidence for the role of COOH terminus of mouse mammary tumor virus superantigen in determining T cell receptor Vβ specificity. *J. Exp. Med.* **178:**737–741.
109. **Zhang, J., and H. M. Temin.** 1994. Retrovirus recombination depends on the length of sequence identity and is not error prone. *J. Virol.* **68:**2409–2414.

PART IV

HIV, AIDS, AND THE IMMUNE SYSTEM

The DNA Provirus: Howard Temin's Scientific Legacy
Edited by G. M. Cooper, R. Greenberg Temin, and B. Sugden
© 1995 American Society for Microbiology, Washington, DC 20005

Chapter 17

Human Retroviruses in the Second Decade: Some Pathogenic Mechanisms and Approaches to Their Control

Robert C. Gallo

HOWARD TEMIN'S INFLUENCE ON OUR WORK

I did not have the good fortune of working in Howard Temin's laboratory, but once I began working on virological topics, he was an influence on virtually every stage of my career: as a scientific thinker and leader, as a strong critic, as a man, and ultimately as a friend. At a symposium organized by David Pauza and the University of Wisconsin's primate center during the last year of Howard's life and again at his memorial service, I spoke of the last two of those influences. In this volume, I wish to elaborate on the first two of these. In the mid- to late 1960s, I became interested in making some biochemical comparisons between normal and neoplastic cells, and with my colleague and friend Bob Ting, I was directed toward exploiting some animal transforming virus systems. Before his arrival at the National Institutes of Health (NIH), Bob had been in the same place as Howard (Dulbecco's laboratory at the California Institute of Technology), and it was from Bob that I first learned of Howard and his ideas. As chance would have it, by 1969 or 1970 I became interested in cellular DNA polymerases just prior to Howard's presentation in Houston of the discovery of the DNA polymerase (reverse transcriptase [RT]) of Rous sarcoma virus, soon to be followed by the Mizutani-Temin publication and the independent Baltimore publication in *Nature* of this scientific milestone.

I wish I could say I soon thought of using RT for the molecular biological tool it soon became, but my thoughts were solely on using it in virology. I had become interested in human leukemia and, like others, was impressed by the number and availability of retrovirus-induced leukemia animal models. Even if human leukemias were never caused by such viruses, a relevant pathogenetic mechanism might be learned from such models. It was the characterization of RT from various animal retroviruses and the comparison of them to cellular DNA

Robert C. Gallo • Laboratory of Tumor Cell Biology, National Cancer Institute, Bethesda, Maryland 20892-4255.

polymerases that brought me into my first personal contact with Howard and his attempts by then (1971 to 1972) to extend his provirus hypothesis to a more complex and encompassing protovirus theory. Though not yielding the same clear-cut result as the provirus hypothesis, it still spawned almost a generation of work and ideas. We began looking at the similarities and dissimilarities of various polymerases with notions of understanding their origins and relatedness. These were the topics which formed most of the discussions I had with Howard during the early to mid-1970s.

But there was also the practical side of RT. The RT assay was a new tool, a sensitive, cheap, rapid, easily doable method for finding and quantitating known viruses and for synthesizing radiolabeled viral cDNA that, in turn, made probing and analyses of integrated sequences routine. As you know, this paved the way for most of the subsequent advances and ideas on the molecular basis for retroviral carcinogenesis. The retroviral genome and the major aspects of its replication cycle were soon understood in detail, and no Nobel prize could have been more obvious or more merited than the one given for this work.

Extension of this work to human leukemia and related diseases brought me into contact with Howard the critic. We used familiarity with RT and cellular DNA polymerases in an effort to see if we could find hints of RT in association with any human leukemias—the so-called ''footprint'' of a retrovirus. Rare positive results made this arduous work useful but only as a boost to morale in continuing the work. Howard's arguments generally came back to the point that ultimately it could never become definitive: isolation of a virus, if it was there, was essential. I am uncertain if his criticisms were necessary for our fortunate concomitant attempts to find and use growth factors for culturing primary human blood cells, but no doubt his arguments catalyzed these efforts, which led to some interesting findings in T-cell biology and to the eventual isolation of the human retroviruses human T-cell lymphotrophic leukemia virus types I and II (HTLV-I and HTLV-II). Having a framework for animal retrovirus genomes and the integrated provirus made our subsequent experiments with the HTLVs obvious, including the rapid demonstration of provirus clonality in the leukemic cells of patients with adult T-cell leukemia (ATL) caused by HTLV-I. Though there were some unprecedented findings with the HTLVs (such as their then novel regulatory genes, multiple splicing, in vitro immortalizing effect on T cells in the absence of an oncogene, a *trans*-acting growth-promoting gene product, and transmissibility mostly as a DNA provirus), nonetheless, most of our work could rely upon this framework.

There is no need to belabor these points or what followed with human immunodeficiency virus (HIV) and AIDS (20–23). Though again, some surprises were in store, the stage was set for virus isolation and propagation, for determination of the criteria needed for establishing causality, and for development of a blood test. Of course, not all of these advances came solely or directly from Howard's work or ideas, but once again, the framework was there, and once again, so was Howard the critic, and nowhere was he more interested and more available for discussions or help than in this new public health menace. It is my belief that AIDS slightly changed Howard's views on disease-oriented research. I believe he came to a fuller appreciation of the need for and the difficulty of such research.

The discussions I had with him then became much more frequent, and I was very fortunate that out of this a real friendship began and intensified in these past 10 years. In the end, he became a model, our "gold standard," as a scientist and as a man.

In the rest of this chapter, I will summarize some current concepts on human retroviruses and some of our current work. I will focus on three topics: HTLV-I, some approaches to the inhibition of HIV replication, and the pathogenesis of AIDS-associated Kaposi's sarcoma (KS). I will try to point to questions and problems I think are important for the future.

HTLV-I

HTLV-I is now known to cause some ATLs/lymphomas, tropical spastic paraparesis/HTLV-I-associated myelopathy (TSP/HAM) (25, 28, 29, 48, 51) (a neurological disease resembling multiple sclerosis), and several apparently autoimmune disorders, including some polymyositis, rheumatoid arthritis-like disorders, uveitis, bronchitis, and mild immune impairment associated with bacterial dermatitis in infants. The importance of this virus now can be thought of as fourfold: first, as a public health problem in certain regions (West Indies, some parts of South America, and some portions of and populations in the United States, southern Japan, and Africa); second, as a model system for other human cancers (because we have the cause in hand as well as in vitro and in vivo models); third, as a model for multiple sclerosis; and fourth, for practical production of various cytokines owing to its capacity to immortalize T cells, especially CD4$^+$ T cells, with production of several cytokines.

New Information

Epidemiological

In the past few years, we have learned that there are several subtypes of HTLV-I. The prototype HTLV-I isolates are now termed a "cosmopolitan" type because they are by far the most widely distributed. New subtypes have been found in Equatorial Africa and in the South Pacific (26, 30) (Fig. 1). The important questions are whether these new subtypes also cause disease, whether they cause the same diseases as the cosmopolitan HTLV-I, and whether more subtypes will be found.

Genome and replication cycle

The HTLV genomes have been known since the early analyses to contain an extra 3′ segment (the X region). Two genes in this region were soon defined. They encode nuclear proteins which take part in *trans*-acting transcriptional activation (p40 Tax) of the HTLV long terminal repeat (LTR) and in transport (p27 Rex) of the larger mRNA molecules (for structural proteins) (Fig. 2). In their general mode

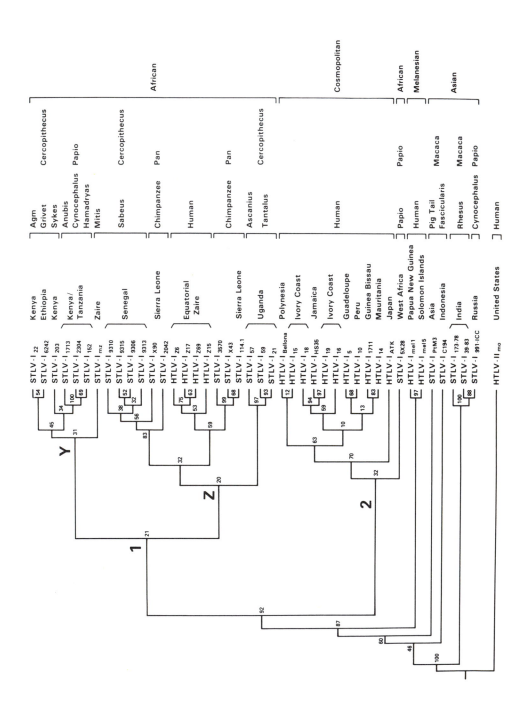

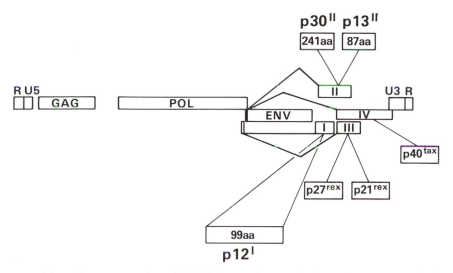

FIGURE 2. Schematic representation of HTLV-I genomic organization. The open boxes in the pX region (located between the end of the envelope gene and the 3' viral LTR) include the open reading frames for the p27^{rex}/p21^{rex} proteins, p40^{tax}, and the proteins p12^{I}, p13^{II}, and p30^{II}, recently demonstrated to be produced through alternative splicing (31, 32). aa, amino acids.

of action, Tax and Rex are the forerunners of HIV Tat and Rev. Tax also induces expression of several cellular genes, e.g., those for interleukin 1 (IL-1), IL-2, IL-3, IL-6, granulocyte-macrophage colony-stimulating factor (GM-CSF), the IL-2 receptor alpha chain (IL-2Rα), c-fos, HLA Cl-I, and others. The ability of HTLV to immortalize T cells is believed to be mediated at least in part by the transcriptional activation of cellular genes by Tax. Many groups contributed to these findings (27, 60), especially M. Yoshida and his group in Tokyo, F. Wong-Staal when she was with our group, W. Haseltine, and T. Waldmann and W. Greene and their colleagues. Recently, we and others found more complex alternative splicing mechanisms giving rise to several previously unidentified mRNAs and proteins (6, 17). One such protein, p12, studied chiefly by G. Franchini with I. Koralnik and J. Mulloy in our group and in collaboration with R. Schlagel and his group at Georgetown (17, 31, 32), is an endomembrane-associated protein which has structural similarities to bovine papillomavirus E5 oncoprotein; enhances the transforming capacity of E5; like E5, binds to a 16-kDa subunit (called ductin) of the H$^+$ vacuolar ATPase of cellular organelles; and binds to cytoplasmic domains

FIGURE 1. Phylogenetic tree of simian T-cell leukemia/lymphotropic virus type I (STLV-I), HTLV-I, and HTLV-II envelope nucleotide (30) DNA sequences (positions 6046 to 6567) are constructed on the outgroup HTLV-II$_{MO}$ as calculated by the DNA Boot program (Phylip V.3:41), which implements the bootstrap method for placing confidence intervals, using parsimony for DNA sequences. The numbers located at the forks indicate the frequency of association of the various virus groups in the bootstrap analysis after 100 replications of the data. Agm, African green monkey.

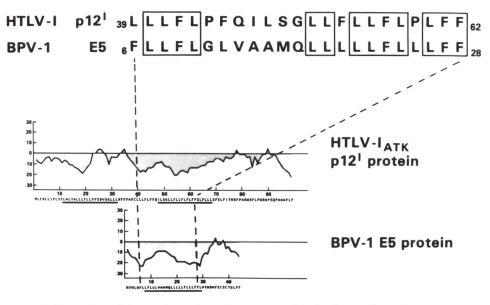

FIGURE 3. Comparison of the putative amino acid sequences of the bovine papillomavirus type 1 (BPV-1) E5 and HTLV-I p12[I] proteins. Single amino acid codes are used, and the lower part of the figure represents the hydrophage profile of both proteins (17).

of the IL-2Rα and IL-Rβ (Fig. 3). Some key questions for the future concern whether p12 is involved in transformation of T cells (after their immortalization), how immortalization in vitro is sustained in cases in which little or no virus expression is detected, what the cell receptor for HTLV-I is, and why CD4[+] T cells are usually the cells immortalized in vitro while many T cells are infected.

ATL is induced in about 2 to 5% of infected people per 70-year lifetime expectancy. Though we know that HTLV-I is clonally integrated into the DNA of these ATL cells, we do not understand the genetic events which allow the immortalized cell to become malignant. To my knowledge, there is as yet no consistent documented role for suppressor genes or oncogenes, nor do we understand how transformation is maintained in the absence of virus expression and a specific integration site.

TSP/HAM

Most studies favor the concept of an autoimmune mechanism for the demyelinating neurological disorder TSP/HAM. This idea is based, I think, chiefly on the work of the late D. McFarlin, which was continued by his colleague S. Jacobson at NIH, and rests on the lack of any clear-cut evidence for the presence of HTLV-I in any central nervous system cell and the presence of cytotoxic T lymphocytes in the central nervous system-targeting cells that express the HTLV Tax and Rex proteins. Moreover, there is evidence that TSP patients have a greater immune response to the virus (and more virus) than ATL patients or healthy carriers.

Recent work by M. Essex and colleagues suggests that a mutation in Tax is present in the HTLV-I endemic in Colombia, South America, where there is a much higher rate of TSP/HAM per infected person than in the United States or Japan. Remarkably, he and his colleagues often find this same variant in association with TSP in Japan and the United States. Other studies in my laboratory by Y. Lunardi-Iskandar in collaboration with J. Bryant of NIH's Institute of Dental Research suggest that this multiple sclerosis-like disease may be cytokine mediated. Though we have not yet identified this factor, our current results indicate that it is released into conditioned medium by a subset of HTLV-I-infected T cells and is able to reproduce a TSP-like disease in immunodeficient mice. The important problems of the immediate future will be to identify this factor and to attempt to reconcile this finding with an autoimmune model. We also need more studies to determine whether any subsets of central nervous system cells are infected by HTLV-I. To date, this possibility remains controversial. Considerable work on TSP is also being carried out in an experimental rat model in Japan as well as in extensive epidemiological studies, e.g., work from Osame's group, Sonoda's group, and the group in Nagasaki (28, 48).

Treatment and Vaccine

Until very recently there was little practical public health progress on HTLV disease. However, infection in Japan has been reduced by stopping breast feeding by infected mothers (the chief mode of spread of HTLV-I). Waldmann and co-workers attacked leukemic cells by using monoclonal antibodies coupled with toxins, and unexpected exciting results were recently obtained by P. Gill and coworkers at the University of Southern California: therapy of ATL with alpha interferon (IFN-α) plus azidothymidine (AZT), but neither alone, induced remissions in some patients. However, in view of the lack of significant virus expression in ATL, neither the rationale for the trial nor the positive outcome is easy to understand.

In a collaborative study with J. Tartaglia and E. Paoletti of Virogenetics, we obtained protection against HTLV-I infection of rabbits by using modified vaccinia vectors carrying the HTLV-I *env* gene (18). In the future, it will be important to learn the mechanism of IFN-α–AZT-induced clinical remission of ATL. We will also need to discuss the value of vaccination of people in areas where HTLV-I is endemic in view of the positive prevention results in rabbits.

HIV

We are at a juncture in AIDS research. It is the 10th anniversary of our understanding of its cause (HIV) and of our access to an accurate blood test, which has prevented infection from contaminated blood transfusions and allowed the epidemic to be properly monitored. As a consequence of those successes, we witnessed enormous progress in the early period (roughly 1983 to 1985), but now there is a consensus among involved scientists, as well as among other interested people, that research results have plateaued despite considerable scientific advances and the addition of innumerable talented scientists to the field. The reason for this current sense of frustration is obvious. It stems from the fact that the remaining major research concerns are curing the infected person (treatment) and preventing infection of others (vaccine), and both of these objectives have proven to be more than elusive. This frustration has fostered many recent discussions on what should be done differently or better. One popular proposal is to greatly broaden our definition of basic research in AIDS: in other words, an expansion of basic research in multiple directions outside the whole area generally considered to be AIDS research. It seems foolish to argue against this notion, since we cannot be certain of which fundamental observations or discoveries in biology will be useful to other fields. However, some others have argued the opposite, namely, that to invigorate the field, we should consider special projects for AIDS, i.e., some accelerated and highly targeted programs. The counterargument for targeted research, of course, is that we have insufficient information, i.e., we do not yet know all the necessary formulae, while the arguments for some targeted programs are partly based on several new openings from recent research on AIDS. For example, steadily (even if slowly) accumulating knowledge about the pathogenic mechanisms by which HIV causes AIDS and abundant information on the replication cycle of HIV provide new ideas for different therapeutic and preventive approaches which some believe have not yet been fully exploited or sufficiently pursued. The only logical approach, it seems to me, is to argue for both approaches but to limit targeted "crash" programs to pilot studies (to determine how well they do) that deal with a few obvious key issues in AIDS research which are ripe for pushing.

Blocking HIV Replication
(the Quest for Inhibitors from Which HIV Will Not Escape)

It is necessary to ask whether blocking HIV will be sufficient to halt progression to AIDS. It is not possible to give a definitive answer to that question, since it is possible that the continued presence of integrated HIV DNA proviruses, even if not expressing virus, will impair immune function. However, the vast majority of clinical and laboratory studies indicate that AIDS progression does correlate with HIV replication. Blocking HIV infection should also reduce the indirect pathological effects of HIV such as extracellular effects from its proteins (such as Tat, Vpr, Env, and possibly Nef) on uninfected cells or their induction of unwanted cytokines (Fig. 4). Therefore, it is reasonable to argue that one sine qua non of all therapy should include attempts to block HIV replication and to do it as early and as completely as possible.

The biggest obstacles to effective therapies are toxicity and the emergence of drug-resistant variants. Most testing for new anti-HIV therapy today is based on targeting the enzymes of HIV (RT, protease, integrase, and RNase H). These programs now seem to be well established in the pharmaceutical industry, and ultimately, the proper combination of some of them will probably make for therapeutic advances. Efforts in our laboratory are aimed at inhibiting HIV for pro-

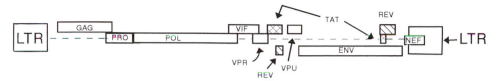

FIGURE 4. Genome structure of HIV-1 with reading frames indicated. Gag and Pol products are translated from unspliced RNA, and Env is translated from singly spliced RNA. More complex, multiple splicing events lead to formation of mRNA for the regulatory proteins.

longed periods, i.e., trying to overcome HIV mutations which escape drug effects, and we are pursuing two major approaches to this end.

Targeting cellular factors

Because viruses require cellular factors for their replication and cellular factors are not hypermutable like HIV, it seems reasonable to add such cellular factors to our list of drug targets. One known key cellular target is the CD4 molecule. In this respect, much has been done with soluble CD4 as a competitive inhibitor, but this approach has not worked well against primary HIV isolates. We are currently exploring another approach: blocking HIV entry into cells by using another virus which binds to the same cell surface receptor (CD4) as HIV. We have recently reported that human herpesvirus 7 (HHV-7) utilizes CD4 as its receptor for infection of CD4$^+$ T cells (work by P. Lusso in our group) (44) (Fig. 5). Consequently, it competes with HIV for this receptor and in so doing inhibits

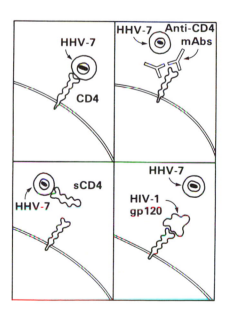

FIGURE 5. Experimental evidence indicating that HHV-7 uses CD4 as a component of its receptor: monoclonal antibodies (mAbs) to human CD4, the soluble form of human CD4 (sCD4), and the soluble form of HIV-1 gp120 all inhibit infection of human CD4$^+$ T cells by HHV-7.

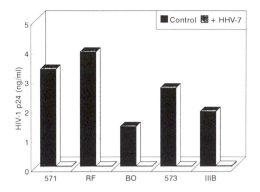

FIGURE 6. Infection of primary human CD4+ T lymphocytes by different HIV-1 isolates is inhibited by preexposure of the cells to HHV-7 for 48 h. Isolates 571, 573, and BO are primary isolates passaged only in primary human peripheral blood mononuclear cells.

infection of every strain of HIV we have tested (Fig. 6). HIV infection of primary T cells, primary macrophages, and T-cell lines is blocked with HHV-7 (44). The current main objective is to identify, purify, and test the effects of the HHV-7 envelope protein responsible for this effect and then to study the toxicity and efficacy of this protein in the simian immunodeficiency virus macaque animal model by delivering active peptide components in liposomes or other vectors.

The chief target of HIV infection most likely is the "resting" CD4+ T cell, but HIV replication is incomplete after entry into these cells (for example, see M. Stevenson and coworkers [57], I. Chen and coworkers [61], and F. Lori in our group [39, 40]). Our results suggest that this lack of completeness is due at least in part to the insufficient amounts of deoxynucleotides in those cells. As a result, viral DNA synthesis is incomplete. The nucleotide pools are increased when T cells are activated, leading to successful completion of HIV replication. The increase in deoxynucleotides is dependent on the induction of an enzyme, ribonucleotide reductase. If this enzyme is inhibited, HIV DNA synthesis remains impaired, and virus replication is inhibited. Hydroxyurea is a well-studied inhibitor of the activity of this enzyme, and our results show that hydroxyurea inhibits HIV replication (Fig. 7). Also, by reducing the pools of deoxynucleotides, more nucleoside analogs such as AZT or dideoxyinosine (ddI) should be incorporated into the viral DNA being synthesized, and as a result, HIV replication should be blocked. This hypothesis was verified in vitro by demonstrating that ddI and AZT synergize with hydroxyurea in inhibiting HIV replication (24, 40). Nonetheless, recent results of W.-Y. Gao and coworkers indicate that the main mechanism for this effect may be more complex. Whatever the mechanism, because hydroxyurea is inexpensive, can be given orally, crosses the blood-brain barrier, and has been safely used in clinical medicine for some chronic leukemias for over 30 years, it may be worthy of clinical trials in combination with nucleoside analogs.

Targeting conserved sequences of HIV

Obviously, there are several possible approaches to designing inhibitors which interact or interfere with HIV conserved sequences. We have pursued two.

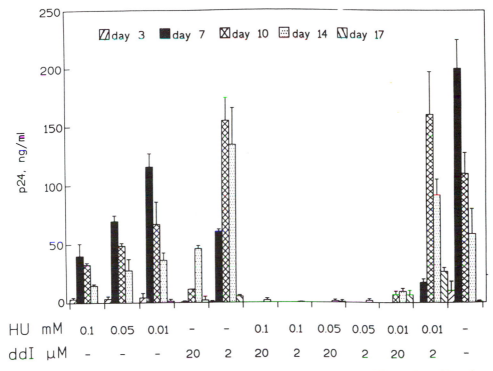

FIGURE 7. Time course of HIV-1 inhibition by hydroxyurea (HU) and/or by ddI in activated lympho-cytes. Lymphocytes were obtained from an HIV-1-infected patient, isolated, and stimulated for 2 days with phytohemagglutinin and IL-2. Subsequently, hydroxyurea and ddI were added at the concentra-tions illustrated, and replication was analyzed as follows: every 3 to 4 days, supernatants were har-vested for p24 analysis, and fresh supernatant and drugs were added. Partial inhibition was obtained when the drugs were used separately, and complete block of HIV-1 replication was achieved when hydroxyurea and ddI were used in combination. Hydroxyurea concentrations necessary to block HIV-1 (in combination with ddI) were not toxic and corresponded to the lower concentrations measured in the sera of cancer patients treated with hydroxyurea (4). (Reprinted with permission from Lori et al., *Science* **266:**801–805, 1994.)

Antisense oligodeoxynucleotides. In collaborative studies with S. Agrawal of the Hybridon Co., P. Zamecnik of the Worcester Foundation for Experimental Biology, and G. Zon of Lynx Co., we have shown that short stretches (20 to 30) of oligodeoxynucleotides complementary to some conserved segments of HIV-1 can inhibit in vitro replication of the virus (34, 35, 38). Hybridon's most thoroughly studied anti-HIV antisense oligodeoxynucleotide, known as GEM 91, complexes to a region of HIV-1 needed to make structural proteins of the virus core (the 5' region of the *gag* gene and sequences just 5' to this) and inhibits replication of all HIV-1 strains tested; it is now in phase one clinical testing in France and the United States (Table 1). The next phase of clinical studies with GEM 91 will determine whether the in vitro cell culture studies, which showed strong inhibition of HIV-1 replication, will be reproduced in vivo. The chief obstacles to antisense therapy are its cost, difficulty in production, problems getting sufficient amounts

TABLE 1
Inhibitory properties of *antitat* gene[a]

- *antitat* gene [poly(TAR)-antisense *tat*] strongly inhibits HIV-1 by sequestering *tat* and affecting Tat mRNA translation.
- Transduction of T-cell lines as well as primary PBMC with *antitat* gene results in suppression of HIV-1 replication.
- Suppressive phenotype of transduced cells is maintained for weeks and months.
- Laboratory and primary isolates of HIV-1 are inhibited.
- Replication of HIV-1 in infected PBMC from AIDS patients is also inhibited upon transduction with *antitat* gene.
- *antitat* gene is also effective against certain strains of HIV-2 and SIV.
- Poly(TAR) decoy inhibits release of Tat from infected cells and may inhibit functions of extracellular Tat that depend on its internalization.
- Expression of *antitat* is autoregulated (HIV-1 LTR), and thus, the gene cannot be overproduced to cause toxic effects by sequestering some essential cellular factor.
- *antitat* is a dual-function gene: it minimizes the generation of HIV escape mutants.

---

[a] PBMC, peripheral blood mononuclear cells; SIV, simian immunodeficiency virus.

across the cell membrane, and difficulty in determining whether apparent specificity is in reality a specific antisense effect. These problems may not be insurmountable.

Gene therapy. We wish to deliver to uninfected cells genes which inhibit HIV infection. Preferably, we will target stem cells. A protected stem cell will produce progeny, like $CD4^+$ T cells, that also contain the inhibitory gene and in theory are protected from HIV infection. The stem cells (and T cells) are obtained from bone marrow, adult peripheral blood, or, preferably, umbilical cord blood of a newborn and then cultured, and the inhibitory gene is transduced into those cells and then reinfused back into the HIV-infected patient. Of course, the goal is to reconstitute the immune system by protecting cells from infection. There are numerous ideas for achieving this end. Some are almost ready for or are already in phase one clinical trials, e.g., G. Nabel's *trans*-dominant *rev* inhibitory gene and F. Wong-Staal's ribozyme, both delivered by retroviral vectors. We have been comparing the inhibitory effects of four different genes (Fig. 8). One is a fragment (VH + VL) of the gene for an antibody to the HIV-1 RT (work of F. Weichold in our group [58a]) (Fig. 9). Another utilizes a gene for the sequence that is antisense to the HIV-1 Tat sequence (9, 34, 35). A third is a polymer of the Tat-binding site (*tar*) (Fig. 10) acting as a molecular decoy (9, 33, 36, 37); poly(TAR) is expressed after infection (because it utilizes an HIV LTR) and binds the Tat formed and needed by the infecting virus (work of J. Lisziewicz). The fourth approach is to use sequences of HIV-2. S. K. Arya in our laboratory and J. Rappaport of NIH's Institute of Dental Research discovered that HIV-2 sequences inhibit HIV-1. The mechanism by which genetic or protein components of HIV-2 inhibit HIV-1 is unknown.

We have found that all of the above-mentioned approaches inhibit HIV replication in laboratory culture systems, that inhibition works against all strains of HIV-1 tested (including new field isolates), that a combination of a few approaches

Poly(TAR)	"molecular decoy", sequesters Tat, inhibits viral RNA expression
Transdominant gag	inhibits formation of functional capsid
Transdominant rev	interferes with Rev function, inhibits expression of mRNA for virion proteins
Anti-RT	binds to RT, inhibits appearance of nuclear viral DNA
Antisense gag, rev, tat	inhibits expression of protein, mechanisms not clear but may be by RNAse H, hybrid arrest of translation, or other ?

FIGURE 8. HIV-1 inhibitory genes. Indicated above are some of the HIV-1 inhibitory genes currently being developed by us for possible clinical use. Poly(TAR) is an oligomer containing 30 to 50 repeats of the Tat activation response (TAR) region (33). *trans*-dominant Gag is a mutant with a short deletion encompassing a Gag precursor cleavage site (56, 58). *trans*-dominant *rev* is a mutant of *rev* which interferes with the activity of the wild-type gene product (45). Antisense oligonucleotides are antisense to the indicated genes and have phosphorothioate or phosphodiester backbones.

is better than use of any one alone, that resistance does not occur after 6 months of culture, and that these effects could be obtained against HIV-infected blood cells as well as infected cell lines (36, 37). The chief problem faced in gene therapy for AIDS is the difficulty of obtaining a sufficient number of successfully transduced stem cells. We are also uncertain of the long-term safety of these genes, i.e., whether they can resist the development of escape mutants (though I am not aware that anyone has found emergence of resistant variants; to my knowledge, no one has tested this by using suboptimal expression of the inhibitory gene and

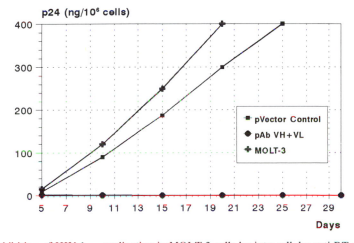

FIGURE 9. Inhibition of HIV-1$_{IIIB}$ replication in MOLT-3 cells by intracellular anti-RT antibody fragments. HIV-1 replication over 30 days in MOLT-3 cell cultures was derived after, minimally, 2 weeks of selection. Values are means of triplicate experiments (standard deviation, ≥15%); multiplicity of infection was 2.0.

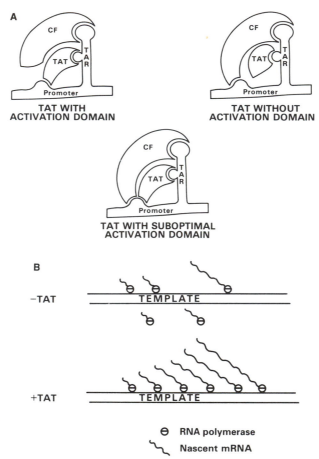

FIGURE 10. (A) Schematic of Tat-Tat activation response (TAR) region interaction leading to transcriptional activation. Tat binds to the bulge of the stem-loop structure of the TAR element and also interacts with a cellular factor(s) (CF) which binds to the loop of the stem-loop. The diagram on the left assumes that Tat has an activation domain that interacts with the promoter element to activate transcription. The diagram on the right on the other band assumes that Tat does not have an activation domain and that its role is to recruit a cellular factor(s) which activates transcription. The lower diagram depicts a combination of the two possibilities. (B) Tat-mediated transcriptional activation. Tat is depicted to promote both transcriptional initiation and elongation. Its major effect may be to enhance the processivity of transcription.

then, after allowing viral replication, retested with optimal levels), and whether a sufficient level of expression of the gene can be maintained over long periods. Nonetheless, it seems reasonable to pursue clinical trials in HIV-1-infected people to obtain information on the in vivo efficacy of these approaches.

HHV-6: A POSSIBLE PROMOTER OF AIDS PROGRESSION

HHV-6 was isolated in my laboratory in 1986 from a patient with AIDS and B-cell lymphoma. Because of its association with B-cell lymphoma, we originally

HHV-6

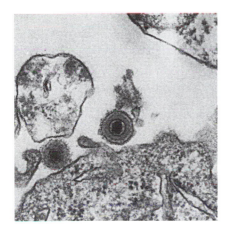

**Pan T-cell tropic
NK-cell tropic**

HHV-7

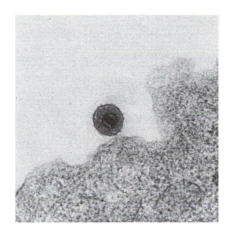

CD4+ T-cell tropic

FIGURE 11. Electron micrograph illustrating the ultrastructures of mature HHV-6 and HHV-7 virions.

called this new herpesvirus herpes B-lymphotropic virus. However, we soon learned it was chiefly a pan-T-cell tropic virus (work by P. Lusso), and we renamed it human herpesvirus 6. In fact, this was the first known T-tropic herpesvirus (Fig. 11). Because HHV-6 has many in vitro detrimental effects on T-cell function, activates HIV expression, and induces de novo expression of CD4, and because active replication has been associated by some investigators with hematopoietic impairment, we have suggested that HHV-6 may catalyze AIDS progression (43). Because of its common presence (most people have antibodies to it), some epidemiologists have argued against this notion. However, this is akin to an argument that cytokines have no bearing on disease pathogenesis because everyone has them. Moreover, antibody assays for herpesviruses have limited meaning and can be misleading. Prospective epidemiological studies need to be carried out with quantitative assays for the virus itself as well as with coinfection experiments in animal models, and such studies are now in progress.

KAPOSI'S SARCOMA

Background and Pathogenesis of Early Stage

KS remains the most important tumor associated with HIV-1 infection in terms of both frequency and the suffering it produces. KS has a greater incidence in males, especially homosexual men (5, 12, 19, 50, 52). In 1986, we began to develop systems for the experimental study of KS. Several concepts and a few conclusions have come from experiments with these new systems. Early KS is a

TABLE 2

Cultured AIDS-KS spindle cells express endothelium-specific markers and activation molecules which are also present in KS spindle cells in vivo[a]

Marker	Specificity	Presence in KS spindle cells[a]	
		Cultured	In situ
FVIII-RA	Vascular endothelium	$-/+$[b]	$-/+$
CD34	Vascular endothelium and hematopoietic cell progenitors	+	+
Cadherin-5	Vascular endothelium	+	+
ICAM-1	Activated endothelial cells and other cell types	+	+
VCAM-1	Activated endothelial cells and other cell types	+	+
ELAM-1	Activated endothelial cells	+	+

[a] Cultured AIDS-KS spindle cells and frozen sections of AIDS-KS lesions were stained by immunohistochemistry for markers expressed by the activated endothelial cells which are listed here. +, Positive expression.
[b] Positive after culture in the absence of HTLV-II CM.

complex tumor consisting of many cell types and resembling an inflammatory response (12, 19, 50, 52). The tumor cell is believed to be a spindle-shaped cell (12, 50). We have been able to culture these cells obtained from KS lesions (15, 47, 53). Cultured KS spindle cells express endothelium-specific cell markers and activation molecules which are also present in KS spindle cells in vivo (13, 16) (Table 2). We recently found these spindle cells in the peripheral blood of KS patients and also in a substantial number of HIV-1-infected homosexual men prior to the appearance of overt KS (8). It is possible that these circulating cells are responsible for the multiple and often widely separated lesions that sometimes develop in the same time frame (multifocal aspect of KS). We showed that these

TABLE 3

Activated T cells release inflammatory cytokines which induce endothelial cells in vitro and in vivo to acquire the phenotypic and functional features of KS spindle cells[a]

Feature	Presence in[b]:		
	Normal endothelial cells		AIDS-KS cells
	Prior to cytokine activation	Cytokine activated	
Spindle shape	−	+	+
Expression of FVIII-RA	+	+/−	+/−
Adhesion molecule expression[c]	−	+	+
Expression of integrin receptors as in KS cells ($\alpha5\beta1$ and $\alpha V\beta3$)	−	+	+
Expression of bFGF, IL-1, IL-6, IL-8, and GM-CSF	−	+	+
Induction of KS-like lesions in nude mice	−	+	+

[a] See text for explanations. Table is modified from a table in reference 12.
[b] −, Negative or activity absent; +, positive or activity present.
[c] Adhesion molecules mediating contact with inflammatory cells (ICAM, VCAM, ELAM).

TABLE 4
Cytokine expression and production by AIDS-KS cells[a]

Cytokine[b]	Expression by mRNA	Protein production
bFGF	+ + + +	+ + + +
aFGF	+	ND
IL-1α	+	+
IL-1β	+ + + +	+ + + +
IL-6	+ + + +	+ + + +
IL-8	ND	+ +
GM-CSF	+ +	+ +
PDGF	+ +	+ +
VEGF	+ +	+ +
TGF-β	+ +	ND

[a] mRNAs from AIDS-KS cells were analyzed for expression of a variety of cytokines. The content of each cytokine was analyzed by radioimmunoprecipitation and/or enzyme-linked immunoassay and is the sum of the intra- and extracellular proteins produced by AIDS-KS cells. Table is modified from a table in reference 12. ND, not done; −, negative; +, detectable levels. + +, + + +, and + + + +, are twofold, threefold, and fourfold the level of +, respectively.
[b] aFGF, acidic FGF; PDGF, platelet-derived growth factor; TGF-β, transforming growth factor beta.

cells have the properties of normal vascular endothelial cells, but ones in a state of chronic activation (13, 16) (Tables 2 and 3). We think inflammatory cytokines like IFN-γ, IL-1, and tumor necrosis factor induce their activation (13, 16). Though the mechanism is uncertain, these cytokines are increased in HIV-infected people and have been reported to be elevated in gay men even prior to HIV-1 infection (16). Paradoxically, our results suggest that KS arises more from chronic immune activation than from immune deficiency. Activated endothelial cells have a propensity for growth, migration, and the production and release of other cytokines (Table 3) (1–3, 10–16, 46, 47, 53, 54). Among the cytokines prominently produced by the early-stage KS spindle cells are platelet-derived growth factor, GM-CSF, IL-1, IL-6, vascular endothelial growth factor (VEGF), IL-8, and especially basic fibroblast growth factor (bFGF) (12–16, 46, 53, 54) (Table 4). For many reasons, among them being the abilities of these factors to promote new blood vessel formation, growth of several cell types, chemotaxis, and edema, we think that these cytokines are critical to the development of early KS lesions. bFGF appears to be the most important among these factors for the development of early KS lesions, because most of the growth in vitro of the spindle cells is inhibited by antisense to bFGF (14) (Fig. 12 and Table 5). Moreover, inoculation of nude mice with

TABLE 5
Evidence for role of bFGF in KS lesion formation

- High expression in KS spindle cells in vitro and in vivo[a]
- Induction of spindle cell growth (autocrine)[a]
- Release by spindle cells in absence of cell death[a] and in biologically active form inducing growth, migration, and invasion of endothelial cells[a]
- Induction in nude mice of vascular lesions closely resembling those of early KS

[a] Increased by inflammatory cytokines.

A

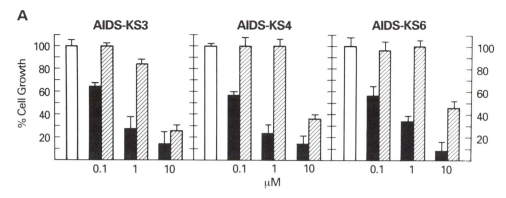

B

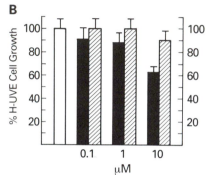

FIGURE 12. Antisense phosphorothioate oligodeoxynucleotides directed against bFGF (AS bFGF) but not sense oligomers (S bFGF) block the growth of AIDS-KS cells derived from different patients (AIDS-KS3, -KS4, and -KS6) (A) but not the proliferation of normal endothelial cells (human umbilical-vein-derived endothelial cells [H-UVE]) grown under standard conditions (aFGF and heparin) (B). □, medium; ■, AS bFGF; ▨-S bFGF. Reproduced from the *Journal of Clinical Investigation*, 1994, **94**:1736–1746, by permission of The Society for Clinical Investigation.

cultured spindle cells produces transient KS-like lesions (14, 53, 54). However, these lesions are of mouse origin. Apparently, the human KS spindle cells produce cytokines, like bFGF, which produce the mouse lesion. We recently found that bFGF alone can mimic this effect (13) and that antisense to bFGF blocks the formation of KS-like lesions induced by inoculated spindle cells in nude mice (work of B. Ensoli) (14) (Table 5). Further, we have shown that the HIV-1 protein, Tat, is released by acutely infected T cells, can enter these spindle cells, and acts synergistically with bFGF in promoting cell growth in vitro and KS-like lesion formation in mice when it acts on activated endothelial cells that have receptors for Tat, the integrins $\alpha V \beta 3$ and $\alpha 5 \beta 1$ (work of B. Ensoli) (1–3, 10–13, 16) (Table 6). These integrins are turned on by bFGF (7, 49). Other new results show that Tat mimics the effect of extracellular matrix molecules such as fibronectin (3, 13). These molecules, which hold cells in a local environment, when "cut" by specific proteases (released with signals that indicate an inflammatory response should occur), can cause cell migration and adhesions as well as induce growth. Tat has the same properties and shares an important region of homology with these extracellular matrix proteins, namely, the RGD domain (3, 13). Moreover, Tat activates collagenase IV, an enzyme involved in basement membrane degradation (13).

It is interesting that HIV-2 is rarely associated with KS and that HIV-2 Tat protein lacks the RGD domain. That these in vitro findings may be operative in

TABLE 6
Effects of HIV-1 Tat protein on KS and endothelial-cell properties

In Vitro
   Induction of spindle cell growth, migration, invasion, and adhesion
   Induction of growth, migration, invasion, tube formation, and adhesion of cytokine-activated endo-
      thelial cells

In Vivo
   Synergistic effects with bFGF in inducing KS-like lesions in nude mice
   Induction of KS-like lesions in presence of inflammatory cytokines

vivo is suggested by our finding by in situ immunohistochemistry assays of high levels of bFGF expression in the spindle cells of KS clinical specimens (13, 59) and detectable Tat in some of these cells which also express the receptors for Tat ($\alpha5\beta1$ and $\alpha V\beta3$) (which must enter by uptake, since these cells are not infected) (13). Finally, it has been shown that HIV-1-infected T cells express new adhesion molecules that facilitate their interactions with endothelial cells and that the latter in turn also express new adhesion molecules induced by inflammatory cytokines as well as by bFGF which facilitate their interactions with T cells (7, 16, 49).

To summarize, we believe that most of the cells of a KS lesion are hyperplastic, not malignant, cells and that their origin lies in the chronic overproduction of inflammatory cytokines. These cytokines activate endothelial cells, inducing them in turn to produce large amounts of other cytokines, in particular, bFGF (12, 16). The activated spindle cells also develop receptors for extracellular matrix molecules (3, 13), but these receptors are also recognized and used by the HIV protein, Tat, and Tat combines synergistically with bFGF in promoting new blood vessel formation and continued proliferation of these cells (13). Control of the growth of the hyperplastic cells, then, seems to lie in inhibition of the effects of bFGF and Tat and/or in reversing the overproduction of inflammatory cytokines like IFN-γ.

KS as a Malignancy

Until very recently, we had no evidence that HIV-associated KS contained any neoplastic cells; i.e., we did not know if the hyperplastic cells ever transformed into malignant cells. One recent result from Levinton-Kriss in Israel indicated that iatrogenic KS cells (KS SLK) can sometimes become neoplastic: she and her coworkers obtained an immortalized cell line from one KS patient (55). The first such cell line (KS Y-1) from HIV-1-associated KS, which we developed, induces metastatic malignant sarcoma in immunodeficient mice, contains abnormal (in both number and type) tetraploid chromosomes, and grows continuously in the absence of any growth factor (42). In short, these are malignant cells. These results were obtained with the first passage of KS Y-1 cells in culture. Therefore, these characteristics did not develop while the cells were in culture in the laboratory but were present in the primary KS tumor. We conclude from these results that

KS can evolve into a true malignancy. It is possible that KS even begins with such malignant cells but that these cells are masked by an abundance of the hyperplastic spindle cells that respond for unknown reasons to the neoplastic clone. In turn, these hyperplastic cells promote angiogenesis.

Control of KS Malignant Cells by a Hormone of Pregnancy, hCG

KS Y-1 cells produce very little bFGF but do not release it and are not responsive to it or to Tat. Inhibiting bFGF or Tat or blocking inflammatory cytokines should have no effect on these cells. Instead, they produce high levels of IL-8, VEGF, and IL-6, but we have no evidence that these cytokines are needed in an auto- or paracrine manner for growth of these cells. However, we have found that these malignant cells are totally controlled by a hormone of pregnancy. We have reported (i) that there is little or no tumor formation in pregnant mice; (ii) that sera of pregnant humans and mice block growth of these cells in vitro and that if the cells are pretreated before inoculation, tumor formation in mice is also blocked; (iii) that the active factor is the β chain of hCG (βhCG); (iv) that the mechanism involves binding of hCG to a cell surface receptor followed by apoptotic death of the tumor cells; and (v) that these findings should not be limited to this single case, because assays of biopsy clinical specimens from tumor lesions of other KS patients show evidence for cell surface receptors for βhCG (41). Most of this work was carried out by Y. Lunardi-Iskandar of my laboratory and J. Bryant of the NIH Institute of Dental Research. To our knowledge, this is the first evidence of a potent antitumor effect of hCG. Since βhCG shares strong homology with the β chains of luteinizing hormone and less but significant homology with follicle-stimulating hormone, and since these hormones rise during a phase of the menstrual cycle to high levels which are far above the levels in males, these results may also provide clues for the greater frequency of KS in males.

ACKNOWLEDGMENTS. I thank Suresh Arya, Barbara Ensoli, Genoveffa Franchini, Julianna Lisziewicz, Franco Lori, Yanto Lunardi-Iskandar, Paolo Lusso, Marvin S. Reitz, Jr., and Frank Weichold, staff members of my laboratory, who helped assemble and place references, figures, and tables.

REFERENCES

1. **Albini, A., G. Barillari, R. Benelli, R. C. Gallo, and B. Ensoli.** Tat, the human immunodeficiency virus type 1 regulatory protein, has angiogenic properties. *Proc. Natl. Acad. Sci. USA*, in press.
2. **Barillari, G., L. Buonaguro, V. Fiorelli, J. Hoffman, F. Michaels, R. C. Gallo, and B. Ensoli.** 1992. Effects of cytokines from activated immune cells on vascular cell growth and HIV-1 gene expression. *J. Immunol.* **149**:3727–3734.
3. **Barillari, G., R. Gendelman, R. C. Gallo, and B. Ensoli.** 1993. The Tat protein of human immunodeficiency virus type 1, a growth factor for AIDS Kaposi sarcoma and cytokine-activated vascular cells, induces adhesion of the same cell types by using integrin receptors recognizing RGF amino acid sequence. *Proc. Natl. Acad. Sci. USA* **90**:7941–7945.
4. **Belt, R. J., C. D. Haas, J. Kennedy, and S. Taylor.** 1980. Studies of hydroxyurea administered by continuous infusion. *Cancer* **46**:455–462.
5. **Beral, V., T. A. Peterman, R. L. Berkelman, and H. W. Jaffe.** 1990. Kaposi's sarcoma among persons with AIDS: a sexually transmitted infection? *Lancet* **335**:123–128.

6. **Berneman, Z. N., R. B. Gartenhaus, M. S. Reitz, Jr., W. A. Blattner, A. Manns, B. Hanchard, O. Ikehara, R. C. Gallo, and M. E. Klotman.** 1992. Expression of alternatively spliced human T-lymphotropic virus type I pX mRNA in infected cell lines and in primary uncultured cells from patients with adult T-cell leukemia/lymphoma and healthy carriers. *Proc. Natl. Acad. Sci. USA* **89:**3005–3009.

7. **Brooks, P. C., R. A. F. Clark, and D. A. Cheresh.** 1994. Requirement of vascular integrin $\alpha v \beta 3$ for angiogenesis. *Science* **264:**569–571.

8. **Browning, P. J., J. M. G. Sechler, M. Kaplan, R. H. Washington, R. Gendelman, R. Yarchoan, B. Ensoli, and R. C. Gallo.** 1994. Identification and culture of Kaposi's sarcoma-like spindle cells from the peripheral blood of human immunodeficiency virus-1-infected individuals and normal controls. *Blood* **84:**2711–2720.

9. **Chang, H.-K., R. Gendelman, J. Lisziewicz, R. C. Gallo, and B. Ensoli.** 1994. Block of HIV-1 infection by a combination of antisense *tat* RNA and TAR decoys: a strategy for control of HIV-1. *Gene Ther.* **1:**208–216.

10. **Ensoli, B., G. Barillari, S. Z. Salahuddin, R. C. Gallo, and F. Wong-Staal.** 1990. Tat protein of HIV-1 stimulates growth of cells derived from Kaposi's sarcoma lesions of AIDS patients. *Nature* (London) **344:**84–86.

11. **Ensoli, B., L. Buonaguro, G. Barillari, V. Fiorelli, R. Gendelman, R. A. Morgan, P. Wingfield, and R. C. Gallo.** 1993. Release, uptake, and effects of extracellular human immunodeficiency virus type 1 Tat protein on cell growth and viral transactivation. *J. Virol.* **67:**277–287.

12. **Ensoli, B., and R. C. Gallo.** 1994. Growth factors in AIDS-associated Kaposi's sarcoma: cytokines and HIV-1 Tat protein, p. 1–12. *In* V. T. DeVita, Jr., S. Hellman, and S. A. Rosenberg (ed.), *AIDS Updates,* vol. 7. J. B. Lippincott, Philadelphia.

13. **Ensoli, B., R. Gendelman, P. Markham, V. Fiorelli, S. Colombini, M. Raffeld, A. Cafaro, H.-K. Chang, J. N. Brady, and R. C. Gallo.** 1994. Synergy between basic fibroblast growth factor and HIV-1 Tat protein in induction of Kaposi's sarcoma. *Nature* (London) **371:**674–680.

14. **Ensoli, B., P. Markham, V. Kao, G. Barillari, V. Fiorelli, R. Gendelman, M. Raffeld, G. Zon, and R. C. Gallo.** 1994. Block of AIDS-Kaposi's sarcoma (KS) cell growth, angiogenesis, and lesion formation in nude mice by antisense oligonucleotide targeting basic fibroblast growth factor. *J. Clin. Invest.* **94:**1736–1746.

15. **Ensoli, B., S. Nakamura, S. Z. Salahuddin, P. Biberfeld, L. Larsson, B. Beaver, F. Wong-Staal, and R. C. Gallo.** 1989. AIDS-Kaposi's sarcoma-derived cells express cytokines with autocrine and paracrine growth effects. *Science* **243:**223–226.

16. **Fiorelli, V., R. Gendelman, F. Samaniego, P. D. Markham, and B. Ensoli.** 1995. Cytokines from activated T cells induce normal endothelial cells to acquire the phenotypic and functional features of AIDS-Kaposi's sarcoma spindle cells. *J. Clin. Invest.* **95:**1723–1734.

17. **Franchini, G., J. C. Mulloy, I. J. Koralnik, A. LoMonico, J. J. Sparkowski, T. Andersson, D. J. Goldstein, and R. Schlegel.** 1993. The human T-cell leukemia/lymphotropic virus type I p12^I protein cooperates with the E5 oncoprotein of bovine papillomavirus in cell transformation and binds the 16-kilodalton subunit of vacuolar H^+ ATPase. *J. Virol.* **67:**7701–7704.

18. **Franchini, G., J. Tartaglia, P. Markham, J. Benson, J. Fullen, M. Wills, J. Arp, G. Dekaban, E. Paoletti, and R. C. Gallo.** 1995. Highly attenuated HTLV type I$_{env}$ poxvirus vaccines induce protection against a cell-associated HTLV-I challenge in rabbits. *AIDS Res. Hum. Retroviruses* **11:**307–313.

19. **Friedman-Kien, A. E.** 1981. Disseminated Kaposi's sarcoma syndrome in young homosexual men. *J. Am. Acad. Dermatol.* **5:**468–471.

20. **Gallo, R.** 1986. The first human retrovirus. *Sci. Am.* **244:**88–89.

21. **Gallo, R.** 1987. The AIDS virus. *Sci. Am.* **256:**47–56.

22. **Gallo, R.** 1991. *Virus Hunting,* p. 352. Basic Books, New York.

23. **Gallo, R., and L. Montagnier.** 1988. AIDS in 1988. *Sci. Am.* **259:**41–48.

24. **Gao, W.-Y., A. Cara, R. C. Gallo, and F. Lori.** 1993. Low levels of deoxynucleotides in peripheral blood lymphocytes: a strategy to inhibit human immunodeficiency virus type 1 replication. *Proc. Natl. Acad. Sci. USA* **90:**8925–8928.

25. **Gessain, A., F. Barin, J. C. Vernant, O. Gout, L. Maurs, A. Calender, and G. deThé.** 1985. Antibodies to HTLV-I in patients with tropical spastic paraparesis. *Lancet* **2:**407–409.

26. **Gessain, A., E. Boeri, R. Yanagihara, R. C. Gallo, and G. Franchini.** 1993. Complete nucleotide sequence of a highly divergent human T-cell leukemia (lymphotropic) virus type I (HTLV-I) variant from Melanesia: genetic and phylogenetic relationship to HTLV-I strains from other geographical regions. *J. Virol.* **67**:1015–1023.

27. **Haseltine, W. A., and F. Wong-Staal.** 1988. The molecular biology of the AIDS virus. *Sci. Am.* **159**:52–62.

28. **Hirose, S., Y. Vemura, M. Fujishita, T. Kitagawa, M. Yamashita, J. Imamura, K. Ohtsuki, H. Taguchi, and I. Miyoshi.** 1986. Isolation of HTLV-I from cerebrospinal fluid of patients with myelopathy. *Lancet* **2**:297.

29. **Jacobson, S., C. S. Raine, E. S. Minglioli, and D. E. McFarlin.** Isolation of an HTLV-I like retrovirus from a patient with tropical spastic paraparesis. *Nature* (London) **331**:540–543.

30. **Koralnik, I. J., E. Boeri, W. C. Saxinger, A. LoMonico, J. Fullen, A. Gessain, H. G. Guo, R. C. Gallo, P. Markham, V. Kalyanaraman, V. Hirsch, J. Allan, K. Murthy, P. Alford, J. P. Slattery, S. J. O'Brien, and G. Franchini.** 1994. Phylogenetic associations of human and simian T-cell leukemia/lymphotropic virus type I strains: evidence for interspecies transmission. *J. Virol.* **68**:2693–2707.

31. **Koralnik, I. J., J. Fullen, and G. Franchini.** 1993. The p12^I, p13^{II}, and p30^{II} proteins encoded by human T-cell leukemia/lymphotropic virus type I open reading frames I and II are localized in three different cellular compartments. *J. Virol.* **67**:2360–2366.

32. **Koralnik, I. J., A. Gessain, M. E. Klotman, A. LoMonico, Z. N. Berneman, and G. Franchini.** 1992. Protein isoforms encoded by the pX region of human T-cell leukemia/lymphotropic virus type I. *Proc. Natl. Acad. Sci. USA* **89**:8813–8817.

33. **Lisziewicz, J., J. Rappaport, and R. Dhar.** 1991. Tat regulated production of multimerized TAR RNA inhibits HIV-1 gene expression. *New Biol.* **3**:1–8.

34. **Lisziewicz, J., D. Sun, M. Klotman, S. Agrawal, P. Zamecnik, and R. C. Gallo.** 1992. Specific inhibition of human immunodeficiency virus type 1 replication by antisense oligonucleotides: an *in vitro* model for treatment. *Proc. Natl. Acad. Sci. USA* **89**:11209–11213.

35. **Lisziewicz, J., D. Sun, V. Metelev, P. Zamecnik, R. C. Gallo, and S. Agrawal.** 1993. Long-term treatment of human immunodeficiency virus-infected cells with antisense oligonucleotide phosphorothioates. *Proc. Natl. Acad. Sci. USA* **90**:3860–3864.

36. **Lisziewicz, J., D. Sun, J. Smythe, P. Lusso, F. Lori, A. Louie, P. Markham, J. Rossi, M. Reitz, and R. C. Gallo.** 1993. Inhibition of human immunodeficiency virus type 1 replication by regulated expression of a polymeric Tat activation response RNA decoy as a strategy for gene therapy in AIDS. *Proc. Natl. Acad. Sci. USA* **90**:8000–8004.

37. **Lisziewicz, J., D. Sun, B. Trapnell, M. Thomson, H.-K. Chang, B. Ensoli, and B. Peng.** 1995. An autoregulated dual-function *antitat* gene for human immunodeficiency virus type 1 gene therapy. *J. Virol.* **69**:206–212.

38. **Lisziewicz, J., D. Sun, F. F. Weichold, A. R. Thierry, P. Lusso, J. Tang, R. C. Gallo, and S. Agrawal.** 1994. Antisense oligodeoxynucleotide phosphorothioate complementary to Gag mRNA blocks replication of human immunodeficiency virus type 1 in human peripheral blood cells. *Proc. Natl. Acad. Sci. USA* **91**:7942–7946.

39. **Lori, F., F. di Marzo Veronese, A. L. DeVico, P. Lusso, M. S. Reitz, Jr., and R. C. Gallo.** 1992. Viral DNA carried by human immunodeficiency virus type 1 virions. *J. Virol.* **66**:5067–5074.

40. **Lori, F., A. Malykh, A. Cara, D. Sun, J. N. Weinstein, J. Lisziewicz, and R. C. Gallo.** 1995. Hydroxyurea as an inhibitor of human immunodeficiency virus type 1 replication. *Science* **266**:801–805.

41. **Lunardi-Iskandar, Y., J. L. Bryant, R. A. Zeman, V. H. Lam, F. Samaniego, J. M. Besnier, P. Hermans, A. R. Thierry, P. Gill, and R. C. Gallo.** Tumorigenesis and metastasis of neoplastic Kaposi's sarcoma cell line (KS Y-1) in immunodeficiency mice blocked by a human pregnancy hormone. *Nature* (London), in press.

42. **Lunardi-Iskandar, Y., P. Gill, V. Lam, R. A. Zeman, F. Michaels, D. L. Mann, M. S. Reitz, Jr., M. Kaplan, Z. N. Berneman, D. Carter, J. L. Bryant, and R. C. Gallo.** A neoplastic cell line (KS Y-1) from HIV-1-associated Kaposi's sarcoma (KS) as a malignancy. *J. Natl. Cancer Inst.*, in press.

43. **Lusso, P., and R. C. Gallo.** 1995. Human herpesvirus 6 in AIDS. *Immunol. Today* **16**:67–71.

44. **Lusso, P., P. Secchiero, R. W. Crowley, A. Garzino-Demo, Z. N. Berneman, and R. C. Gallo.** 1994. CD4 is a critical component of the receptor for human herpesvirus 7: interference with human immunodeficiency virus. *Proc. Natl. Acad. Sci. USA* **91:**3872–3876.

45. **Malim, M. H., S. Bohnlein, J. Hauber, and B. R. Cullen.** 1989. Functional dissection of the HIV-1 Rev trans-activator-derivation of a trans-dominant repressor of Rev function. *Cell* **58:**205–214.

46. **Miles, S. A., A. R. Rezai, J. F. Salazar-Gonzalez, M. Vander Meyden, R. H. Stevens, R. T. Mitsuyasu, T. Taga, T. Hirano, and T. Kishimoto.** 1990. AIDS Kaposi sarcoma-derived cells produce and respond to interleukin 6. *Proc. Natl. Acad. Sci. USA* **87:**4068–4072.

47. **Nakamura, S., S. Z. Salahuddin, P. Biberfeld, B. Ensoli, P. D. Markham, F. Wong-Staal, and R. C. Gallo.** 1988. Kaposi's sarcoma cells: long-term culture with growth factor from retrovirus-infected CD4$^+$ T cells. *Science* **242:**426–430.

48. **Osame, M., K. Usuku, S. Izumo, N. Izichi, H. Amitani, A. Igara, M. Matsumoto, and M. Tara.** 1986. HTLV-I associated myelopathy: a new clinical entity. *Lancet* **i:**1031.

49. **Pober, J. S., and R. S. Cotran.** 1990. Cytokines and endothelial cell biology. *Physiol. Rev.* **70:**427–451.

50. **Regezi, S. A., L. A. MacPhail, T. E. Daniels, J. G. DeSouza, J. S. Greenspan, and D. Greenspan.** 1993. Human immunodeficiency virus-associated oral Kaposi's sarcoma: a heterogeneous cell population dominated by spindle-shaped endothelial cells. *Am. J. Pathol.* **43:**240–249.

51. **Rodgers-Johnson, P., D. C. Gadjusek, O. Morgan, V. Zaninovic, P. S. Sarin, and D. S. Graham.** 1985. HTLV-I and HTLV-III antibodies and tropical spastic paraparesis. *Lancet* **ii:**1247–1248.

52. **Safai, B., K. G. Johnson, P. L. Myskowski, B. Koziner, S. Y. Yang, S. Cunningham-Rundles, J. H. Godbold, and B. Dupont.** 1985. The natural history of Kaposi's sarcoma in the acquired immunodeficiency syndrome. *Ann. Intern. Med.* **103:**744–750.

53. **Salahuddin, S. Z., S. Nakamura, P. Biberfeld, M. H. Kaplan, P. D. Markham, L. Larsson, and R. C. Gallo.** 1988. Angiogenic properties of Kaposi's sarcoma-derived cells after long-term culture *in vitro*. *Science* **242:**430–433.

54. **Samaniego, F., P. D. Markham, R. C. Gallo, and B. Ensoli.** 1995. Inflammatory cytokines induce AIDS-Kaposi's sarcoma-derived spindle cells to produce and release basic fibroblast growth factor and enhance Kaposi's sarcoma-like lesion formation in nude mice. *J. Immunol.* **154:**3582–3592.

55. **Siegal, B., S. Levinton-Kriss, A. Schiffer, J. Sayar, I. Engelberg, A. Vonsover, Y. Ramon, and E. Rubinstein.** 1990. Kaposi's sarcoma in immunosuppression. Possibly the result of a dual viral infection. *Cancer* **65:**492–498.

56. **Smythe, J. A., D. Sun, M. Thomson, P. D. Markham, M. S. Reitz, R. C. Gallo, and J. Lisziewicz.** 1994. A rev-inducible mutant gag gene stably transferred into T-lymphocytes: an approach to gene therapy against HIV-1 infection. *Proc. Natl. Acad. Sci. USA* **91:**3657–3661.

57. **Stevenson, M., T. L. Stanwick, M. P. Dempsey, and C. A. Lamonica.** 1990. HIV-1 replication is controlled at the level of T cell activation and proviral integration. *EMBO J.* **9:**1551–1560.

58. **Trono, D., M. B. Feinberg, and D. Baltimore.** 1989. HIV-1 Gag mutants can dominantly interfere with the replication of the wild-type virus. *Cell* **59:**113–120.

58a.**Weichold, F.** Unpublished results.

59. **Xerri, L., H. J. Hausson, J. Planche, V. Guigou, J. J. Grobb, P. Parc, D. Birnbaum, and O. Delapexriere.** 1991. Fibroblast growth factor gene expression in AIDS-Kaposi's sarcoma detected by *in situ* hybridization. *Am. J. Pathol.* **138:**9–15.

60. **Yoshida, M.** 1994. Host-HTLV type I interaction at the molecular level. *AIDS Res. Hum. Retroviruses* **10:**1193–1197.

61. **Zack, J. A., S. J. Arrigo, S. R. Weitsman, A. S. Go, A. Haislip, and I. S. Y. Chen.** 1990. HIV-1 entry into quiescent primary lymphocytes: molecular analysis reveals a labile, latent viral structure. *Cell* **61:**213–222.

The DNA Provirus: Howard Temin's Scientific Legacy
Edited by G. M. Cooper, R. Greenberg Temin, and B. Sugden
© 1995 American Society for Microbiology, Washington, DC 20005

Chapter 18

Cherish An Idea That Does Not Attach Itself to Anything[1]

Tak Wah Mak

Smiling, sincere, incorruptible—
His body disciplined and limber.
A man who had become what he could,
And was what he was—
Ready at any moment to gather everything
Into one simple sacrifice.

—Dag Hammarskjold, *Markings*

MADISON, WISCONSIN

In the mid-1960s, I left Hong Kong for the United States, en route to Madison, Wisconsin. It was an exciting time. I was entering an undergraduate program in chemical engineering at the University of Wisconsin and was ultimately looking forward to a rewarding and stable career. Within a few months of my arrival, however, it became frightfully apparent that I was most unsuited to the profession. So it came as little surprise that after my repeated refusals to complete assignments, which consisted of seemingly unending and repetitive calculations, my professor declared that he had never met anyone less suited to be an engineer.

I was very disappointed that what had begun so auspiciously and with such anticipation had ended so abruptly, but I was also quite relieved. Determined that I would pursue an academic career, without delay I ventured across University Avenue and joined the Department of Biochemistry at Henry Mall. The move seemed somewhat arbitrary at the time, but quitting the engineering program had unnerved me, and I needed to quickly regain some confidence that I was indeed suited for some kind of career. I thought laboratory work was as good a place to start as any.

---

[1] From the Vajracchedika prajnaparamita sutra (Diamond Sutra).

Tak Wah Mak • The Amgen Institute, The Ontario Cancer Institute, and the Departments of Medical Biophysics and Immunology, University of Toronto, 500 Sherbourne Street, Toronto, Ontario, Canada M4X 1K9.

The timing was perfect. Roland Rueckert had just joined the faculty in the Department of Biochemistry and was at that time heading a group in the McArdle Laboratory for Cancer Research, since his space in the biophysics building on Linden Drive was still under construction. I enrolled in the undergraduate research program, working on nucleic acid chemistry and virology in Roland's laboratory. Although it was Roland's congeniality and kindness that drew me into his laboratory, his quick and analytical mind, as well as his unyielding enthusiasm, quickly became apparent. I was intrigued with the workings of viruses and was soon convinced that I had made the right decision. After graduation, I started a graduate program in his laboratory, working on infectious RNA viruses and pursuing the study of picornavirus.

It was in Rueckert's laboratory that I first encountered Howard Temin.

Every Tuesday noon, the virologists on the faculty got together (as they still do) for lunch. Paul Kaesberg, John Anderegg, Roland, and Howard would meet at Roland's laboratory before convening at the Pine Room or some other lunch spot. I never spoke with Howard during his rendezvous at our laboratory, but in his presence, it was impossible not to notice his calm and his firm confidence.

Toward the end of the 1960s, Roland and Howard became very close friends. Even though we did not often see Howard in our laboratory, it was obvious to me that something important was going to happen. Roland was a molecular virologist by training; he was, in fact, one of the leading figures involved in the purification and characterization of virions and their protein components. As a result, Roland had a natural and keen interest in Howard's quest to identify reverse transcriptase in Rous sarcoma virus and seemed almost alone in his support of Howard and his attempts to discover a mechanism for DNA synthesis from an RNA template. Howard pursued the isolation of an enzyme for what he believed would catalyze DNA synthesis from RNA despite enormous pressures from the scientific community to abandon his thinking; it was almost heresy to break with the central dogma of the times, which offered no exceptions to the rule that the information flow was only from DNA to RNA. Thus, Roland's expertise in molecular virology and his open mind regarding an RNA template provided Howard with an excellent source of molecular expertise. However, in 1970, only Howard could have been in any way sure that he was about to make the most exciting biological find of the decade: the discovery of reverse transcriptase (42). David Baltimore independently made the same discovery that same year (3).

I left Wisconsin for Canada without ever really knowing Howard. This did not reflect any lack of desire to know him, since I regarded him with the highest esteem: it was simply that I had been intimidated enough not to brave engaging him in any scientific discussions. Nevertheless, while in Canada, I managed to keep quite closely in touch with Howard's work through occasional phone talks with Roland and through reading the literature that Howard was publishing.

I spent the mid-1970s in Toronto, working in E. A. McCulloch's laboratory. There I learned much about hematopoiesis. I also discovered that Friend virus was an excellent system for studying the interactions between retroviruses and the hematopoietic system. A few years later, Roland and Rex Risser invited me to present a seminar at the University of Wisconsin. It was my first opportunity to give a seminar there since my days as a student; armed with some modest success and excited about returning to the warm, friendly enclaves of Madison, I jumped at the invitation.

TEMIN'S WORLD

I was eager to return to Wisconsin and make a favorable impression on Dr. Temin. I was therefore highly delighted, albeit anxious, when he showed up at the lecture theater and actually sat in the first row. Despite being intensely nervous, I did manage to finish my presentation, and immediately afterwards, I was asked to meet with Dr. Temin. At that point I was overwhelmed by a peculiar set of emotions, finding myself filled with pride and apprehension simultaneously: I simply did not know what to expect. But after almost 10 years, I would finally meet Howard Temin.

It was one of my most unforgettable and humbling experiences. For over an hour, Dr. Temin dissected both my research project and my presentation. During this time, he managed to reveal many aspects of my research that had not occurred to me; he even made recommendations on improving my slides and style of presentation. Even today, I do not know whether he would have been so frank and critical if I had not been a student under his very good friend Roland.

I immediately began wondering what the possibility of working under Dr. Temin might be. The prospect of being exposed to Temin's world was very exciting. I felt that if I had the opportunity to observe his brilliance and learn how he operated the laboratory, some of his success might rub off on me. At the same time, I believed I would benefit from further exposure to retrovirus research, since the field was attracting many new researchers from all areas of virology and it appeared that there might be important things I should learn. Soon after our meeting, I arranged through Roland to join the Temin laboratory for more apprenticeship and for work on reticuloendotheliosis virus (REV-A) and its transforming counterpart (REV-T). Together with Irvin Chen, we eventually cloned and sequenced the REV-T-specific sequences (9) (Fig. 1). We subsequently gave the name *rel* (reticuloendotheliosis) to a REV-T sequence that was similar to a cellular host gene.

I have to confess that I was slightly disappointed when I was given the REV-T project. To me, REV was a virus of no demonstrable significance: the cells it transformed were of undefined lineages, and at that time, only a few laboratories seemed interested enough to study it. I was quietly hoping for a project that would distinguish me from the thousands of scientists who had jumped on the retrovirology bandwagon. In retrospect, I am sure that Dr. Temin realized the significance of both the virus and *rel*; nevertheless, I left Madison and the Temin laboratory without appreciating the importance of *rel*.

However, I was sure that being a part of the Temin team would be an invaluable experience. I was able to observe his unwavering dedication to scientific research and to issues involving science policy. I was also witness to the regimented and businesslike manner in which he operated his laboratory, where we were

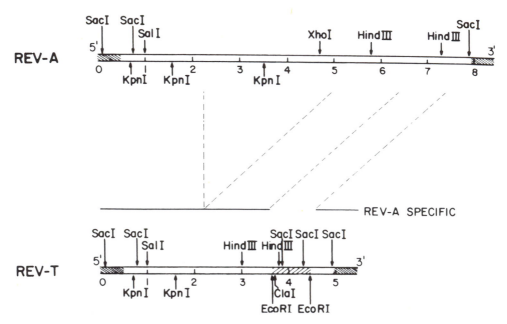

FIGURE 1. Schematic map of restriction sites for REV-A and REV-T unintegrated linear cDNAs. Numbers show distances in kilobases from the 5′ end of the viral DNA. Shaded areas denote *rel* sequences.

expected to work tirelessly through the week and on weekends and to report our progress to him each Friday. The Friday one-on-one discussion sessions (confessions) were excruciatingly detailed. He demanded to know the exact manner by which our experiments were performed and even the chemical compositions of solutions and buffers. He often recalculated and replotted our data, sometimes arriving at additional or slightly different conclusions. If we did not have new experiments to report during the week, he would make us reiterate progress reports of the previous week. These "confessions" were such a ritual that sometimes between out-of-town meetings, Dr. Temin would return to Madison simply to conduct them.

A VIROLOGIST INFECTED BY T CELLS

The Temin laboratory had in many ways been very challenging; it charged me with renewed confidence and enthusiasm. I returned to Canada and redoubled my research efforts on Friend virus, unaware that our work on viruses was soon going to give way to new efforts in T-cell biology. In collaboration with Alan Bernstein, we cloned the spleen focus-forming virus (SFFV) (47), and Stephen Clark, a graduate student in our laboratory, determined the entire sequence of the provirus (11). We found that, unlike the other rapidly transforming viruses that encode homologs of eukaryotic genes involved in control of cellular proliferation,

SFFV transforming sequences consist of only an ecotropic-xenotropic viral envelope fusion protein (gp55) (11) (Fig. 2). It was not until 7 years later that it was understood that the envelope fusion protein in SFFV induces erythroproliferation by interacting with erythropoietin receptor (22). While the role of gp55 in the transformation process may be better understood today, the pathophysiology of virus-host interactions continues to be unresolved.

While Stephen Clark was busy studying the molecular biology of the viruses, Tsunefumi Shibuya was active in analyzing the genetics of the hosts. He uncovered two additional host genes that affect cellular responses to viral infection. Expression of FV-5 in an infected cell was found to play an important role in determining the nature of ensuing disease. For example, hematocrit levels (erythropoiesis) vary dramatically after exposure to the polycythemia strain of SFFV (36). This suggests that the envelope fusion protein could stimulate different degrees of proliferation and differentiation depending on the host target cell. We also uncovered a host gene (*Fv-6*) that affected Friend virus infection and appeared to control the type of leukemia that is induced after infection of newborn mice with the replication competent Friend murine leukemia virus (37). At the time, these results were most exciting.

It appeared that the momentum I had gained in Howard's laboratory was continuing to gather speed. In fact, after my return to Toronto from Madison, our work went so well that I became convinced that it had something to do with my experience in Howard's laboratory. The progress we were making in virology may even have provided me the courage to hazard our new experimental program on T cells. These studies were aimed at cloning genes involved in T-cell differentiation and antigen recognition and were motivated intellectually by the knowledge that many T-cell-specific gene products are expressed at various stages of T-cell development and that these molecules are undoubtedly involved in T-cell antigen recognition and selection. Thus, we believed that the differential expression of these genes would in fact permit the elucidation of T-cell genetics.

Pre-T cells migrate from the bone marrow into the thymus, where they mature. The most primitive T cells begin as double-negative thymocytes with no CD4 or CD8 on the cell surface. Mature effector cells, on the other hand, express either CD4 or CD8, accessory molecules associated with helper and cytotoxic-T-cell

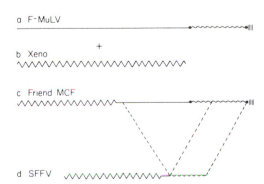

FIGURE 2. Mechanism proposed in 1983 for the production of SFFV from murine leukemia virus. F-MuLV, Friend murine leukemia virus; Xeno, xenotropic virus; MCF, mink cell focus virus.

(CTL) functions, respectively. The largest population of thymocytes, however, expresses both CD4 and CD8 (Fig. 3). Other surface molecules known to be on T lymphocytes include CD2, CD3, CD28, and CD45, which participate in cell-cell interactions and activation. In addition to these surface molecules, intracellular molecules are involved in T-cell activation and development. In the thymus, T-cell recognition of antigens in the context of self-major histocompatibility complex MHC molecules is an obligatory step in the selection process. Selection enriches for self-restricted T cells and silences those with potential autoreactivity, while at the same time producing a T-cell repertoire capable of coping with a broad spectrum of pathogens. Signals delivered to T cells during the selection process are the burden of the T-cell antigen receptor. While characterization of the recep-

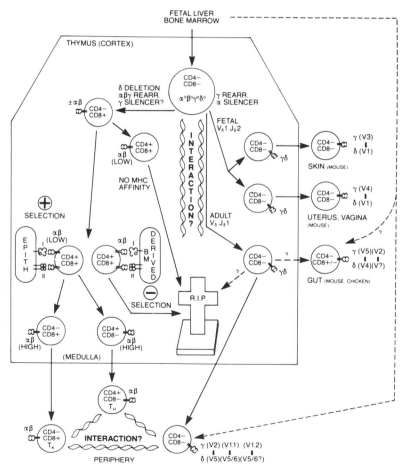

FIGURE 3. Cartoon depicting steps in T-cell development in the murine thymus. Positive selection is by stromal epithelial (EPITH) cells. Negative selection occurs on bone marrow (BM)-derived elements.

tor genes is clearly important to understanding the selection process and other T-cell functions, the genes coding for the T-cell receptor (TcR) had eluded isolation.

For over a decade, the molecular basis of the T-cell recognition structure remained the object of intense speculation. However, by using monoclonal antibodies raised against idiotype-bearing T-cell-specific surface proteins, it had been shown that on the surface of a T cell is a heterodimer that is probably part of a larger cell surface complex, a system that was expected to be analogous to the B-cell receptor (2, 15, 25). In order to sequence T-cell-specific genes and deduce their primary structures, we implemented a series of strategies that were aimed at exploiting differential gene expression between T and B cells and that we hoped would provide us with T-cell-specific genes. We used a technique of differential screening and molecular subtraction, favoring a subtraction approach partly because of our previous use of the protocol to isolate virus-specific sequences (23). Our decision to use this method was opposed by the granting agencies and criticized by some colleagues. It was thought that the subtraction of genes between two different cells was not feasible, considering that the kinetics of hybridization between the cDNA from one cell type and the cellular RNA (or genomic DNA) from the second source may not be adequate. It was asserted that cDNA representing low-frequency cellular RNA would fail to hybridize with adequate efficiency before degradation occurred.

Since we were working in a relatively small laboratory with no prior achievements in immunology, establishing a research project that had as its sole objective the cloning of the TcR genes would have been foolhardy and indeed difficult to justify. We simply set out to isolate T-cell-specific genes involved in differentiation and function. If this approach was to yield the TcR genes, we had to assume that the mRNA was T cell specific, a premise that was problematic, since the prevailing hypothesis was that the TcR genes were composed of T-cell-variable segments linked to immunoglobulin-constant regions; in addition, we realized that RNA would have to be present at levels high enough to survive the degradation that would occur with the subtraction method. With these considerations, we certainly could not count on our strategy delivering us the TcR. However, we expected that over 90% of the genes expressed in T and B cells would be similar and that it would not be an impossible task to analyze the small number of T-cell-specific genes that, with a little luck, we might happen to stumble on. We were, however, reasonably optimistic that genes we might manage to isolate would correspond to proteins involved in either T-cell development or antigen recognition, and we thought that discovery of any one of these genes would be significant to the understanding of T-cell biology.

To this end, a series of differential screening and molecular subtraction experiments were performed by Yusuke Yanagi and myself in 1982 and 1983. In one such experiment, about 1,500 T-cell-specific clones were isolated after three rounds of subtraction of T-cell cDNA from RNA of a B-cell line and classified into over 25 different groups on the basis of sequence similarity. To our astonishment, one of these groups of cDNA actually showed properties similar to that hypothesized for the TcR (49), excluding the conjectured immunoglobulin-constant region. The protein sequence deduced from these clones showed a variable (V) and a constant

(C) region linked by joining (J) and diversity (D) segments. Moreover, the sequences of this cDNA rearranged differently in various T-cell lines. This was, of course, exactly what was expected if the TcR genes were analogous to the immunoglobulin molecules. Independent of our efforts, the group of Mark Davis at Stanford University, using almost identical methodology, also cloned a TcR gene from mice (18).

T-CELL ANTIGEN RECEPTOR

Thus, although it is now paradigmatic, unraveling the genetic nature of the TcR genes provided the first evidence for the generation of T-cell diversity by somatic-gene recombination. Yanagi et al. had in fact cloned the beta chain of the TcR (49), as the amino terminus encoded by the gene was identical to the partial sequence reported for a beta-chain protein (1). Within a year of the initial cloning of the beta chain, several groups had isolated the alpha chain of the TcR (10, 17, 38, 48). In the headlong rush to isolate the alpha chain by molecular subtraction, a T-cell-specific cDNA that encoded an immunoglobulin-like cell surface protein and appeared to be capable of rearrangement was also identified (31). Although it was thought that this cDNA encoded a TcR alpha chain, closer examination revealed discrepancies between the biochemical and functional properties of this chain and those expected of the alpha chain. Subsequently, it was discovered that a second rearranging TcR system existed. These were dubbed gamma-delta heterodimers, and they defined a new minor subset of T cells (7). The role of these gamma-delta T cells remains enigmatic: many investigators favor a hypothesis that they represent a primitive defense system, probably recognizing a limited repertoire of bacterial or parasitic components presented on nonconventional MHC molecules. However, recent findings suggest that the gamma-delta receptors may not have to recognize combinations of peptide and MHC-like molecules in all cases.

The genomic organization for sequences encoding the TcR is similar to immunoglobulin gene clusters, which contain multiple V-region gene segments upstream of the C-region genes. Nested between these regions are the D and J gene segments, which are in place to enhance germ line diversity as well as to provide added joints whereby terminal transferase can generate additional N-region diversity (13, 14, 16, 39, 43, 46, 50). There is increasing evidence that this location of the TcR codes for the CDR3 region, which would require hyperdiversity, since it encodes amino acid residues recognizing the peptides bound within the binding cleft of the MHC. It is also clear from the genomic organization of these genes (Fig. 4) and the number of known V, D, and J gene segments that the germ line diversity of the alpha-beta genes is considerably greater than that of the gamma-delta genes. Another interesting aspect of the design of these genes was the finding that the TcR delta locus was situated within the TcR alpha genes, a design that could minimize the chance of both alpha-beta and gamma-delta receptors being expressed on the same T cell.

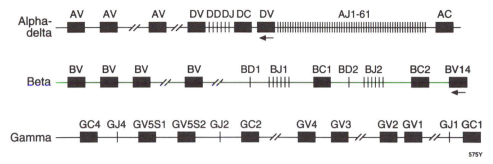

FIGURE 4. Genomic arrangement of alpha-beta and gamma-delta TcR genes.

An important step was the establishment of a one-receptor model for T cells. Pamela Ohashi, a graduate student in my laboratory, first demonstrated that the TcR was part of a signal transduction complex with CD3 (27), while investigators in Switzerland (12) and at the National Institutes of Health (32) further demonstrated that specificities to both antigen and MHC were reconstituted with the transfer of genes coding for an alpha-beta heterodimer. These studies, together with those of the crystal structures of peptide-MHC complexes (6), strongly asserted the validity of the one-receptor model for T-cell antigen recognition.

Perhaps the most exciting employment of the new knowledge of TcR genes was in the delineation of thymic selection events. Through the use of transgenic mice expressing specific TcR genes against defined combinations of peptide and class I and II MHC gene products, we learned that T cells have three fates during thymic selection: cells carrying TcR with no affinity to self-peptide–MHC complexes perish by programmed cell death; thymocytes expressing TcR with strong reactivity to self-peptide–MHC complexes are deleted, as they could become autoreactive; and finally, thymocytes that have low-affinity TcR reactivity with self-peptide–MHC complexes are chosen, or positively selected, to be exported to peripheral organs, where these cells become a clonotypic part of the peripheral repertoire. Thus, the employment of TcR transgenic mice helped resolve a key issue in immunology: the mechanism of immune tolerance and thymic selection (4, 19, 21, 34, 41, 45). Now TcR genes are used as standard tools to study T-cell recognition, development, and malignancies.

Thus, what started as a side project for cloning T-cell-specific sequences resulted in the cloning of the TcR and permanently altered the course of my laboratory. After the beta chain had been identified, I discussed with Howard on several occasions the implications the T-cell work might have for my laboratory. He avoided giving me any advice and was careful not to influence my decisions on whether or not to pursue the study of immunogenetics exclusively. However, he did remark that if I pursued T-cell biology, it would put me in the fast lane for a while.

THE ERA OF KNOCKOUT MICE

Toward the end of the 1980s, our laboratory became involved in a theme that has preoccupied us ever since. This development was made possible by new advances in the manipulation of embryonic stem cells in vitro and the discovery that these cells injected into blastocytes contribute to germ cell embryogenesis and allow transmission of genetic material. When these technologies were combined with the new technique of gene-disruptive homologous recombination (8), it was believed that they would make it possible to generate animals with specific genetic mutations transmittable through their germ lines. If these animal were produced, they might provide extremely valuable animal models for studying specific gene functions in vivo. Targeted gene disruption has, in fact, become especially useful to immunological studies, as different stages of lymphocyte activation and differentiation are processes that involve discrete genetic elements, whereas disruption does not usually result in lethal mutations.

By the summer of 1988, our laboratory had become convinced that the combination of these technologies would successfully yield gene-targeted mice. As a result, Marco Schilham, Wai-Ping Fung-Leung, and Amin Rahemtulla, two postdoctoral fellows and a graduate student, respectively, combined their efforts to develop the techniques in our laboratory. Our first two targets were the CD4 and CD8 genes. CD4 and CD8 molecules, expressed on mutually exclusive subsets of peripheral T cells, interact with nonpolymorphic regions of the MHC class II and class I molecules, respectively, and are thought to stabilize interactions between the T cell and its target. Wai-Ping was the first to succeed in producing mice lacking CD8 on the cell surface (13). She found that these mutant animals continued to produce differentiated $CD4^+$ T cells, demonstrating that the collateral expression of CD4 and CD8 molecules on the cell surface is not required for development of mature T cells. However, the mice were unable to mount detectable responses against viral antigens and were thus defective in the cell mediated arm of the immune response. Mice deficient in CD4 expression maintained development of $CD8^+$ T cells and generated normal CTL responses to viral challenge (30). Furthermore, the humoral-response arm of the immune system, normally regulated by $CD4^+$ T cells, was only partially defective.

Encouraged by the results with the CD4 and CD8 knockout mice, we proceeded to generate mice lacking genes engaged in TcR-mediated signal transduction. Mice lacking an intact *lck* gene, which codes for a tyrosine kinase associated with the cytoplasmic tails of CD4 and CD8, were shown to have a very early block in T-cell development in the thymus (26). The few lymphocytes found in the peripheral blood of these mice are also functionally inactive, as no detectable CTLs or immunoglobulin against specific viral targets can be found. Thymic development in mice with disruptions at the genetic loci of another *src* family protein tyrosine kinase (PTK) gene, *fyn,* is normal compared to that in *lck*-deficient mice.

The CD45 molecule is a large transmembrane protein tyrosine phosphatase expressed on all cells of hematopoietic origin except erythrocytes and platelets. TcR-mediated PTK activation and subsequent signal transduction are blocked in T- and B-cell lineages deficient in CD45 expression (44). Since the earliest signal transduction event taking place after TcR ligation is the activation of PTK activity, it is suggested that TcR-mediated PTK activity is regulated by CD45 dephosphorylation of substrates involved in early signal transduction. One potential substrate of CD45 is *lck,* whose activity is negatively regulated by phosphorylation of a C-terminal tyrosine residue. After TcR ligation, CD45 may facilitate signal transduction through dephosphorylation of the *lck* C-terminal tyrosine residue (44). However, in mice, a disruption at exon 6 of the CD45 gene does not impair development of the hematopoietic system (20). In addition, even though wild-type mice express CD45 during early B-cell development, B-cell ontogeny appears to be normal in these mice. Failure of these cells to respond to stimulation indicates a requirement for CD45 in B-lymphocyte activation. T-cell development is also arrested, and few mature T lymphocytes are found in these mice, which also lack the abilities to generate CTL and to mount humoral responses against specific antigens. Recently, it was demonstrated that mast cells deficient for CD45 do not degranulate as a result of immunoglobulin E receptor activation (5), a finding of great interest to the study of allergy.

One of the most exciting advances in immunology during the last several years was the finding that T lymphocytes require costimulatory molecules for full activation (33). While it is still not known how many kinds of these costimulatory molecules exist, the cell surface molecule CD28 and a related molecule, CTLA-4, appear to be important for this function. Furthermore, these molecules are known to interact with B7-1 or B7-2, found mainly on B cells, macrophages, and certain dendritic cells (33). To determine whether CD28 is an essential costimulating molecule in vivo, Arda Shahinian generated a mouse strain with disruptions at the genetic loci for CD28 (35).

These animals are defective in their abilities to respond to the potent T-lymphocyte mitogen concanavalin A and are dramatically impaired in their abilities to mediate a specific humoral response. However, while levels of specific immunoglobulin G produced against vesicular stomatitis virus are greatly reduced in these mutant mice, animals challenged with lymphocytic choriomeningitis virus (LCMV) generated normal numbers of functional CTL, and delayed-type hypersensitivity responses to LCMV were unaltered. These studies reveal that certain pathogens are capable of initiating immune responses without the involvement of CD28 costimulation. There are at least two possible explanations for these results. The ability of the immune system to circumvent the need for CD28 costimulation may reflect the existence of redundant molecules subserving the CD28-B7 costimulatory function. On the other hand, since signaling through CD28 after engagement with B7 is thought to be mainly responsible for production of interleukin-2 (IL-2), infection by certain pathogens (i.e., LCMV) might bypass the need for IL-2 by inducing the production of other lymphokines with functions that overlap those of IL-2. Either one or both of these scenarios could explain why mice lacking CD28 generate a normal LCMV-specific CTL response.

While many surface and intracellular proteins may be involved in the initial activation of immune cells, cytokines and their receptors participate significantly in maintaining and attenuating the responses. One of the most important of these factors is tumor necrosis factor (TNF). There are two TNFs, p55 and p75, each with high affinities for TNF-α and TNF-β. To determine the importance of each of these receptors, Klaus Pfeffer in our laboratory generated a mouse strain lacking TNF-Rp55. These animals maintained normal expression of TNF-Rp75, supporting previous reports that the two TNF receptors are regulated independently (29). Nevertheless, most phenomena known to be associated with TNF are altered in mice lacking TNF-Rp55, which is consistent with earlier thinking that effects associated with TNF are mediated via the p55 receptor. These animals also succumbed easily to infection by certain intracytoplasmic bacteria, such as listeriae. These studies helped demonstrate the importance of TNF-Rp55 and highlight the beneficial and harmful effects of TNF.

To ascertain the importance of type 1 interferons (alpha and beta [IFN-α and IFN-β]), in controlling immune responses and cell growth, we also examined the consequences of disruptions in genes coding for interferon regulatory factors 1 and 2 (IRF-1, IRF-2), which are transcriptional factors involved in type 1 IFN production. In collaboration with Tatagutsu Taniguchi, Toshifumi Matsuyama and Toru Kimura generated mice with disruptions in these genes. Data from these animals indicated that IRF-1 is involved in type 1 IFN production and suggested that IRF-1-independent and IRF-1-dependent pathways for IFN induction exist.

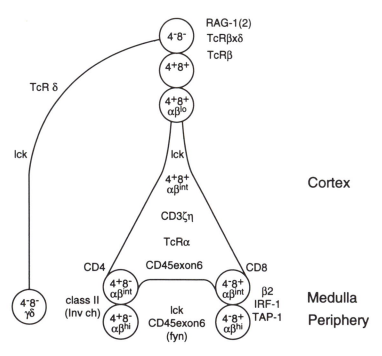

No effect: TNFRp55, IRF-2, TGF-β, TGF-α, (IFN-γ) R, IL-2, CD28, IRF-2, CD2, IL-4, Il-2Rβ

FIGURE 5. Locations of blocks in T-cell development in gene-targeted mutant mice. For example, alpha-beta T cells in the CD3$\zeta\eta$ knockout mice are blocked at the double-positive 4^+8^+ stage, while ontogeny of the gamma-delta lineage is unimpeded. TGF-β, TGF-α, transforming growth factors β and α, respectively; TAP-1, transporter associated with antigen processing; Inv ch, invariant chain.

Curiously, these mutant animals are defective in promoting the development of $CD8^+$ CTL, while the generation of $CD4^+$ T cells is normal (24). In addition, it was demonstrated that aspects of oncogene-induced transformation or apoptosis are dependent on IRF-1 (40). In contrast, mice lacking IRF-2 produce larger amounts of IFNs, indicating that this transcriptional factor might act to suppress IFN production. While these mutant animals show normal T-cell development, they do display a defect in early B-cell differentiation such that the number of pre-B cells is reduced in the bone marrow (24).

Many other laboratories have also been active in generating these gene-targeted mice in the study of immunology. A summary of these findings can be found in a recent review by Klaus Pfeffer and myself (28). Genetic checkpoints for these genes are illustrated in Fig. 5.

THE LOTUS

However muddy the water is, the lotus retains its purity; indeed it blooms beautifully just because it grows out of mud.

—Old Chinese Proverb

Looking back, it seems almost unreal that I started out to study engineering in Wisconsin and ended up developing a scientific career under the influence of a great scientist such as Howard Temin. Of course, there were other invaluable mentors, such as Roland Rueckert and Ernest McCulloch, along the way. I will always be especially indebted to Roland not only for giving me a start in science but also for introducing me to Howard. I regret not having kept more closely in touch with Howard after I left Wisconsin (the second time). I am sure that I would have been more in touch with him if I had stayed in the field of virology. There was, however, the occasional visit to Madison, mainly to present seminars and visit old friends. Howard always remained fatherly and provided invaluable advice on numerous scientific issues as well as personal matters.

My last visit with Howard was in January 1993. Having learned of his illness, I arranged through Roland to call on him. Although he was obviously frail from chemotherapy, he took his ailment with great courage, and we spent much of our time together discussing in detail his treatment protocols and what he thought was happening to him clinically. We also discussed extensively a subject of profound concern to him: AIDS. He was very worried about the lack of progress in the treatment and prevention of this disease. His thesis was that the war on AIDS will be a long haul and that we should initiate clinical trials (especially vaccines) even if we are not confident that the existing approaches will be efficacious. His belief was that what we would learn from these experiments and that experiences derived from these clinical trials would help us design better strategies in the future. He was active until his last days in advising different agencies on these and other scientific matters. One cannot help but have the greatest admiration for his unyielding devotion to science. In the face of a great many lucrative offers, he took extraordinary pains to avoid benefiting personally from industry.

I will always remember Howard as a kind, pure, gentle human being with the utmost integrity and as one of the most dedicated and original of scientists.

As for myself, I will, of course, never really know how much impact Howard had on my thinking and career. For the most part, it was probably subtle; on the other hand, his influence will be with me forever. From Howard you learned that science has its own integrity and that you must work tirelessly, be innovative, and, above all, occasionally cherish an idea that does not attach itself to anything.

REFERENCES

1. **Acuto, O., M. Fabbi, J. Smart, C. B. Poole, S. F. Schlossman, and E. L. Reinherz.** 1984. Purification and NH$_2$-terminal amino acid sequencing of the beta subunit of a human T-cell antigen receptor. *Proc. Natl. Acad. Sci. USA* **81:**3851–3855.
2. **Allison, J. P., R. W. McIntyre, and D. Bloch.** 1982. Tumor specific antigen and murine T lymphoma detected with monoclonal antibody. *J. Immunol.* **129:**2293–2300.
3. **Baltimore, D.** 1970. RNA-dependent DNA polymerase in virions of RNA tumor viruses. *Nature* (London) **226:**1209–1211.
4. **Berg, L. J., B. F. Groth, A. M. Pullen, and M. M. Davis.** 1989. Phenotypic differences between alpha beta versus beta T-cell receptor transgenic mice. *Nature* (London) **340:**559–562.
5. **Berger, S. A., T. W. Mak, and C. Paige.** 1994. Leukocyte common antigen (CD45) is required for immunoglobulin E-mediated degranulation of mast cells. *J. Exp. Med.* **180:**471–476.
6. **Bjorkman, P. J., M. A. Saper, B. Samraoui, W. S. Bennett, J. L. Strominger, and D. C. Wiley.** 1987. Structure of the human class I histocompatibility antigen, HLA-2. *Nature* (London) **329:**506–512.
7. **Brenner, M. B., J. McLean, D. P. Dialynas, J. L. Strominger, J. A. Smith, F. L. Owen, J. G. Seidman, S. Ip, F. Rosen, and M. S. Krangel.** 1986. Identification of a putative second T cell receptor. *Nature* (London) **322:**145–149.
8. **Capecchi, M. R.** 1989. Altering the genome by homologous recombination. *Science* **244:**1288–1292.
9. **Chen, I. S., T. W. Mak, J. J. O'Rear, and H. M. Temin.** 1981. Characterization of reticuloendotheliosis virus strain T DNA and isolation of a novel variant of reticuloendotheliosis virus strain T by molecular cloning. *J. Virol.* **40:**800–811.
10. **Chien, Y., D. Becker, T. Lindsten, M. Okamura, D. Cohen, and M. M. Davis.** 1984. A third type of murine T cell receptor gene. *Nature* (London) **312:**36–40.
11. **Clark, S. P., and T. W. Mak.** 1983. Complete nucleotide sequence of an infectious clone of Friend spleen focus-forming provirus: gp55 is an envelope fusion glycoprotein. *Proc. Natl. Acad. Sci. USA* **80:**5037–5041.
12. **Dembic, Z., W. Haas, S. Weiss, J. McCubrey, J. Keifer, H. von Boehmer, and M. Steinmetz.** 1986. Transfer of specificity by murine α and β T cell receptor genes. *Nature* (London) **320:**232–238.
13. **Fung-Leung, W. P., M. W. Schilham, A. Rahemtulla, T. M. Kundig, M. Vollenweider, and J. Potter.** 1991. CD8 is needed for development of cytotoxic T cells but not helper T cells. *Cell* **65:**443–449.
14. **Gascoigne, N., Y. Chien, D. Becker, J. Kavaler, and M. Davis.** 1984. Genomic organization and sequence of T cell receptor β-chain constant and joining regions. *Nature* (London) **310:**387–391.
15. **Haskins, K., R. Kubo, J. White, M. Pigeon, J. Kappler, and P. Marrack.** 1983. The major histocompatibility complex-restricted antigen receptor on T cells. I. Isolation with a monoclonal antibody. *J. Exp. Med.* **157:**1149–1169.
16. **Hayday, A. C., D. J. Diamond, G. Tanigawa, J. S. Heilig, V. Folson, H. Saito, and S. Tonegawa.** 1983. Unusual organization and diversity of T cell receptor α chain genes. *Nature* (London) **316:**828–832.
17. **Hayday, A. C., H. Eisen, and S. Tonegawa.** 1984. A third rearranged and expressed gene in a clone of cytotoxic T lymphocytes. *Nature* (London) **312:**36–40.
18. **Hedrick, S. M., D. I. Cohen, E. A. Nielsen, and M. M. Davis.** 1984. Isolation of cDNA clones encoding T cell-specific membrane-associated proteins. *Nature* (London) **308:**149–153.

19. **Kaye, J., M. L. Hsu, M. E. Sauron, S. C. Jameson, M. R. Gascoigne, and S. M. Hendrik.** 1989. Selective development of CD4[+] T cells in transgenic mice expressing a class II MHC-restricted antigen receptor. *Nature* (London) **341**:746–749.

20. **Kishihara, K., J. Penninger, V. A. Wallace, T. M. Kundig, K. Kawai, and A. Wakeham.** 1993. Normal B lymphocyte development but impaired T cell maturation in CD45-exon 6 protein tyrosine phosphatase-deficient mice. *Cell* **74**:143–156.

21. **Kisielow, P., H. S. Teh, H. Bluthmann, and H. von Boehmer.** 1988. Positive selection of antigen-specific T cells in thymus by restricting MHC molecules. *Nature* (London) **335**:730–733.

22. **Li, J. P., A. D. D'andrea, H. F. Lodish, and D. Baltimore.** 1990. Activation of cell growth by binding of Friend spleen focus-forming virus gp55 glycoprotein to the erythropoietin receptor. *Nature* (London) **343**:762–764.

23. **Mak, T. W., D. Penrose, C. Gamble, and A. Bernstein.** 1978. The Friend spleen focus-forming virus (SFFV) genome: fractionation and analysis of SFFV and helper virus-related sequences. *Virology* **87**:73–80.

24. **Matsuyama, T., T. Kimura, M. Kitagawa, K. Pfeffer, T. Kawakami, and N. Watanabe.** 1993. Targeted disruption of IRF-1 or IRF-2 results in abnormal type I IFN gene induction and aberrant lymphocyte development. *Cell* **75**:1–20.

25. **Meuer, S. C., K. A. Fitzgerald, R. E. Hussey, J. C. Hodgdon, S. F. Schlossman, and E. L. Reinherz.** 1983. Clonotypic structures involved in antigen-specific human T cell function. Relationship to the T3 molecular complex. *J. Exp. Med.* **157**:705–719.

26. **Molina, T. J., K. Kishihara, D. P. Siderowski, W. van Ewijk, A. Narendran, and E. Timms.** 1992. Profound block in early thymocyte development in mice lacking p56lck. *Nature* (London) **357**:161–164.

27. **Ohashi, P., T. W. Mak, P. van den Elsen, Y. Yanagi, Y. Yoshikai, A. F. Calman, C. Tehorst, J. D. Stobo, and A. Weiss.** 1985. Reconstitution of an active surface T3/T cell antigen receptor by DNA transfer. *Nature* (London) **316**:606–609.

28. **Pfeffer, K., and T. W. Mak.** 1994. Lymphocyte ontogeny and activation in gene targeted mutant mice. *Annu. Rev. Immunol.* **12**:367–411.

29. **Pfeffer, K., T. Matsuyama, T. M. Kundig, A. Wakeham, K. Kishihara, and A. Shahinian.** 1993. Mice deficient for the 55kD tumor necrosis factor receptor are resistant to endotoxic shock yet succumb to *Listeria monocytogenes* infection. *Cell* **73**:457–467.

30. **Rahematulla, A., W. P. Fung-Leung, M. W. Schilman, T. M. Kundig, S. R. Sambhara, and A. Narendran.** 1991. Normal development and functional CD8[+] cells but markedly decreased helper cell activity in mice lacking CD4. *Nature* (London) **353**:180–184.

31. **Saito, H., D. M. Kranz, Y. Takagaki, A. Hayday, H. Eisen, and S. Tonegawa.** 1984. Complete primary structure of heterodimeric T cell receptor deduced from cDNA sequences. *Nature* (London) **309**:757–762.

32. **Saito, T., A. Weiss, J. Miller, M. A. Norcross, and R. N. Germain.** 1987. Specific antigen-Ia activation of transfected human T cells expressing murine Ti alpha beta-human T3 receptor complexes. *Nature* (London) **325**:125–130.

33. **Schwartz, R. H.** 1992. Costimulation of T lymphocytes: the role of CD28, CTLA-4, and B7/BB1 in interleukin-2 production and immunotherapy. *Cell* **71**:1065–1068.

34. **Sha, W. C., C. A. Nelson, R. D. Newberry, D. M. Krantz, L. H. Russell, and D. Y. Loh.** 1988. Positive and negative selection of an antigen receptor on T cells in transgenic mice. *Nature* (London) **336**:73–76.

35. **Shahinian, A., K. Pfeffer, K. P. Lee, T. M. Kundig, K. Kishihara, A. Wakeham, K. Kawai, P. S. Ohashi, C. B. Thompson, and T. W. Mak.** 1993. Differential T cell costimulatory requirements in CD28-deficient mice. *Science* **261**:609–612.

36. **Shibuya, T., and T. W. Mak.** 1982. Host control of susceptibility to erythroleukemia and to the types of leukemia induced by Friend murine leukemia virus: initial and late stages. *Cell* **31**:483–493.

37. **Shibuya, T., and T. W. Mak.** 1982. A host gene controlling early anaemia or polycythaemia induced by Friend erythroleukaemia virus. *Nature* (London) **296**:577–579.

38. **Sim, G. K., J. Yague, J. Nelson, P. Marrack, E. Palmer, A. Augustin, and J. Kappler.** 1984. Primary structure of human T-cell receptor alpha-chain. *Nature* (London) **312**:771–775.

39. **Takihara, Y., D. Tkachuk, E. Michalopoulos, E. Champagne, J. Reimann, M. Minden, and T. W.**

Mak. 1988. Sequence and organization of the diversity, joining and constant region genes of the human T cell δ chain locus. *Proc. Natl. Acad. Sci. USA* **85:**6097–6101.

40. **Tanaka, N., M. Ishihara, M. Kitagawa, H. Harada, T. Kimura, T. Matsuyama, M. S. Lamphier, S. Aizawa, T. W. Mak, and T. Taniguchi.** 1994. Cellular commitment to oncogene-induced transformation or apoptosis is dependent on the transcription factor IRF-1. *Cell* **77:**829–839.

41. **Teh, H. S., P. Kisielow, B. Scott, H. Kishi, Y. Uematsu, H. Bluthmann, and H. von Boehmer.** 1988. Thymic major histocompatibility complex antigens and the alpha beta T-cell receptor determine the CD4/CD8 phenotype of T cells. *Nature* (London) **335:**229–233.

42. **Temin, H. M., and S. Mizutani.** 1970. RNA dependent DNA polymerase in virions of Rous sarcoma virus. *Nature* (London) **226:**1211–1213.

43. **Toyonaga, B., Y. Yoshikai, V. Vadasz, B. Chin, and T. W. Mak.** 1985. Organization and sequences of the diversity, joining and constant region of genes of the human T cell receptor β chain. *Proc. Natl. Acad. Sci. USA* **82:**8624–8628.

44. **Trowbridge, I. S., and M. L. Thomas.** 1994. CD45: an emerging role as a protein tyrosine phosphatase required for lymphocyte activation and development. *Annu. Rev. Immunol.* **12:**85–116.

45. **von Boehmer, H.** 1990. Developmental biology of T cells in T cell receptor transgenic mice. *Annu. Rev. Immunol.* **8:**531–556.

46. **Winoto, A., S. Mjolsness, and L. Hood.** 1985. Genomic organization of genes encoding mouse T cell receptor α chain. *Nature* (London) **316:**832–836.

47. **Yamamoto, Y., C. L. Gamble, S. P. Clark, A. Joyner, T. Shibuya, M. E. MacDonald, D. Mager, A. Bernstein, and T. W. Mak.** 1981. Clonal analysis of early and late stages of erythroleukemia induced by molecular clones of integrated spleen focus-forming virus. *Proc. Natl. Acad. Sci. USA* **78:**6893–6897.

48. **Yanagi, Y., A. Chan, B. Chin, M. Minden, and T. W. Mak.** 1985. Analysis of cDNA clones specific for human T cells and the alpha and beta chains of the T-cell receptor heterodimer from a human T-cell line. *Proc. Natl. Acad. Sci. USA* **82:**3430–3434.

49. **Yanagi, Y., Y. Yoshikai, K. Leggett, S. P. Clark, I. Aleksander, and T. W. Mak.** 1984. A human T cell-specific cDNA clone encodes a protein having extensive homology to immunoglobulin chains. *Nature* (London) **308:**145–149.

50. **Yoshikai, Y., S. P. Clark, S. Taylor, U. Sohn, M. D. Minden, and T. W. Mak.** 1985. Organization and sequences of the variable, joining and constant region genes of the human T cell receptor α chain. *Nature* (London) **316:**837–840.

The DNA Provirus: Howard Temin's Scientific Legacy
Edited by G. M. Cooper, R. Greenberg Temin, and B. Sugden
© 1995 American Society for Microbiology, Washington, DC 20005

Chapter 19

Gene Therapy in CD4$^+$ T Lymphocytes in SCID-hu Mice

Elizabeth S. Withers-Ward and Irvin S. Y. Chen

AIDS is a candidate for treatment by gene transfer therapy. Since human immunodeficiency virus type 1 (HIV-1) infects multiple hematopoietic cell lineages, pluri- or multipotent stem cells are potential targets for introducing therapeutic genes (28). Such an introduction would result in subsequent expression of the therapeutic gene in functional hematopoietic cells of all lineages following proliferation and differentiation of the transduced stem cells. In this way, all of the hematolymphoid organs would be targeted. Various gene therapy strategies involving the transfer of genes that would inhibit viral replication have been proposed in the treatment of AIDS. Two gene therapeutic approaches currently being considered for the treatment of AIDS are (i) a protein-based strategy that uses a Rev M10 transdominant mutant (3, 22) and (ii) an RNA-based approach that uses a ribozyme that has catalytic activity against the 5′ leader sequence of HIV-1 (33, 35, 39, 40). The data on their mechanisms of action as well as on their efficacy in inhibition of HIV-1 replication in vitro are substantial.

Development of gene therapy strategies for the treatment of human T-lymphocyte disorders, including AIDS, requires an in vivo system in which transduced human hematopoietic stem cells can be used to reconstitute the T-lymphoid compartment. The severe combined immunodeficient (SCID) mouse reconstituted with human tissue (SCID-hu) has been developed as an in vivo system for modeling gene therapy strategies in the T-lymphoid compartment and testing for protection against pathology following infection with HIV-1 (2, 4, 23, 31, 36). The combination of HIV-1 pathology and CD34$^+$ progenitor cell transduction taking place within a relatively short time in the SCID-hu mouse provides an experimental system in which to address many of the issues that are critical to successful gene therapy approaches for AIDS but are not easily tested in clinical trials. Here, we review recent progress in the use of the SCID-hu system to further our understand-

Elizabeth S. Withers-Ward • Department of Microbiology and Immunology, UCLA School of Medicine and Jonsson Comprehensive Cancer Center, Los Angeles, California 90024. *Irvin S. Y. Chen* • Department of Microbiology and Immunology and Division of Hematology-Oncology, Department of Medicine, UCLA School of Medicine and Jonsson Comprehensive Cancer Center, Los Angeles, California 90024.

ing of vector transduction and expression in modeling potential therapeutic approaches in stem cells.

HIV-1 INFECTION AND PATHOGENESIS IN SCID-hu MICE

The SCID-hu mouse is generated by transplantation of human fetal liver and thymus fragments under the kidney capsule of the SCID mouse, resulting in growth of a conjoint thymus-liver organ (Thy-Liv) that appears to behave like a normal human thymus (23). This model system mimics many of the hematopoietic differentiation processes of human T cells. The SCID-hu mouse system also allows us to monitor acute HIV-1 infection of a human lymphoid organ (2, 4, 13, 16, 31, 36). This is the first animal model to display pathology following HIV-1 infection, and it is thus a logical system in which to model novel therapeutic approaches such as gene therapy in vivo. Direct injection of HIV-1 into the human Thy-Liv implants in SCID-hu mice results in a reproducible infection and severe depletion of human cells bearing the CD4 molecule resembling that seen in humans (Fig. 1) (2). In addition, histological analysis of HIV-infected implants shows a disruption of the thymic epithelial cell, which supports the framework of the thymus. Virus-induced thymocyte death in the SCID-hu system requires relatively high virus loads in the infected implants, and the kinetics of cell depletion varies with the virus strain used. Virus load increases for the first 3 weeks and then decreases. The decrease in virus load coincides with the decrease in $CD4^+$ cells in infected implants and is probably due to the loss of infected target cells. Thus, decreasing virus replication and, consequently, virus load by therapeutic approaches such as gene therapy may avert thymocyte depletion in this system.

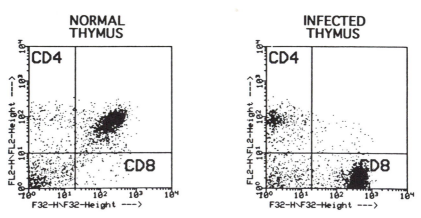

FIGURE 1. Two-color flow cytometric analysis of Thy-Liv implants. At 6 weeks postinfection, implants were biopsied and stained with monoclonal antibodies to human CD8 and CD4 directly conjugated with fluorescein isothiocyanate or phycoerythrin. (Left) Cells from a mock-infected implant; (right) cells from an HIV-1-infected implant.

The following observations in the SCID-hu model lend further support to the idea that the mechanisms of pathogenesis are likely to be related to those seen in an adult: (i) only primary isolates of HIV-1 will replicate in SCID-hu mice, whereas laboratory strains such as HTLV-IIIB will not; (ii) mutation of the *nef* gene of HIV-1$_{JR-CSF}$ and HIV-1$_{NL4-3}$ results in a decrease in viral load by 2 to 3 orders of magnitude (15), similar to what is observed in macaques infected with simian immunodeficiency virus *nef* mutants but is not seen in tissue culture (9, 18); and (iii) a similar correlation of HIV-1 variants with disease course is seen. Syncytium-inducing strains of HIV-1 result in pathology in the SCID-hu mouse, whereas non-syncytium-inducing strains show either delayed or no pathology (5a, 17).

Most anti-AIDS gene therapy strategies are based on the assumption that direct cell killing is the primary pathogenic mechanism for HIV-1 CD4$^+$ cell destruction. If this were indeed the case, gene therapy approaches for AIDS would also have the additional advantage that cells protected from HIV-1 infection might have a growth advantage in vivo. However, CD4$^+$ cell destruction may involve indirect effects upon uninfected cells (11). In this case, gene therapy strategies could deliver genes that would target the sequence of events leading to immunodeficiency.

USE OF THE SCID-hu SYSTEM AS A MODEL FOR IMMUNE RECONSTITUTION AND GENE THERAPY

Studies in our and other laboratories have established that the SCID-hu system is amenable to reconstitution with exogenous CD34$^+$ cells (1, 34). Human CD34$^+$ progenitor cells transduced with a retroviral vector containing the neomycin resistance gene (*neo*) were able to engraft and reconstitute irradiated Thy-Liv implants in the SCID-hu mouse (1). The retroviral vector used in this study, LNL6, contained *neo* as a selectable marker located between the murine sarcoma virus and the murine leukemia virus (MLV) long terminal repeats (LTRs) (27). CD34$^+$ cells were transduced in vitro by cocultivation with the amphotropic PA317 clone 8 producer line (25). The efficiency of retroviral transduction was determined by quantitative PCR analysis to be approximately 10% (Fig. 2) (42, 43). To ensure that the thymocytes were of donor origin, female thymic graft recipients and male donor stem cells were used to normalize total donor DNA for the Y-chromosome-specific sequences (30).

For optimal engraftment of the in vitro-retrovirus-transduced CD34$^+$ cells, it was necessary to reduce the endogenous stem cells present in the Thy-Liv implant by using total-body irradiation (200 rads) (37). For engraftment and reconstitution, the Thy-Liv implants of irradiated SCID mice were directly injected with 5×10^5 LNL6-transduced CD34$^+$ cells within 24 h of irradiation. To determine whether the transduced CD34$^+$ cells successfully engrafted in these mice, sequential biopsies were taken at 4 and 6 weeks posttransplantation and analyzed by flow cytometry. Irradiated animals that received exogenous CD34$^+$ cells had normal distributions of CD4$^+$, CD8$^+$, and CD4$^+$/CD8$^+$ cells, whereas in the two-

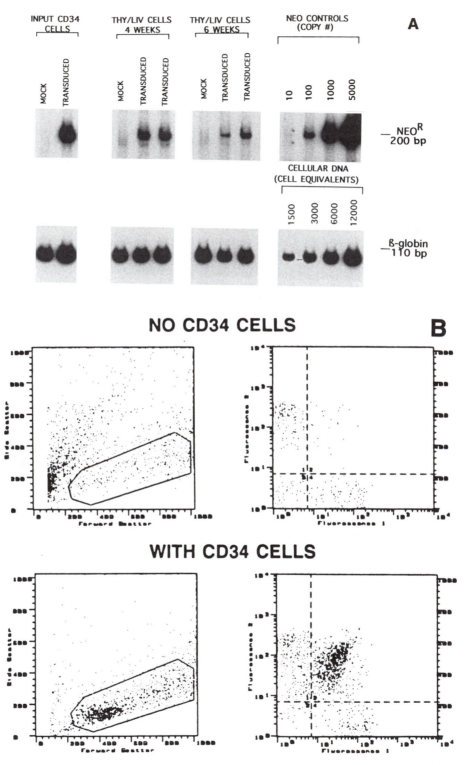

FIGURE 2. *Continued on next page*

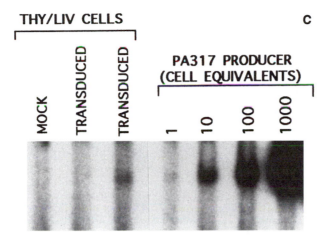

FIGURE 2. Reconstitution of SCID-hu Thy-Liv implants with LNL6-transduced CD34$^+$ cells. (A) Quantitative PCR analysis of transduced *neo* sequences in input CD34$^+$ cells and in reconstituted irradiated implants. CD34$^+$ cells were transduced in vitro by cocultivation with the amphotropic PA317 clone 8 producer line (25). For engraftment and reconstitution, the Thy-Liv implants of irradiated (200 rads) SCID mice were directly injected with 5×10^5 LNL6-transduced CD34$^+$ cells within 24 h of irradiation. Sequential biopsies of injected implants were taken at 4 and 6 weeks posttransplantation and analyzed by quantitative PCR and flow cytometry (see panel B). Analysis of the input CD34$^+$ cells and two representative reconstituted implants are shown. In each case, PCR analysis for *neo* sequences (upper panels) was performed in parallel with PCR analysis for β-globin sequences (lower panels). The copy number was determined by comparison with standard reactions containing known amounts of cell or plasmid DNA, which are shown to the right of panel A. Mock control implants were irradiated but did not receive transduced CD34$^+$ cells. (B) Representive flow cytometric analysis of control and reconstituted irradiated implants. Panels on the left show forward versus side scatter plots and gating of the live thymocyte population, and panels on the right show the flow cytometric analysis for human CD4 (horizontal axis) and human CD8 (vertical axis) antigens. (C) RT-PCR analysis of control and LNL6-transduced implants. A longer exposure of the gel shown in this panel revealed *neo* mRNA expression in both of the reconstituted Thy-Liv implants (shown in panel A) at 6 weeks postreconstitution. RT-PCR analysis of serial dilutions of mRNA isolated from the PA317 producer line is shown at the right. This figure was reprinted from Akkina et al. (1) with permission from the publisher.

thirds of the animals that did not receive CD34$^+$ cells, implants were depleted of thymocytes (Fig. 2). Using quantitative PCR, it was determined that up to 4% of the thymocyte population contained *neo* gene sequences (Fig. 2). Also, at least 5×10^4 CD34$^+$ cells were required to reconstitute an animal (Fig. 3). *neo* gene sequences were not detected by PCR in animals that received 10-fold-fewer cells. About 0.1% of the CD34$^+$ cell population is believed to constitute true stem cells. On the basis of this estimate, it was concluded that about 50 stem cells are required for reconstitution of the irradiated Thy-Liv implant.

    neo mRNA expression was detected in reconstituted Thy-Liv implants at 6 weeks postreconstitution by reverse transcription PCR (RT-PCR) (Fig. 2). However, on a per-infected-cell basis, *neo* gene expression was about 200-fold less than that in the PA317 producer line. Thus, it appears that transduction of the *neo* gene into CD34$^+$ stem cells results in expression of the transduced *neo* gene

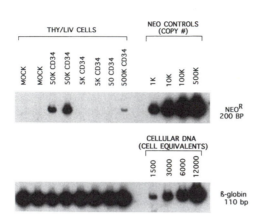

FIGURE 3. Determination of the number of LNL6-transduced stem cells required for reconstitution of irradiated Thy-Liv implants. Irradiated implants were injected with various amounts of LNL6-transduced CD34⁺ cells as indicated. Reconstituted implants were biopsied at 6 weeks posttransplantation and analyzed for the presence of *neo* sequences by quantitative PCR. In each case, PCR analysis for *neo* sequences (upper panel) was performed in parallel with PCR analysis for β-globin sequences (lower panels). The copy number was determined by comparison with standard reactions containing known amounts of cell or plasmid DNA, which are shown to the right. Mock control implants were irradiated but did not receive transduced CD34⁺ cells. Each lane represents a separate implant. This figure was reprinted from Akkina et al. (1) with permission from the publisher.

upon differentiation of thymocytes. It is not clear whether all transduced cells express low amounts of vector mRNA or whether only a subset of cells expresses the *neo* gene.

Cells obtained from mice at 4 weeks postreconstitution were sorted into CD4⁺/CD8⁻, CD4⁻/CD8⁺, and CD4⁺/CD8⁺ subsets to determine whether the transduced gene was present in the various thymocyte populations as they matured into end-stage cells. Approximately 2% of the cells in each of the three populations contained the *neo* gene, as determined by quantitative PCR analysis (Fig. 4). Thus, the percentage of cells harboring the *neo* signal was similar on a per-cell basis,

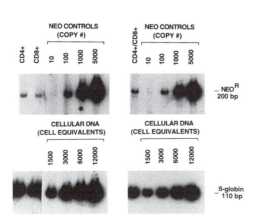

FIGURE 4. Detection of the transduced *neo* gene in the various thymocyte populations of reconstituted implants as they mature into end-stage cells. Cells were obtained from an implant reconstituted with 50,000 LNL6-transduced CD34⁺ cells at 4 weeks postreconstitution (see Fig. 3) and were sorted into CD4⁺/CD8⁻, CD4⁻/CD8⁺, and CD4⁺/CD8⁺ subsets after staining with fluoresceinated monoclonal antibodies specific for the human CD4 and CD8 surface markers. Each population was analyzed by quantitative PCR to determine the percentage of cells in each subgroup that contained the *neo* gene. Upper panels show PCR analysis for *neo* sequences, which was performed in parallel with PCR analysis for β-globin sequences (lower panels). The copy number was determined by comparison with standard reactions, which are shown to the right in each case. This figure was reprinted from Akkina et al. (1) with permission from the publisher.

suggesting that the vector sequences were not deleterious to stem cells undergoing lineage-specific expansion and differentiation.

OPTIMIZATION OF TRANSDUCED GENE EXPRESSION
IN HUMAN T CELLS

One of the key issues in human gene therapy has been the development of retroviral vectors with optimal gene expression in differentiated hematopoietic cells (7, 29). The study described above suggests that retroviral vectors that rely on the murine sarcoma virus LTR for transcription of inserted sequences direct gene expression in human CD4$^+$ cells and may lead to sustained gene expression in transduced cells. However, the level of transduced gene expression is rather low relative to that in the PA317 producer line. Similar levels of expression have been achieved in human hematopoietic cells transduced with a retroviral vector that relies on the MLV LTR for transcription in vivo (7, 8). Clearly, a better understanding of the factors that control expression of genes introduced into hematopoietic cells is required to provide a solid rationale for extending these types of studies to clinical trials. The SCID-hu mouse is an ideal model system for evaluating therapeutic vectors because it allows examination of gene expression in undifferentiated precursors to mature T lymphocytes.

Gene therapy strategies aimed at inhibiting HIV infection or replication require vectors that are expressed in cells infected by HIV, the human CD4$^+$ T lymphocytes. To address the issue of improving gene expression in T lymphocytes, vectors that would be expressed specifically in T cells could be designed. They would be MLV-based SIN vectors containing self-inactivating LTRs (41). After reverse transcription and integration, the 5' LTR is no longer functional. Elimination of viral transcriptional control elements from the LTRs creates vectors that, when transduced, should be innocuous in target cells. The viral enhancer and promoter elements could then be replaced with cis-acting elements that regulate tissue-specific expression in T cells (e.g., enhancers from the alpha chain of the interleukin-2 receptor [20] and the Tα3 element of the T-cell receptor gene [14]), resulting in tissue-specific expression of the therapeutic gene. Insertion of the HIV-1 enhancer, including the TAR element, would result in vectors with HIV-inducible expression. This method of inserting heterologous enhancers, resulting in chimeric MLV LTRs, has been successful in altering MLV transcriptional properties in tissue culture and in mice (12, 21). Use of the promoter/enhancer elements within the LTR to drive expression should also result in more efficient expression by eliminating transcriptional interference between competitive promoters (10).

Vectors would then be tested in SCID-hu mice, starting from initial infection of precursor cells through expression in differentiated mature T cells. These studies would allow identification of control elements that lead to efficient expression in the cell types infected by HIV. It would also be possible to assess whether constitutively expressed foreign genes introduced with retroviral vectors have deleterious effects on stem cell reconstitution, survival, multipotency, or differentiation.

EFFICIENCY OF INFECTION OF CD34⁺ CELLS

Since the titer of the virus used in the study described above was approximately 5×10^6 CFU/ml, the inability to infect a greater percentage of cells reflects a restriction at the level of infection rather than a limitation in infectious virus. The restriction is likely to be at or before the stage of reverse transcription because of either failure to fully reverse transcribe or inefficient entry of the virus into cells. The restriction by failure to fully reverse transcribe occurs because the cell needs to be actively proliferating for retroviral infection to occur (6, 42). The restriction by inefficient entry into cells is possibly due to low levels of the amphotropic receptor on the surface and/or relatively poor infectivity of human cells compared with murine cells, events that are not mutually exclusive or well defined.

Optimization of the efficiency of viral entry into CD34⁺ cells can be approached in three ways: (i) producing pseudotyped virions with an extended host range, (ii) increasing the packaging efficiency of virions in producer cells, and (iii) increasing virus titers. The most promising strategies include replacing the murine amphotropic envelope with the gibbon ape leukemia virus envelope (26, 38) and/ or the G glycoprotein of the vesicular stomatitis virus (VSV-G) (5). The gibbon ape leukemia virus envelope offers the advantages of an increased packaging function that is up to 10-fold greater than that in the PA317 cells and a broader host range. The VSV-G protein has an extremely broad host range, and VSV-G virions can be concentrated by ultracentrifugation to titers of $>10^9$ CFU/ml.

One of the challenges in optimizing the efficiency of infection of cells is to utilize conditions in which cells undergo division for self-renewal but do not commit to a differentiated lineage (7, 32). This is achieved by using conditions that allow maximal infection of cells within the shortest time to minimize cell division and associated loss of self-renewing capacity. This hypothesis can be tested directly with the SCID-hu mouse by infecting cells for various periods and testing the efficiency of reconstitution with various amounts of CD34⁺ cells. Also, further characterization of the pluripotent stem cell and the factors and conditions required for stem cell proliferation and survival (37) are critical for the optimization, maintenance, and selection of pluripotent precursor cells and subsequent infection by vectors.

MODELING STEM CELL GENE THERAPY STRATEGIES DESIGNED TO TARGET HIV-1-INFECTED CD4⁺ T LYMPHOCYTES

Important issues to be considered in the development of antiviral approaches to gene therapy include (i) the dosage of the antiviral reagent relative to the levels of wild-type target that are required to block HIV-1 replication in an in vivo setting, (ii) the possible toxicity to hematopoietic differentiation and/or T-cell function, and (iii) the possibility of immune recognition of the transdominant mutant as a foreign protein. The SCID-hu mouse model allows us to address the first two issues. It is not possible to address the third issue with this model, since the extent of immune system function has not been well characterized in the SCID-hu mouse and is thought to be minimal. Although SCID mice are severely immune deficient for T and B cells, they have intact natural killer cells and nonlymphoid resistance systems such as macrophages (19). Thus, any exposure to antigen can stimulate these resistance mechanisms. No immune response (either murine or human) to viral antigens has been described in this system. The absence of an immune response in SCID-hu mice has two advantages. First, viral pathogenesis can be studied in the absence of an immune response, thereby shortening the time frame for onset of disease. Second, while it is likely that introduction of a foreign protein will elicit a host immune response in patients, the effects of therapeutic reagents can be assessed independently of the immune response. This issue is not a problem with ribozymes, which are nonimmunogenic.

If they prove to be successful, these approaches will have broad general applications in targeting specific viral proteins or RNA sequences encoding viral genes, for which the mechanisms of action and interaction with other host or viral proteins are understood.

POTENTIAL LIMITATIONS OF THE SCID-hu MODEL FOR GENE THERAPY

As with many model systems, the SCID-hu mouse model is not a perfect representation of HIV-1 infection or hematopoietic differentiation in humans and may not accurately mimic HIV-1 pathogenesis in humans. The SCID-hu model specifically addresses infection of the human thymus and not the entire immune system. Although AIDS is primarily a T-cell disease, other types of cells may be infected as well, including monocyte/macrophages and dendritic, brain, microglial, and endothelial cells. It is important to recognize that this is principally a model for T-cell differentiation and that other aspects relevant to HIV-1 gene therapy, such as infection of other cell types (monocyte/macrophages, dendritic cells, etc.), may not be accurately mimicked. Also, if aberrant T-cell function has deleterious effects on cells other than T cells, these effects may not be manifested in the SCID-hu system. This model is also limited in that the effect of host-derived immune responses on the gene therapeutic approach may not be seen. However, it is advantageous in that detailed studies of thymopoiesis and the kinetics of HIV-induced T-cell depletion have already been performed, so the effects of stem cell gene therapy on T-cell differentiation and on viral replication and pathogenesis in vivo can be assessed.

SUMMARY

Since traditional therapies have not been effective in the treatment of HIV-1-infection individuals, it is imperative that novel approaches be considered in the treatment of AIDS. Effective treatment of AIDS will probably require both antiviral and immune restoration approaches. Gene therapy, if successful, would accomplish both. HIV-1-induced pathogenesis in SCID-hu mice, which closely mimics many aspects of disease in humans, is certainly more representative of what happens in humans than tissue culture systems are and provides an ideal model for the optimization of retroviral transduction and stable, constitutive gene expression in human T cells. This model will provide an important preclinical assessment of stem cell gene therapies on HIV-1 infection, stem cell function, and T-cell development. The insight gained from these studies will have broad general applications in human stem cell gene therapy.

RECOLLECTIONS OF HOWARD TEMIN AS AN ADVISOR AND A FRIEND

The very first correspondence I (I.S.Y.C.) ever received from Howard Temin was after my acceptance to the graduate program of the Department of Oncology and McArdle Laboratory for Cancer Research at the University of Wisconsin in 1977. After my meeting with Dr. Temin (as he was known by all of the students), he indicated that he would accept me in his laboratory, and I inquired whether there were any materials I could read prior to joining the laboratory. He wrote to me on 5 August 1977:

> Thank you for your letter. I'm looking forward to a mutually profitable association. It is unnecessary for you to do any reading before you come. However, if you like, you can look at the Cold Spring Harbor book on the Molecular Biology of Tumor Viruses or read in the Cold Spring Harbor Symposium Volume 39, 1974.

This letter was important to me in a number of ways: first, because this marked the beginning of my formal studies in molecular biology and retrovirology, and second, because the tone of this letter was so much different from my years of training under him as a graduate student. Indeed, I cannot recall a single other instance in which it was "unnecessary" for me to do any additional work. Graduate school under the tutelage of Howard Temin was a constant series of challenges and expectations that none of the students or postdoctoral fellows could ever fulfill completely.

During my time as a student, his approach to teaching was to ask probing questions and expect precise, articulate answers. He challenged us to think critically and imaginatively; cajoling and holding hands were certainly not his style. I can remember many occasions during the legendary Friday conferences when each member of the laboratory would show Dr. Temin results obtained during the week. Rarely at the end of the discussion would he say, "You've had a good week" or "Good job"; more often, it would be, "Is that all?" We all understood that this was not meant to be condescending or insulting but rather was his way of challenging us, no matter what our accomplishments were, to achieve more. He showed his appreciation of our work in other ways. After the success of important experiments, he would become animated, pose more questions, and suggest more experiments. At other times, he would reflect, as if we were not present, on the implications or consequences of our findings. It was often a struggle to keep up with his flow of thoughts. This was his way of teaching, and this is one of the important lessons I carried forward into all of my endeavors, both as a student and thereafter.

Clearly, the years of tutelage under Dr. Temin were the most important few years of my research career. I learned not only about how to perform scientific research in general but also about the many fascinating aspects of retroviruses, which have become the basis of my academic studies. Since leaving Dr. Temin's laboratory, I have worked with human retroviruses, first human T-cell leukemia viruses and then HIV, the causative agents of various T-cell leukemias/lymphomas and AIDS, respectively. It is likely that I will continue to work with retroviruses in some manner for the rest of my academic life. This direction is because Dr. Temin himself was so enthralled with and instilled in me an appreciation for the evolutionary simplicity yet potentially deadly nature of these viruses. They have provided biological insights ranging from an understanding of their unique replicative properties, including reverse transcription and integration, to the knowledge that they have contributed to the evolution of genetic materials that are important in such diverse areas as genetic variation of HIV-1, resistance to drugs, and incorporation into the viral genomes of oncogenes, now known to be fundamental to the evolution of cancer.

My relationship with Dr. Temin changed completely after I left his laboratory. He changed from a demanding advisor to a kinder and gentler man who was anxious to know how other members of the laboratory and I were doing. Our conversations often included discussions of personal matters, including family and friends. At the same time, his role as an advisor was a constant, and even several years later, when I visited McArdle Laboratory, he would comment on my oral presentations and critique the work. As before, he sometimes became animated with enthusiasm at our accomplishments, but now he also smiled broadly as a proud teacher in appreciation of his students.

It is evident that he always viewed himself first and foremost as a teacher, and I believe that he felt that different stages of being a student required different means and approaches to teaching. Whether he did this consciously or subconsciously I do not know, but it was nevertheless a highly successful technique in all stages of my learning. In retrospect, the rather uncharacteristic tone of the first letter I ever received from Howard Temin may not have been so surprising; after all, I was not yet his student. A few months following that letter, I entered his laboratory as a graduate student and have been his student, student-friend, and student-colleague ever since.

REFERENCES

1. Akkina, R. K., J. D. Rosenblatt, A. G. Campbell, I. S. Y. Chen, and J. A. Zack. 1994. Modeling human stem cell gene therapy in the SCID-hu mouse. *Blood* **84:**1393–1398.
2. Aldrovandi, G. M., G. Feuer, L. Gao, M. Kristeva, I. S. Y. Chen, B. Jamieson, and J. A. Zack. 1993. HIV-1 infection of the SCID-hu mouse: an animal model for virus pathogenesis. *Nature* (London) **363:**732–736.
3. Bahner, I., C. Zhou, X.-J. Yu, Q.-L. Hao, J. C. Guatelli, and D. B. Kohn. 1993. Comparison of *trans*-dominant inhibitory human immunodeficiency virus type 1 genes expressed by retroviral vectors in human T lymphocytes. *J. Virol.* **67:**3199–3207.

4. **Bonyhadi, M. L., L. Rabin, S., Salimi, D. A. Brown, J. Kosek, J. M. McCune, and H. Kaneshima.** 1993. HIV induces thymus depletion *in vivo. Nature* (London) **363**:728–736.

5. **Burns, J. C., T. Friedmann, W. Driever, M. Burrascano, and J.-K. Yee.** 1993. Vesicular stomatitis virus G glycoprotein pseudotyped retroviral vectors: concentration to very high titer and efficient gene transfer into mammalian and nonmammalian cells. *Proc. Natl. Acad. Sci. USA* **90**:8033–8087.

5a. **Camerini, D., and I. Chen.** Unpublished data.

6. **Chen, I. S. Y., and H. M. Temin.** 1982. Establishment of infection by spleen necrosis virus: inhibition in stationary cells and the role of secondary infection. *J. Virol.* **41**:183–191.

7. **Cournoyer, D., and C. T. Caskey.** 1993. Gene therapy of the immune system. *Annu. Rev. Immunol.* **11**:297–329.

8. **Cournoyer, D., M. Scarpa, K. Mitani, K. A. Moore, D. Markovitz, A. Bank, J. W. Belmont, and C. T. Caskey.** 1993. Gene transfer of adenosine deaminase into primitive human hematopoietic progenitor cells. *Hum. Gene Ther.* **2**:203–213.

9. **Daniel, M. D., F. Kirchoff, S. C. Czajak, P. K. Sehgal, and R. C. Desrosiers.** 1992. Protective effects of a live attenuated SIV vaccine with a deletion in nef gene. *Science* **258**:1938–1941.

10. **Emerman, M., and H. M. Temin.** 1984. Genes with promoters in retrovirus vectors can be independently suppressed by an epigenetic mechanism. *Cell* **39**:459–467.

11. **Fauci, A. S.** 1993. Multifactorial nature of human immunodeficiency virus disease: implications for therapy. *Science* **262**:1011–1018.

12. **Feuer, G., and H. Fan.** 1990. Substitution of murine transthyretin (Prealbumin) regulatory sequences into Moloney murine leukemia virus long terminal repeat yields infectious virus with altered biological properties. *J. Virol.* **64**:6130–6140.

13. **Grody, W., S. Fligiel, and F. Naeim.** 1985. Thymus involution in the acquired immunodeficiency syndrome. *Am. J. Clin. Pathol.* **84**:85–95.

14. **Ho, I.-C., P. Vorhees, N. Marin, B. K. Oakley, S. F. Tsai, S. H. Orkin, and J. M. Leiden.** 1991. Human GATA-3: a lineage-restricted transcription factor that regulates the expression of the T cell receptor α gene. *EMBO J.* **10**:1187–1192.

15. **Jamieson, B. D., G. M. Aldrovandi, V. Planelles, J. B. M. Jowett, L. Gao, L. M. Bloch, I. S. Y. Chen, and J. A. Zack.** 1994. Requirement of human immunodeficiency virus type 1 *nef* for in vivo replication and pathogenicity. *J. Virol.* **68**:3478–3485.

16. **Joshi, V. V., J. M. Oleske, S. Saad, C. Gadol, E. Conner, R. Bobila, and A. B. Minnefor.** 1986. Thymus biopsy in children with acquired immunodeficiency syndrome. *Arch. Pathol. Lab. Med.* **110**:837–842.

17. **Kaneshima, H., L. Su, M. Bonyhadi, R. I. Connor, D. D. Ho, and J. M. McCune.** 1994. Rapid-high, syncytium-inducing isolates of human immunodeficiency virus type 1 induce cytopathicity in the human thymus of the SCID-hu mouse. *J. Virol.* **68**:8188–8192.

18. **Kestler, H. W., III, D. J. Ringler, K. Mori, D. L. Panicali, P. K. Sehgal, M. D. Daniel, and R. C. Desrosiers.** 1991. Importance of the *nef* gene for maintenance of high virus loads and for development of AIDS. *Cell* **65**:651–662.

19. **Krowka, J. F., S. Sarin, R. Namikawa, J. M. McCune, and H. Kaneshima.** 1991. Human T cells in the SCID-hu mouse are phenotypically normal and functually competent. *J. Immunol.* **146**:3751–3756.

20. **Kuang, A. A., K. D. Novak, S.-M. Kang, K. Bruhn, and M. J. Lenardo.** 1993. Interaction between NF-κB- and serum response factor-binding elements activates an interleukin-2 receptor a-chain enhancer specifically in T lymphocytes. *Mol. Cell. Biol.* **13**:2536–2545.

21. **Li, Y., E. Golemis, J. W. Hartley, and N. Hopkins.** 1987. Disease specificity of nondefective Friend and Moloney murine leukemia viruses is controlled by a small number of nucleotides. *J. Virol.* **61**:693–700.

22. **Malim, M. H., S. Bohnlein, J. Hauber, and B. R. Cullen.** 1989. Functional dissection of the HIV-1 Rev transactivator—derivation of a *trans*-dominant repressor of Rev function. *Cell* **58**:205–214.

23. **McCune, J. M., R. Namikawa, H. Kaneshima, L. D. Shultz, M. Lieberman, and I. L. Weissman.** 1988. The SCID-hu mouse: murine model for the analysis of human hematolymphoid differentiation and function. *Science* **241**:1632–1639.

24. **Miller, A. D.** 1992. Human gene therapy comes of age. *Nature* (London) **257**:455–460.

25. **Miller, A. D., and C. Buttimore.** 1986. Redesign of retrovirus packaging cell lines to avoid recombination leading to helper virus production. *Mol. Cell. Biol.* **6:**2895–2902.

26. **Miller, A. D., J. V. Garcia, N. von Suhr, C. M. Lynch, C. Wilson, and M. V. Eiden.** 1991. Construction and properties of retrovirus packaging cells based on gibbon ape leukemia virus. *J. Virol.* **65:**2220–2224.

27. **Miller, A. D., and G. J. Rosman.** 1989. Improved retroviral vectors for gene transfer and expression. *BioTechniques* **7:**980–987.

28. **Morgan, R. A., and W. F. Anderson.** 1993. Human gene therapy. *Annu. Rev. Biochem.* **62:**191–217.

29. **Mulligan, R. C.** 1993. The basic science of gene therapy. *Science* **260:**926–932.

30. **Nakahori, Y., K. Hamano, M. Iwaga, and Y. Nakagome.** 1991. Sex identification polymerase chain reaction using X-Y homologous primer. *Am. J. Med. Gen.* **39:**472–473.

31. **Namikawa, R., H. Kaneshima, M. Lieberman, I. L. Weissman, and J. M. McCune.** 1988. Infection of SCID-hu mouse by HIV-1. *Science* **242:**1684–1686.

32. **Nolta, J. A., and D. B. Kohn.** 1990. Comparison of the effects of growth factors on retroviral vector mediated gene transfer and proliferative status of human hematopoietic progenitor cells. *Hum. Gene Ther.* **1:**257–268.

33. **Ojwang, J. O., A. Hampel, D. J. Looney, F. Wong-Staal, and J. Rappaport.** 1992. Inhibition of human immunodeficiency virus type 1 expression by a hairpin ribozyme. *Proc. Natl. Acad. Sci. USA* **89:**10802–10806.

34. **Peault, B., I. L. Weissman, C. Baum, J. M. McCune, and A. Tsukamoto.** 1991. Lymphoid reconstitution of the human fetal thymus in SCID mice with CD34 precursor genes. *J. Exp. Med.* **174:**1283–1286.

35. **Rossi, J. J., E. M. Cantin, J. A. Zaia, P. A. Ladne, J. Chen, D. A. Stephens, N. Sarver, and P. S. Chang.** 1990. Ribozymes as therapies for AIDS. *Ann. N.Y. Acad. Sci.* **616:**184–200.

36. **Stanley, S. K., J. M. McCune, H. Kaneshima, J. S. Justement, M. Sullivan, E. Boone, M. Baseler, J. Adelsberger, M. Bonyhadi, J. Orenstein, C. H. Fox, and A. Fauci.** 1993. Human immunodeficiency virus infection of the human thymus and disruption of the thymic microenvironment in the SCID-hu mouse. *J. Exp. Med.* **178:**1151–1163.

37. **Tsukamoto, A., C. Baum, B. Peult, I. Weissman, S. Chen, B. Chen, and A.-M. Buckle.** 1993. Characterization and enrichment of candidate human hematopoietic stem cells. *J. Hematol. Ther.* **2:**117–119.

38. **von Kalle, C., H. P. Keim, S. Goehle, B. Darovsky, S. Heimfeld, B. Torok-Storb, R. Strob, and F. G. Schuening.** 1994. Increased gene transfer into human hematopoietic progenitor cells by extended in vitro exposure to a pseudotyped retroviral vector. *Blood* **84:**2890–2897.

39. **Weerasinghe, M., S. E. Liem, S. Asad, S. E. Read, and S. Joshi.** 1991. Resistance to human immunodeficiency virus type 1 (HIV-1) infection in human CD4$^+$ lymphocyte-derived cell lines conferred by using retroviral vectors expressing an HIV-1 RNA-specific ribozyme. *J. Virol.* **65:**5531–5534.

40. **Yu, M., J. Ojwang, O. Yamada, A. Hampel, J. Rapapport, D. Looney, and F. Wong-Staal.** 1993. A hairpin ribozyme inhibits expression of diverse strains of human immunodeficiency virus type 1. *Proc. Natl. Acad. Sci. USA* **90:**6340–6344.

41. **Yu, S. F., T. V. Ruden, P. W. Kantoff, C. Garber, M. Seiberg, U. Ruther, W. F. Anderson, E. F. Wagner, and E. Gilboa.** 1986. Self-inactivating retroviral vectors designed for transfer of whole genes into mammalian cells. *Proc. Natl. Acad. Sci. USA* **83:**3194–3198.

42. **Zack, J. A., S. J. Arrigo, S. R. Weitsman, A. S. Go, A. Haislip, and I. S. Y. Chen.** 1990. HIV-1 entry into quiescent primary lymphocytes: molecular analysis reveals a labile, latent viral structure. *Cell* **61:**213–222.

43. **Zack, J. A., A. Haislip, P. Krogstad, and I. S. Y. Chen.** 1992. Incompletely reverse transcribed human immunodeficiency virus type 1 genomes in quiescent cells can function as intermediates in the retrovirus life cycle. *J. Virol.* **66:**1717–1725.

The DNA Provirus: Howard Temin's Scientific Legacy
Edited by G. M. Cooper, R. Greenberg Temin, and B. Sugden
© 1995 American Society for Microbiology, Washington, DC 20005

Chapter 20

The Dilemma of Developing and Testing AIDS Vaccines

Dani P. Bolognesi

On 17 June 1994, the National Institute of Allergy and Infectious Diseases AIDS Research Advisory Committee recommended to the Institute that it continue but not expand current vaccine trials with the two recombinant gp120 subunit vaccines furthest along in development and proceed with expanded clinical trial evaluation only when more compelling information becomes available with these or other candidates. To some, particularly vaccine developers, this action was viewed as a major setback, given the extensive efforts and resources that have been expended for developing and testing these products. The valuable scientific and practical information that could emerge from the trials even if the vaccines were partly successful is now further from reach. To others, who felt that testing products with questionable promise for efficacy constituted a risky and unjustifiably costly undertaking that would ultimately have a negative impact on present and future vaccine trials, the recommendations of this body were more satisfying. Between these two extremes lies a sizable gulf of uncertainty as to what the best way to proceed might be.

These events illustrate the hurdles that human immunodeficiency virus (HIV) vaccines must face in order to proceed to large-scale trials. They also highlight the continuing struggle to establish standards that a vaccine must meet that are acceptable to scientists, vaccine developers, government officials, and representatives of the communities affected by such trials. Why these are such difficult issues and what needs to be achieved in order to maintain the momentum toward an effective vaccine against HIV are the focus of this discussion.

GENERAL PRINCIPLES OF DEVELOPMENT OF VACCINES AGAINST VIRUSES

From the standpoint of vaccine development, it is becoming more and more apparent that HIV is like no other virus. Key features that have led to successful

Dani P. Bolognesi • Duke University Medical Center, LaSalle Street Extension, SORF Building, Room 204, North Carolina P.O. Box 2926, Durham, North Carolina 27710.

vaccines with other viruses appear to be missing and are replaced by ones that are not conducive to vaccine development. First and foremost is the issue of natural immunity, or the spontaneous resolution of the infection and disease caused by a virus as a consequence of an effective immune response mounted by the host. This signals that the pathogen harbors targets against which successful immune defenses can be mounted. Correlates of immunity (or protection) can be derived and used to guide vaccine development. The ability of host defenses to effectively clear the pathogen is also important for another reason, namely, that it provides the rationale for developing live attenuated forms of the organism as vaccines. Like the pathogen itself, these would be eliminated after the establishment of protective immunity, thus providing an important measure of safety. Indeed, the most successful vaccines are live attenuated forms, although whole inactivated preparations have also been effective. Moreover, development of vaccines against such viruses has not had to deal with features such as variability, latency, and immunopathogenicity, which have prevented the development of successful vaccines with other viruses (e.g., influenza virus, members of the herpesvirus family) and are the trademarks of HIV.

In the absence of natural immunity, one must confront several serious obstacles: (i) correlates of immunity become difficult to establish, (ii) the rationale for live attenuated or even whole inactivate vaccines is weakened because of concerns for safety, (iii) the specter that all immune responses to the pathogen may not necessarily be salutary must be resolved, and (iv) the need to better understand virulence and how to overcome it becomes paramount. Thus, the empiricism that historically has been so dominant in development of vaccines against viruses gives way to a concerted effort to understand the fine details of infection and pathogenesis and how these are balanced with the ensuing host responses.

ANIMAL MODELS FOR HIV VACCINE DEVELOPMENT

When faced with such obstacles, vaccine developers have sometimes turned to animal models. However, once again, HIV presents a major barrier, since animal models have been unable to provide uniform guiding principles or correlates for protection easily translatable to humans. This lack is largely due to the fact that the requirements for protection in various animal models have been rather uneven. The differences between the models and the vaccine outcomes may reflect the respective virulence of the virus in a particular host (Fig. 1). Acute disease models induced by strains such as simian immunodeficiency virus (SIV) SIV-mac251 isolate are refractory to most vaccination attempts with the exception of those using live attenuated viruses and, to a lesser extent, whole inactivated virus (5). On the other hand, a more moderate but nonetheless lethal disease course occurs in pigtailed macaques infected with SIVmne, and vaccine approaches that are not successful with acute disease models do show efficacy in this model (12). Possibly even less stringent are criteria for vaccine success in models such as HIV-2 infections in macaques and HIV-1 infections in chimpanzees, neither of which produces disease. This may reflect the measure of host control of the virus

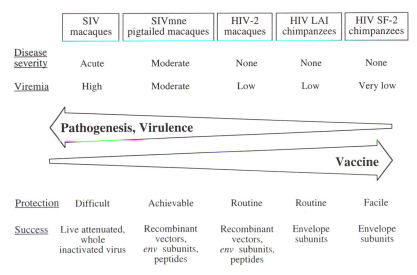

	SIV macaques	SIVmne pigtailed macaques	HIV-2 macaques	HIV LAI chimpanzees	HIV SF-2 chimpanzees
Disease severity	Acute	Moderate	None	None	None
Viremia	High	Moderate	Low	Low	Very low

Pathogenesis, Virulence

Vaccine

	SIV macaques	SIVmne pigtailed macaques	HIV-2 macaques	HIV LAI chimpanzees	HIV SF-2 chimpanzees
Protection	Difficult	Achievable	Routine	Routine	Facile
Success	Live attenuated, whole inactivated virus	Recombinant vectors, *env* subunits, peptides	Recombinant vectors, *env* subunits, peptides	Envelope subunits	Envelope subunits

FIGURE 1. Relative vaccine efficacy in nonhuman primates infected with SIV, HIV-2, and HIV-1 suggests that viral pathogenesis and virulence correlate inversely with ease of vaccination.

or some form of natural immunity; the more effective this natural immunity is, the more likely it is that a vaccine will be efficacious. The virulence of the challenge virus itself and its ability to expand and establish high levels of viremia are related factors, since in models like SIVmne, the virus loads can be between 10- and 100-fold lower than those in animals infected with SIVmac251.

In aggregate, these observations may be related to the question of natural immunity cited earlier, in that even degrees of host control that fall short of complete clearance of the virus appear to be important. The overriding issues to be resolved are which, if any, of these models are the most representative of HIV infection and disease in humans and how best to use them. Perhaps each one represents a segment of the overall spectrum of HIV infection in people defined now as rapid (SIVmac251), intermediate (SIVmne), and long-term (chimpanzees/HIV-1) nonprogressors. It is therefore likely that animal models will continue to be an indispensable component for development of vaccines against HIV, and a better understanding of the lessons they teach us will make them even more valuable.

PRECLINICAL AND CLINICAL STUDIES WITH ENVELOPE SUBUNIT VACCINE CANDIDATES

These issues and uncertainties notwithstanding, vaccine developers initially focused on vaccine approaches based on the virus envelope. Several independent studies had demonstrated that recombinant envelope products were effective in preventing HIV infection in chimpanzees and that antibodies were the best correlate of protection (for a review, see reference 3). The hypothesis that threshold levels of neutralizing antibodies might protect people against HIV infection became plausible, and a number of clinical trials were initiated to evaluate the safety and immunogenicity of envelope-based candidate vaccines. The best performance was achieved by two recombinant gp120 vaccines prepared in mammalian cells. In terms of magnitude, breadth, and duration of neutralizing-antibody responses to several laboratory strains, the results in humans surpassed even those achieved in the chimpanzee model (for a review, see reference 25). In both low- and high-risk volunteers, these vaccines also proved to be very well tolerated (25).

It was at this point that the question arose of proceeding to larger trials with high-risk volunteers in order to evaluate efficacy. Guidelines to the features a vaccine would have to exhibit in order to enter such trials began to be fashioned, but in the absence of defined correlates of immunity, these guidelines were kept very general (Table 1). When the two envelope vaccine candidates were measured against such standards in the spring of 1993, they had essentially met all but one, that being an indication of how well they matched the target viruses in the population virologically, immunologically, and genetically. The initial approach used to determine this was to evaluate their ability to induce antibodies capable of neutralizing fresh patient isolates by use of peripheral blood mononuclear cells (PBMCs) as targets (Fig. 2). Although quite effective in their ability to induce neutralizing antibodies to HIV isolates that are adapted to T-cell lines, which actually overlap those found in HIV-infected individuals, the immune responses elicited by these vaccines have failed to neutralize fresh patient isolates on PBMCs (10, 14). A flurry of studies ensued to determine whether this was an assay problem or whether it reflected a fundamental difference between primary and laboratory isolates. The outcome of these and continuing efforts points heavily away from this being an assay problem, a point that will be revisited later.

In the meantime, a new challenge virus for chimpanzees, one that had been propagated only on PBMCs from the time of its isolation from a patient, became available and made it possible to determine whether the absence of in vitro neutralization correlated with lack of protection in vivo (19). The results were quite surprising in that protection was achieved with both gp120 products despite the absence of significant levels of neutralizing antibodies to the challenge virus (7,

TABLE 1
Guidelines for entry into efficacy trials[a]

Guideline	Response of vaccine in:			
	Spring 1993	Fall 1993	Spring 1994	Summer 1994
Primate protection studies done	√	√	√√	√√
Safety in phase I/II	√	√	√	?
Immunogenic for:				
Vaccine strain	√	√	√	√
Strains circulating in target population	?	×	×	×

[a] √, Meets criteria; √√, exceeds criteria; ?, not determined; ×, fails to meet criteria based on virus neutralization in vitro.

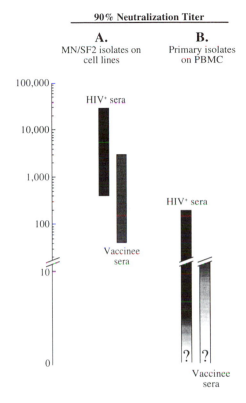

FIGURE 2. Neutralization of laboratory strains versus primary isolates. Bars indicate the range of neutralization titers of sera from HIV-1-infected individuals and uninfected volunteers who were vaccinated with MN and SF-2 envelope glycoproteins performed with laboratory isolates grown on T-cell lines (MN, SF-2) (A) and primary isolates (panel of 10) representative of clade B (origin of MN and SF-2) passaged only on PBMCs (B). Note that sensitivity of primary isolates to neutralization is much lower with both HIV-positive sera and sera from vaccines, but while a fraction of sera from HIV-1-infected individuals can still neutralize some of the isolates, none of the vaccinee sera tested registered positive readings. The minimum positive reading in our assay systems was a titer of 1:10.

20). A possible explanation of why protection against this virus is easily achieved in these and other studies reflects a point made earlier (Fig. 1) about the relationship between virulence and vaccine efficacy. The virus load in HIV SF-2-challenged chimpanzees is considerably lower than in HIV LAI-challenged animals, in which a threshold of neutralizing antibodies is a requirement for protection. As the relative significance of the neutralization results and protection studies was being debated, another important set of information became available, namely, detection of infections in volunteers participating in both phase I and phase II vaccine trials with these products (2).

How each of these developments affected the perception of the suitability of these vaccines for expanded clinical testing is illustrated in Table 1. In the spring of 1993, the two envelope candidates appeared to be on their way to meeting all of the requirements, but by the fall of 1993, they fell somewhat short on measures of breadth of activity as determined by neutralization of primary isolates. In the spring of 1994, the chimpanzee data became available, casting considerable doubt on the significance of the in vitro neutralization and leading many observers to turn to further clinical studies for definitive answers. However, by the summer of 1994, the issue of infections in vaccinees emerged and decreased confidence in the potential efficacy of these vaccines. This perception also exacerbated concerns that an increase in risk behavior might occur as a result of participation in such a trial, which could place volunteers at a higher risk for HIV infection. Such concerns for safety, coupled with the desire to preserve precious human and material resources for future trials, led to the recommendation not to go forward until more could be learned about the unresolved scientific and safety issues. This action thus overrode the recognized value of performing an efficacy trial in order to obtain definitive answers about vaccine performance as well as possibly some of the scientific questions.

However, one element that featured heavily in the decision to delay the trials, namely, infections in vaccinees, must be placed in perspective (2). It is important to remember that breakthrough infections can be associated even with vaccines such as those for hepatitis B, which are more than 90% efficacious (8, 22). Most occur before the vaccine regimen is completed (intercurrent infections), but they also occur in individuals who are poor responders to the vaccine or when vaccine immunity wanes. Therefore, it should be no surprise if they are found in HIV vaccine trials; rather, it is a matter of defining the circumstances surrounding the infections as they relate to the vaccine regimen. In fact, in all but one of the vaccinees who became infected, the immunization schedule had not been completed (2). Therefore, until more is known about the conditions that define these infections and until the phenomenon of enhancement of infection and disease can be excluded to the best of our ability, these infections do not in and of themselves indicate a lack of efficacy and thus should not be a barrier for entry of a vaccine into efficacy trials.

On the other hand, valuable insights about vaccine efficacy as well as HIV infection and pathogenesis might be obtained by careful studies of these infections. For instance, how vaccine failure is related to the immunization schedule or the responder status of the vaccinee could be determined. Does a vaccine fail against certain but not other isolates? Does vaccination influence the course of infection and disease progression in individuals who become infected? And finally, can such information be used to help guide decisions as to when a vaccine is ready for efficacy trial testing?

THE DEBACLE OF PRIMARY ISOLATE NEUTRALIZATION

Returning to the question of the primary isolate neutralization dilemma, the problem, as mentioned earlier, is not likely to lie with the assay, because a number of human monoclonal antibodies as well as a limited number of sera from acutely infected individuals and long-term nonprogressors who have been studied effectively neutralize such isolates. Examples of monoclonal antibodies to both gp120 and gp41 that have impressive neutralization potency on field isolates come from studies in several laboratories. As originally shown by Eva Maria Fenyo and colleagues (1) and more recently in studies from David Ho's group (13), polyclonal sera derived within a 6-month window after acute infection effectively neutralize autologous primary isolates, a property that appears to be eventually lost as the virus diversifies. However, there are also examples of polyclonal sera from a subset of long-term nonprogressors that exhibit broad and effective neutralization of panels of field isolates (11). In all of these instances, the immune response has recognized sites that are shared by the complex virus populations that comprise primary isolates. To date, as noted earlier, immunization with gp120 or gp160 has not produced antibodies with this capability.

One explanation might be that in all of these instances, the immune system is exposed to the dynamics of virus infection, replication, and diversification, while immune responses to static molecules are not. We also now know from studies in several laboratories, including our own, that the envelope of the virus exists in the form of an oligomer rather than the monomer form, gp120 (for a review, see reference 15). Moreover, subsequent to binding to the CD4 receptor, this complex, which is held together by the gp41 transmembrane protein, undergoes dramatic conformational changes that trigger the processes of fusion and entry into the cell (Fig. 3). These changes involve intermolecular interactions between gp120 and gp41 as well as intramolecular rearrangements within each molecule, particularly gp41, because it forms a membrane attack complex. This is a process that is common to many fusogenic viruses, the prototype being the

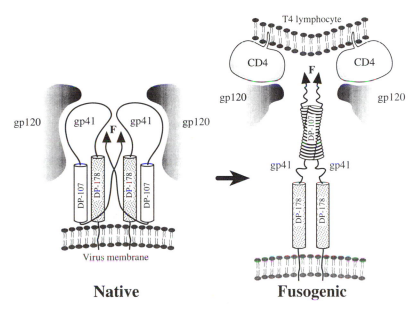

Native   **Fusogenic**

FIGURE 3. Model for structural transition in the HIV-1 TM protein leading to an active fusogenic attack complex. A model that indicates a structural transition from a native oligomer to a fusogenic state following a trigger event (possibly gp120 binding to CD4) is proposed. Features include (i) a native state held together by noncovalent protein-protein interactions to form the heterodimer of gp120/41 and other interactions, principally through gp41 interactive sites (DP-178 and DP-107), to form homo-oligomers on the virus surfaces of the gp120/41 complexes; (ii) shielding of the hydrophobic fusogenic peptide at the N terminus (F) in the native state; and (iii) a leucine zipper domain (DP-107) that exists as a homo-oligomer coiled coil only in the fusogenic state. When triggered, the fusion complex is generated through formation of coiled-coil interactions in homologous DP-107 domains, resulting in an extended α-helix similar to that in the model for influenza. This conformational change positions the fusion peptide for interaction with the cell membrane.

influenza virus, as brought to light by recent elegant studies by Carr and Kim (4), and our model for HIV derives in part from that example. The immune system may thus ''see'' these transitional forms and mount responses to what otherwise might be inaccessible (hidden) epitopes. Defining such target epitopes, which are essential for the virus and probably represent conserved domains shared by most or all HIV species, may contribute to the design of more effective vaccines.

There is indeed new evidence from several laboratories that the envelopes of primary isolates are configured differently from those of T-cell-line-adapted viruses in that they shield certain targets for neutralization by both recombinant CD4 and antibodies (18, 21). This has prompted suggestions that envelope vaccines based on primary isolates and exhibiting an oligomeric form of the envelope may be more effective for induction of neutralizing antibodies to field isolates. However, results of studies in animal models and in humans indicate that the development of antibodies that effectively cross-neutralize primary isolates not only is rare but also occurs only after long-term exposure to the infecting virus. It is thus difficult to envision how immunization with nonreplicating oligomeric envelope would overcome this problem. Actual experiments in which envelope oligomers have been used to immunize animals have indeed not produced neutralizing antibodies effective against primary isolates (14a). Therefore, other strategies for devising immunogens with such properties will be necessary. Encouraging in this regard are approaches based on identifying target epitopes of human monoclonal antibodies that effectively neutralize primary isolates and then reconfiguring these as effective immunogens.

DISCUSSION

In summary, we have witnessed unexpected developments and perhaps learned valuable lessons as the first wave of vaccines approached the all-important milestone of efficacy trial testing. These vaccines were based on the hypothesis that neutralizing antibodies could represent a correlate of protection. For the present, the definitive clinical studies to answer this question have been put on hold until more information becomes available. Other hypotheses, such as cellular immunity or mucosal immunity, could either stand alone or, as is much more appealing to many, be combined through imaginative vaccine approaches in order to elicit a more comprehensive immunity. These are currently in development and could be ready for large-scale trials over the next several years (Table 2). They include complex recombinant pox vectors that, in addition to the envelope, include other structural as well as regulatory gene products of the virus that have proven capable of inducing both neutralizing antibodies and cytotoxic T lymphocytes (CTLs). The added benefit of immunity at mucosal sites can be achieved with other viral and bacterial vectors such as adenovirus, poliovirus, mengovirus, and salmonellae. Other approaches include vaccination with DNA and the live attenuated virus vaccines. The latter, spearheaded by Ron Desrosiers (6), is thus far the most successful vaccine strategy, and now a major research topic is to determine

Table 2
HIV vaccine candidates in development[a]

Second generation (being considered for clinical trials)
- Complex pox vectors (vaccinia, avipox) consisting of multiple HIV gene products (*gag, pol, env, nef*)
- VLP[b]
- Pseudovirions[c]
- Peptide cocktails (including oral formulations)

Third generation (in research)
- Live attenuated viruses
- Whole inactivated viruses
- Naked DNA
- Viral vectors (adenovirus, poliovirus, mengovirus, rhinovirus)
- Bacterial vectors (BCG, *Salmonella* spp., *Shigella* spp.)
- Chimeric proteins[d]

---

[a] For review, see reference 25.
[b] VLP, nucleic acid-free, noninfectious, viruslike particles that self-assemble in yeast, insect, and mammalian expression systems involving *env, gag,* and *pol* gene products of HIV.
[c] Nonreplicating multiply mutated HIV.
[d] Immunogenic protein sequences from other organisms coupled to HIV proteins or peptides.

whether it can be made safe and what mechanism is responsible for its superior efficacy.

Thus, despite substantial efforts to develop a vaccine against HIV, it is evident that many important questions still remain unanswered. Continued efforts are needed to understand basic features of HIV infection and pathogenesis, to improve animal models, and to identify correlates of protection. In addition to searching for correlates of immunity in cases of vaccine breakthrough infections or in long-term nonprogressors, one must also consider the significance of a growing number of instances in which protection against HIV infection occurs without recognizable correlates of immunity. For instance, there are the examples with the SF-2 challenge virus in chimpanzees, in which protection occurs in the absence of neutralizing antibodies even against laboratory strains (9, 16). Similarly, studies with several different vaccine approaches against HIV-2 reveal few or no neutralizing antibodies or CTLs at the time of challenge, and yet protection against infection is achieved (23). Finally, from studies in cohort HIV-exposed nonseroconverters (26) and in monkeys exposed to low-dose rectal challenges (17), one again finds indications of possible protection in the absence of classic immune correlates.

What are these situations telling us about host control of virus replication and pathogenesis? Are there host defense mechanisms that are not easily detectable as classic effector mechanisms (e.g., neutralizing antibodies and CTLs) but that when supplemented with vaccination can effectively block infection? Investigation of these and related examples of protection in the absence of identifiable correlates seems worthwhile, and such studies may eventually show that even the measurable correlates that are familiar to us may be only markers but not the actual mechanism of protection.

To conclude, the biomedical establishment cannot become passive or discouraged over recent developments. It has little alternative but to redouble its efforts and be prepared to maintain a long-term solid commitment until an effective vaccine against this devastating pathogen is achieved. This commitment must not only emphasize basic research but also support the properly justified preclinical and clinical studies of a more empirical nature that have historically proven so valuable in vaccine development. Regardless of how much progress one can achieve scientifically, it will never be enough to guarantee that a vaccine will or will not succeed; such a guarantee requires well-designed clinical trials that in turn will identify the elusive correlates of protection that can guide development of improved vaccines.

In any event, it is now evident that a great deal of momentum is required to drive an HIV vaccine to the all-important milestone of an efficacy trial. Badly needed are acceptable guidelines designed to best forecast the likelihood that vaccine candidates will prove efficacious. In the absence of immune correlates, more emphasis might be placed on vaccine efficacy in several animal models with various degrees of virus virulence and disease. Similarly, appropriately designed phase II studies in individuals at risk for HIV infection may help guide entry into large, definitive efficacy trials. Thus, whereas the recent experience has been a trying one for the vaccine field, one can take some comfort in the fact that when we are next faced with such decisions, the foundation for going forward will be more solid and justifiable.

HOWARD TEMIN'S CONTRIBUTION TO WORK ON HIV VACCINES

Throughout much of the tumultuous course of HIV vaccine evolution, Howard Temin was both a constructive proponent and a critical appraiser. While he clearly recognized the need for a vaccine and the urgency to undertake definitive clinical trials, he was concerned about the safety of the volunteers, and he insisted that informed consent be unambiguous as to benefits and risks. When the two envelope vaccines revealed good immunogenicity and safety in several trials in 1992 and 1993, he was generally positive about moving ahead to expanded trials on the principle that it was reasonable to move forward with what scientists perceived to be the most promising candidates currently available. "Don't let the best be the enemy of the good," he once wrote to me in late 1992. However, as the problem of neutralization of primary isolates revealed itself, he was just as quick to recommend delaying going forward until the significance of these disappointing results was understood.

In the meantime, Howard was actively working on his own strategies for HIV vaccines, about which he was genuinely excited. He wrote of this in a perspective entitled "A Proposal for a New Approach to a Preventive Vaccine against Human Immunodeficiency Virus Type 1" (24). He believed that replicating viruses were the only candidates likely to be effective in preventing HIV infection, but he was not prepared to introduce an attenuated HIV into humans. Instead, he proposed inclusion of structural HIV genes within simple retroviral genomes in the belief that humans would mount a protective response against such chimeras. His students are continuing to work on this unique approach that stemmed from Howard's exquisite knowledge of retrovirus replication and recombination and his intuitive sense of certain features of virus evolution that could be applied to vaccines against viral diseases. The reader is encouraged to read this wonderful and insightful treatise.

REFERENCES

1. **Albert, J., B. Abrahamsson, K. Nagy, E. Aurelius, H. Gaines, G. Nystrom, and E. M. Fenyo.** 1990. Rapid development of isolate-specific neutralizing antibodies after primary HIV-1 infection and consequent emergence of virus variants which resist neutralization by autologous sera. *AIDS* **4:** 107–112.
2. **Belshe, R. B., D. P. Bolognesi, M. L. Clements, L. Corey, R. Dolin, J. Mestecky, M. Mulligan, D. Stablein, and P. Wright.** 1994. HIV infection in vaccinated volunteers. *JAMA* **272:**431. (Letter.)
3. **Bolognesi, D. P.** 1993. Human immunodeficiency virus vaccines. *Adv. Virus Res.* **42:**103–148.
4. **Carr, C. M., and P. S. Kim.** 1993. A spring-loaded mechanism for the conformational change of influenza hemagglutinin. *Cell* **73:**823–832.
5. **Daniel, M. D., G. P. Mazzara, M. A. Simon, P. K. Sehgal, T. Kodama, D. L. Panicali, and R. C. Desrosiers.** 1994. High-titer immune responses elicited by recombinant vaccinia virus priming and particle boosting are ineffective in preventing virulent SIV infection. *AIDS Res. Hum. Retroviruses* **10:**839–851.
6. **Desrosiers, R. C.** 1992. HIV with multiple gene deletions as a live attenuated vaccine for AIDS. *AIDS Res. Hum. Retroviruses* **8:**1457.
7. **Francis, D. P., P. Fast, S. Harkonen, M. J. McElrath, R. Belshe, P. Berman, T. Gregory, and the AIDS Vaccine Evaluation Group.** 1994. MN rgp120 (Genentech) vaccine safe and immunogenic—but will it protect humans?, abstr. 314A, p. 90. *In Tenth International Conference on AIDS; 1994 Aug. 7–12; Yokohama, Japan. Abstract Book,* vol. 1.
8. **Francis, D. P., S. C. Hadler, S. E. Thompson, J. E. Maynard, D. G. Ostrow, N. Altman, E. H. Braff, P. O'Malley, D. Hawkins, F. N. Judson, K. Penley, T. Nylund, G. Christie, F. Meyers, J. N. Moore, Jr., A. Gardner, I. L. Doto, J. H. Miller, G. H. Reynolds, B. L. Murphy, C. A. Schable, B. T. Clark, J. W. Curran, and A. G. Redeker.** 1982. The prevention of hepatitis B with vaccine. Report of the Centers for Disease Control multi-center efficacy trial among homosexual men. *Ann. Intern. Med.* **97:**362–366.
9. **Girard, M.** 1994. Further studies on HIV-1 vaccine protection in chimpanzees. Paper presented at Tenth International Conference on AIDS, 7–12 August 1994, Yokohama, Japan.
10. **Hanson, C. V.** 1994. Measuring vaccine-induced HIV neutralization: report of a workshop. *AIDS Res. Hum. Retroviruses* **10:**645–648.
11. **Ho, D. D.** 1994. Long-term non-progressors, abstr. PS10, p. 50. *In Tenth International Conference on AIDS; 1994 Aug. 7–12; Yokohama, Japan. Abstract Book,* vol. 1.
12. **Hu, S. L., K. Abrams, G. N. Barber, P. Moran, J. M. Zarling, A. J. Langlois, L. Kuller, W. R. Morton, and R. E. Benveniste.** 1992. Protection of macaques against SIV infection by subunit vaccines of SIV envelope glycoprotein gp160. *Science* **255:**456–459.
13. **Koup, R. A., J. T. Safrit, Y. Cao, C. A. Andrews, G. McLeod, W. Borkowsky, C. Farthing, and D. D. Ho.** 1994. Temporal association of cellular immune responses with the initial control of viremia in primary human immunodeficiency virus type 1 syndrome. *J. Virol.* **68:**4650–4655.
14. **Matthews, T. J.** 1994. Dilemma of neutralization resistance of HIV-1 field isolates and vaccine development. *AIDS Res. Hum. Retroviruses* **10:**631–632.
14a. **Matthews, T. J., P. Berman, and B. Moss.** Personal communication.
15. **Matthews, T. J., C. Wild, C. H. Chen, D. P. Bolognesi, and M. L. Greenberg.** 1994. Structural rearrangements in the transmembrane glycoprotein after receptor binding. *Immunol. Rev.* **140:** 93–104.

16. **Natuk, R., M. Robert-Guroff, M. Lubeck, K. Steimer, R. Gallo, and J. Eichberg.** 1994. Adeno-HIV priming and subunit boost: 2nd generation AIDS vaccines, abstr. 248A, p. 74. *In Tenth International Conference on AIDS; 1994 Aug. 7–12; Yokohama, Japan. Abstract Book,* vol. 1.

17. **Pauza, D., P. Trivedi, E. Johnson, K. K. Meyer, D. N. Streblow, M. Malkovsky, P. Emau, K. T. Schultz, and M. S. Salvato.** 1993. Acquired resistance to mucosal SIV infection after low dose intrarectal inoculation: the roles of virus selection and CD8-mediated T cell immunity, p. 151–156. *In* M. Girard and L. Valette (ed.), *Retroviruses of Human A.I.D.S. and Related Animal Diseases. 8ᵉ Colloque des Cent Gardes, 1993 Oct. 25–27.* Fondation Marcel Mérieux, Lyon.

18. **Sattentau, Q.** 1994. Studies with monomeric and oligomeric forms of the HIV envelope from primary and cell line adapted viruses. Paper presented at Tenth International Conference on AIDS, 7–12 August 1994, Yokohama, Japan.

19. **Schultz, A. M., and S. L. Hu.** 1993. Primate models for HIV vaccines. *AIDS* 7:S161–S170.

20. **Steimer, K. S.** 1994. Status of gp120 subunit vaccine development. Paper presented at Tenth International Conference on AIDS, 7–12 August 1994, Yokohama, Japan.

21. **Sullivan, N., R. Wyatt, U. Olshevsky, J. Moore, and J. Sodroski.** 1994. Neutralizing antibodies directed against the HIV-1 envelope glycoproteins. *AIDS Res. Hum. Retroviruses* 10(Suppl 3): S110.

22. **Szmuness, W., C. E. Stevens, E. A. Zang, E. J. Harley, and A. Kellner.** 1981. A controlled clinical trial of the efficacy of the hepatitis B vaccine (Heptavax B): a final report. *Hepatology* 1:377–385.

23. **Tartaglia, J., G. Franchini, M. Robert-Guroff, A. Abimuku, J. Benson, K. Limbach, M. Wills, R. C. Gallo, and E. Paoletti.** 1993. Highly attenuated poxvirus vector strains, NYVAC and ALVAC, in retrovirus vaccine development, p. 293–298. *In* M. Girard and L. Valette (ed.), *Retroviruses of Human A.I.D.S. and Related Animal Diseases. 8ᵉ Colloque des Cent Gardes, 1993 Oct. 25–27.* Fondation Marcel Mérieux, Lyon.

24. **Temin, H. A** 1993. A proposal for a new approach to a preventive vaccine against human immunodeficiency virus type 1. *Proc. Natl. Acad. Sci. USA* 90:4419–4420.

25. **Walker, M. C., and P. E. Fast.** 1993. Human trials of experimental AIDS vaccines. *AIDS* 7: S147–S159.

26. **Willerford, D. M., J. J. Bwayo, M. Hensel, W. Emonyi, F. A. Plummer, E. N. Ngugi, N. Negalkerke, W. M. Gallatin, and J. Kreiss.** 1993. Human immunodeficiency virus infection among high-risk seronegative prostitutes in Nairobi. *J. Infect. Dis.* 167:1414–1417.

The DNA Provirus: Howard Temin's Scientific Legacy
Edited by G. M. Cooper, R. Greenberg Temin, and B. Sugden
© 1995 American Society for Microbiology, Washington, DC 20005

Chapter 21

Reprint of Temin's 1993 Paper Proposing a New Approach to Developing a Human Immunodeficiency Virus Vaccine

From the mid-1980s, Howard Temin not only was keenly aware of the threat to public health posed by the expanding AIDS epidemic but also acted to limit that threat. He tried to minimize the personal and scientific conflicts among early workers on human immunodeficiency virus (HIV) so as not to obscure the national and international problem this virus presents. He worked to focus public support of research on HIV and to be rigorous and generous. He also sought to contribute scientifically to the resolution of this problem, as evidenced by his "Proposal for a New Approach to a Preventive Vaccine against Human Immunodeficiency Virus Type 1" and his experimental pursuit of that proposal.

Most human vaccines have been generated either serendipitously, as was the case for the smallpox virus vaccine, or by scientific insight coupled with ignorance. Virologists have grown to appreciate that by adapting virus strains isolated from human hosts to propagation in cells in culture, viral variants that retain infectivity but lose their pathogenicity in vivo may arise. This forced propagation in vitro can therefore yield viral mutants that can serve as live vaccines. However, the phenotypes being selected in vitro are not understood, and the mutations that affect the phenotypes both in vitro and in vivo have often not been identified.

In this article, Howard Temin built on his appreciation of simple and complex retroviruses to formulate a hypothesis, to develop a test of that hypothesis, and to propose an application of the hypothesis, if it were proven correct, in developing a vaccine against HIV. The hypothesis is based on the observation that simple retroviruses (those containing *gag, pol,* and *env* as structural genes) today frequently infect chickens and mice but are not common pathogens in some other animals, including humans. On the other hand, complex retroviruses (those containing additional regulatory genes not present in simple retroviruses) are pathogens in people today but have not been isolated from chickens and mice. Howard proposed that this dichotomy reflects the immune responses of the species. In his hypothesis, our immediate ancestors' immune responses succeeded in eliminating simple retroviruses as human pathogens. Vestiges of these simple retroviruses are present as related proviruses in our germ line DNA. If his hypothesis is correct, then developing an infectious, replication-competent derivative of HIV that is a simple retrovirus could elicit an immune response protective against subsequent infections by HIV. There is evidence that a derivative of simian immunodeficiency virus lacking a single auxiliary regulatory gene is no longer pathogenic. Such a vaccine might also have the valuable characteristic of evolving in vivo, as does HIV, to yield all possible replicating variants as potential immunogens for the swarms of relatives generated by HIV in vivo.

Howard's proposal is not conventional. It does not build on forced passage of HIV in cells in culture, an approach that seems unwise in view of this virus's capacity to generate variants in vivo. It is based on his novel biological insight. More important, it provides an approach that can be tested in animal models and thereby yields one rational approach to the development of a vaccine for HIV and eventual prevention of AIDS.

Bill Sugden

Proc. Natl. Acad. Sci. USA
Vol. 90, pp. 4419–4420, May 1993
Medical Sciences

A proposal for a new approach to a preventive vaccine against human immunodeficiency virus type 1

(simpler retrovirus/more complex retrovirus/co-virus)

HOWARD M. TEMIN

McArdle Laboratory, 1400 University Avenue, Madison, WI 53706

Contributed by Howard M. Temin, February 22, 1993

ABSTRACT   Human immunodeficiency virus type 1 (HIV-1) is a more complex retrovirus, coding for several accessory proteins in addition to the structural proteins (Gag, Pol, and Env) that are found in all retroviruses. More complex retroviruses have not been isolated from birds, and simpler retroviruses have not been isolated from humans. However, the proviruses of many endogenous simpler retroviruses are present in the human genome. These observations suggest that humans can mount a successful protective response against simpler retroviruses, whereas birds cannot. Thus, humans might be able to mount a successful protective response to infection with a simpler HIV-1. As a model, a simpler bovine leukemia virus which is capable of replicating has been constructed; a simpler HIV-1 could be constructed in a similar fashion. I suggest that such a simpler HIV-1 would be a safe and effective vaccine against HIV-1.

There is as yet no safe and effective vaccine against human immunodeficiency type 1 (HIV-1), the causative agent of AIDS (1–4). Probably the primary reason for this failure is that the nature of a protective immune response against HIV-1 is not known. Clearly, the usual immune response in persons after infection with HIV-1 is not sufficiently protective. In this article, I shall argue that evolution has provided a natural experiment that might direct us to a safe and effective vaccine against HIV-1.

HIV-1 is a lentiretrovirus, one of the more complex retroviruses (5); Hilleman (2) calls it an extraordinary virus. As a more complex retrovirus, HIV-1, like human T-cell leukemia virus type I (HTLV-I), bovine leukemia virus (BLV), and human spumaretrovirus (HSRV), differs from simpler retroviruses in having many genes in addition to *gag*, *pol*, and *env*, which are common to all retroviruses.

All retroviruses contain the *gag*, *pol*, and *env* genes and the cis-acting sequences acted on by the Gag and Pol proteins, as well as the sequences involved in the control of transcription, splicing, and polyadenylylation. In the case of simpler retroviruses, such as murine leukemia virus, the processes of viral transcription, splicing, and polyadenylylation are controlled by cellular proteins. The additional genes in more complex retroviruses code for proteins that act with cellular proteins to control transcription, splicing, and polyadenylylation and enable more complex retroviruses to have more-complex replication cycles (5, 6).

Simpler retroviruses were first found in chickens and have been much studied in chickens and mice. They are also found in vipers, fish, and cats, and there are even a few isolates from monkeys (7). Gibbon ape leukemia virus is a murine leukemia virus-related simpler retrovirus, and Mason–Pfizer monkey virus is a primate type D simpler retrovirus. Study of the human genome indicates that in the past many simpler

retroviruses infected our ancestors, as shown by the relic proviruses in human DNA (7–16).

More complex retroviruses were first isolated in horses (equine infectious anemia virus). Since then they have been isolated from many other mammals, including at least five from humans (HIV-1, HIV-2, HTLV-I, HTLV-II, and HSRV). So far, more complex retroviruses have not been isolated from vertebrate families other than mammals, and simpler retroviruses have not been isolated from humans or ungulates. However, more complex retroviruses have been isolated from both humans and ungulates.

I propose that this phylogenetic distribution is not an artifact of virus isolation techniques, but that it reflects the ability of humans and ungulates to respond to infection by simpler retroviruses with a protective response, and the inability of birds and mammals other than humans and ungulates to respond to infection by simpler retroviruses with such a complete protective response.

Under this hypothesis, retroviruses infecting birds were under no selective pressure to evolve into more complex retroviruses, since the simpler retroviruses were already quite successful in infecting birds. On the other hand, in some mammals the immune system and perhaps other host characteristics were more successful in controlling simpler retroviruses, providing selective pressure for the evolution of more complex retroviruses. [Welsh *et al.* (17) describe such a mechanism in human serum for controlling simpler retroviruses.] The presence of genomes of simpler retroviruses in the human genome (endogenous viruses) indicates that, earlier in evolution, ancestors of present-day humans were successfully infected by such simpler retroviruses (7–16).

A conclusion from this line of reasoning is that present-day humans are able to mount a protective response against simpler retroviruses. Since HIV-1 is a more complex retrovirus, this conclusion does not yet help direct us toward methods for making a safe and effective HIV-1 vaccine. However, if we were able to construct a simpler retrovirus that expressed only the Gag, Pol, and Env proteins of HIV-1, this virus should induce in humans a protective response against itself, the simpler HIV-1. If such a protective response occurs against infection by a simpler HIV-1, I further propose that this protective response would protect against wild-type HIV-1 infection, and thus that the simpler HIV-1 would form an effective vaccine against HIV-1, either as a live "attenuated" virus or as an inactivated virus.

Such a simpler HIV-1 is fundamentally different from vector- or DNA-expressed HIV-1 structural proteins, which cannot replicate in the vaccine recipient (18–21). The simpler HIV-1 would replicate in the vaccine recipient, and it would continue to replicate until a complete protective response was stimulated in the recipient. Furthermore, since the

Abbreviations: BLV, bovine leukemia virus; HIV, human immunodeficiency virus; HSRV, human spumaretrovirus; HTLV, human T-cell leukemia virus; LTR, long terminal repeat; SNV, spleen necrosis virus.

Proc. Natl. Acad. Sci. USA 90 (1993)

simpler HIV-1 replicates as a retrovirus using the HIV-1 reverse transcriptase, any vaccine preparation of the simpler HIV-1 would be a swarm containing different nucleotide sequences, not a monosequence as occurs from a cloned DNA preparation. In addition, the simpler HIV-1 would vary as it replicates in the vaccine recipient and would induce a polytypic response, not the monotypic response of the vector- or DNA-expressed proteins.

The simpler HIV-1 would not be HIV-1; it would not have the additional HIV-1 genes that make HIV-1 a more complex retrovirus. Thus, the simpler HIV-1 should not cause AIDS (22). Furthermore, the simpler HIV-1 would replicate less well than HIV-1, further reducing its possible pathogenicity (23). [The simpler HIV-1 would also differ fundamentally from partially deleted HIV-1 (4), since there would be many fewer HIV-1 sequences in the simpler HIV-1.] However, it is possible that such a simpler HIV-1 used as a live virus vaccine might result in some low incidence of leukemia as a result of insertional activation of protooncogenes (24). Such leukemogenesis would be rare, if it occurred at all (25), and any leukemias induced by the simpler HIV-1 would appear only after a long latent period. Also, the possibility of leukemia induction by a simpler retrovirus can be drastically reduced by making the promoter/enhancer sequences of the simpler HIV-1 inefficient.

A simpler BLV with spleen necrosis virus (SNV) long terminal repeat (LTR) sequences has been constructed and shown to replicate (K. Boris-Lawrie and H.M.T., unpublished work). As a further safety feature, the BLV *gag–pol* genes and the BLV *env* gene were expressed from different simpler BLV constructs—that is, as co-viruses—and the co-viruses were shown to replicate as a chimeric BLV/SNV virus. Expression from the BLV/SNV LTR allows functional *env* expression in the absence of other BLV proteins.

A simpler HIV-1 could be constructed in a similar fashion by taking the HIV-1 *gag, pol,* and *env* genes and the cis-acting sequences acted on by the Gag and Pol proteins of HIV-1 (*att, pbs, E, ppt*) and placing them in the partially deleted LTRs of a simpler retrovirus. In particular, as for BLV, this means substituting simpler retrovirus LTR sequences for all of the HIV-1 LTR sequences except for the internal *att* sequences. [Such a substitution would also delete the HIV transactivation response (TAR) sequence.]

A simpler HIV-1 construct with only HIV-1 *gag, pol,* and *env* genes might not replicate. Rev protein may be necessary for *env* expression unless its cis-acting sequences are mutated (26). Assembly of Gag, Pol, and Env proteins into virions appears to proceed in the absence of accessory proteins (18–22). If not, Vif or other accessory proteins may also need to be present. They could be added back into the constructs.

The utility of such constructs could first be validated in simpler HIV-1/chimpanzee and simian immunodeficiency virus/macaque model systems (see refs. 4 and 27). If there is concern about insertional activation increasing the possibility of leukemia, a killed vaccine made from the simpler HIV-1 could first be tested. However, if it appeared that only a live virus vaccine would be adequate to prevent HIV-1 infection and insertional activation by the simpler HIV-1 appeared to be a problem, the simpler HIV-1 could be further crippled by mutating promoter and enhancer sequences or adding a gene that can be selected against (a suicide gene) expressed from a picornaviral internal ribosome entry site (28–30).

The seriousness of the AIDS pandemic and the so-far low effectiveness of other immunodeficiency virus vaccines, ex-

cept possibly *nef⁻* simian immunodeficiency virus (27), makes other approaches like this one worthy of consideration.

I thank D. Bolognesi, K. Boris-Lawrie, D. Burns, G. Pulsinelli, and B. Sugden for comments. The research in my laboratory is supported by Public Health Service Grants CA22443 and CA07175 from the National Cancer Institute. I am an American Cancer Society research professor.

1. Sabin, A. B. (1992) *Proc. Natl. Acad. Sci. USA* **89,** 8852–8855.
2. Hilleman, M. R. (1992) *AIDS Res. Hum. Retroviruses* **8,** 1743–1747.
3. Ada, G., Blanden, B. & Mullbacher, A. (1992) *Nature (London)* **359,** 572.
4. Desrosiers, R. (1992) *AIDS Res. Hum. Retroviruses* **8,** 411–421.
5. Temin, H. M. (1992) in *The Retroviridae,* ed. Levy, J. A. (Plenum, New York), Vol. 1, pp. 1–18.
6. Vaishnav, Y. N. & Wong-Staal, F. (1991) *Annu. Rev. Biochem.* **60,** 577–630.
7. Coffin, J. M. (1992) in *The Retroviridae,* ed. Levy, J. A. (Plenum, New York), Vol. 1, pp. 19–50.
8. Larsson, E., Kato, N. & Cohen, M. (1989) *Curr. Top. Microbiol. Immunol.* **148,** 115–132.
9. Leib-Mosch, C., Brack-Weiner, R., Bachmann, M., Faff, O., Erfle, V. & Hehlmann, R. (1990) *Cancer Res.* **50,** 5636s–5642s.
10. Harada, F., Tsukada, N. & Kato, N. (1987) *Nucleic Acids Res.* **15,** 9153–9162.
11. Mager, D. L. & Freeman, J. D. (1987) *J. Virol.* **61,** 4060–4066.
12. Ono, M., Kawakami, M. & Takezawa, T. (1987) *Nucleic Acids Res.* **15,** 8725–8737.
13. Mariani-Costanttini, R., Horn, T. M. & Callahan, R. (1989) *J. Virol.* **63,** 4982–4985.
14. Callahan, R., Chiu, I.-M., Wong, J. F. H., Tronick, S. R., Roe, B. A., Aaronson, S. A. & Schlom, J. (1985) *Science* **228,** 1208–1211.
15. Repaske, R., Steele, P. E., O'Neill, R. R., Rabson, A. B. & Martin, M. A. (1985) *J. Virol.* **54,** 764–772.
16. O'Connell, C., O'Brien, S., Nash, W. G. & Cohen, M. (1984) *Virology* **138,** 225–235.
17. Welsh, R. M., Jr., Cooper, N. R., Jensen, F. C. & Oldstone, M. B. A. (1975) *Nature (London)* **257,** 612–614.
18. Karacostas, V., Nagashima, K., Gonda, M. A. & Moss, B. (1989) *Proc. Natl. Acad. Sci. USA* **86,** 8964–8967.
19. Haffar, O., Garrigues, J., Travis, B., Moran, P., Zarling, J. & Hu, S.-L. (1990) *J. Virol.* **64,** 2653–2659.
20. Vzorov, A. N., Bukrinsky, M. I., Grigoriev, V. B., Tentsov, Y. Y. & Bukrinskaya, A. G. (1991) *AIDS Res. Hum. Retroviruses* **7,** 29–36.
21. Hoshikawa, N., Kojima, A., Yasuda, A., Takayashiki, E., Masuko, S., Chiba, J., Sata, T. & Kurata, T. (1991) *J. Gen. Virol.* **72,** 2509–2517.
22. Sabatier, J.-M., Vives, E., Mabrouk, K., Benjouad, A., Rochat, H., Duval, A., Hue, B. & Bahraoui, E. (1991) *J. Virol.* **65,** 961–967.
23. Nowak, M. A., Anderson, R. M., McLean, A. R., Wolfs, T. F. W., Goudsmit, J. & May, R. M. (1991) *Science* **254,** 963–969.
24. Peters, G. (1990) *Cell Growth Differ.* **1,** 503–510.
25. Moolten, F. L. & Cupples, L. A. (1992) *Hum. Gene Ther.* **3,** 479–486.
26. Schwartz, S., Campbell, M., Nasioulas, G., Harrisin, J., Felber, B. & Pavlakis, G. N. (1992) *J. Virol.* **66,** 7176–7182.
27. Daniel, M. D., Kirchhoff, F., Czajak, S. C., Sehgal, P. K. & Desrosiers, R. C. (1992) *Science* **258,** 1938–1941.
28. Moolten, F. L. & Wells, J. M. (1990) *J. Natl. Cancer Inst.* **82,** 297–305.
29. Mullen, C. A., Kilstrup, M. & Blaese, R. M. (1992) *Proc. Natl. Acad. Sci. USA* **89,** 33–37.
30. Ghattas, I. R., Sanes, J. R. & Majors, J. E. (1991) *Mol. Cell. Biol.* **11,** 5848–5859.

Index